Pathobiology of Cancer Regimen-Related Toxicities

Stephen T. Sonis • Dorothy M. Keefe
Editors

Pathobiology of Cancer Regimen-Related Toxicities

Springer

Editors
Stephen T. Sonis
Brigham and Women's Hospital
Boston, MA, USA
and
Biomodels, LLC
Watertown, MA, USA

Dorothy M. Keefe
Royal Adelaide Hospital
Adelaide, South Australia, Australia

ISBN 978-1-4614-5437-3 ISBN 978-1-4614-5438-0 (eBook)
DOI 10.1007/978-1-4614-5438-0
Springer New York Heidelberg Dordrecht London

Library of Congress Control Number: 2012953271

Printed on acid-free paper

Springer is part of Springer Science+Business Media (www.springer.com)

Preface

Toxicities have been a constant companion of drug and radiation regimens used for the treatment of cancer. Depending on the severity, timing, and tissues affected, regimen-related toxicities (RRTs) vary widely on their clinical impact from being irritating and annoying to causing death or permanent disability. In many cases RRTs impede patients' ability to receive optimum therapy. Their negative impact on quality of life, both during and after treatment, has been exhaustively documented. In today's cost conscience environment, the extended resource use and expenses associated with RRTs are being more closely examined. Aside from the soft costs associated with lost work days, increased need for caregiver time and use of over-the-counter remedies, the hard costs incurred by RRTs are dramatic and are driven by unplanned office and emergency room visits, hospitalizations (or extended hospital length of stay), and increased use of supportive medications for palliation and pain control, among others.

For years toxicities were accepted as being inevitable and simply the price that had to be paid for effective cytotoxic cancer treatment. It is an interesting commentary on health care providers' perspective on their significance that in many manuscripts only those toxicities which are graded as severe are included. A critical examination of the NCI's Common Terminology Criteria, version 4, suggests that the rigor with which toxicities are graded is driven more by tolerability than by actual impact on patients' symptoms and quality of life. This observation reflects, in part, the longstanding disconnect between clinicians and patients relative to toxicity impact and the finding that both the frequency and severity of toxicities are markedly underreported. By way of example, the incidence of oral mucositis was studied in patients receiving standard cycled chemotherapy for colorectal cancer. The literature suggests that clinically significant mucositis is not frequent in this cohort reportedly occurring with a frequency of about 15–20%. Yet when patients were directly queried about significant mouth pain during treatment, fully three-quarters answered in the affirmative. Furthermore, it should not be unanticipated that patients are often reluctant to report toxicities as they realize that the consequence could be a less than optimum (from the standpoint of tumor kill) adjustment in treatment regimen.

There is no anticancer treatment regimen which is immune from toxicities. Virtually every cancer patient who is treated with radiation or anticancer drug therapy will suffer some adverse effect. The list of potential toxicities is extensive; some are tissue-specific like dermatitis, pulmonary fibrosis, or mucositis, while others are more systemically manifested such as cachexia, fatigue, or cognitive dysfunction. As a consequence of their clinical and fiscal impact, interest in preventing and treating RRTs has mushroomed in the past decade. In order to best develop effective interventions, better understanding of the pathobiology of toxicities was critical. As investigations began to yield data, two major themes have emerged: first, the biology underlying toxicity development was markedly more complex than was historically imagined; second, that toxicities do not occur as isolated events. Rather they occur in clusters and that the factors which meld one toxicity to another are attributable to common biology.

Historically, toxicities have been lumped into two buckets—those associated with tissue injury and those associated with behavioral or systemic changes. Since many tissue toxicities occurred in replenishing tissues like the gastrointestinal mucosa, the popular hypothesis was that injury was the consequence of nonspecific chemotherapy or radiotherapy-mediated damage to DNA in mother cells which resulted in their death and consequent atrophy and injury. Results of research conducted over the past decade has demonstrated that the biology of regimen-related tissue injury is more intricate and is largely dependent on the sequential interaction of multiple pathways that culminate in definitive damage to parent cells. Likewise, symptoms of fatigue, depression and cognitive dysfunction have been linked to downstream biological signals mediated by biologically active molecules associated with cancer therapy.

Accumulating data derived from studies examining the biological basis for RRTs has shown consistent themes which are likely to be responsible for the clinical observation of toxicity clustering. We concluded that now was an appropriate time to assemble in a single source, a collection of manuscripts which describe the current status of the science associated with RRTs.

The contents of this book are organized into two sections. The first section defines the scope, impact, and behavior of cancer regimen-related toxicities and frames the issue of balancing treatment success and physiological cost. In the second segment of the book, the most current thinking around the pathobiology of specific, common, and representative toxicities is presented by leading researchers and translational scientists. The conclusion summarizes the common biological relationships between toxicities, bioinformational approaches to analyzing key and common pathways, and strategies for the development of effective interventions.

We recognize that trying to define the pathobiological basis for RRTs is like shooting at a moving target. The oncology therapy pipeline is rich and consists of modifications of existing drugs, totally new classes of drugs, and combinations of new and existing agents. As each evolves, so do toxicities. Nonetheless, the principles delineated in this work should be illustrative of the breadth of the biological underpinnings which drive adverse events associated with cancer treatment. Importantly, the commonality of many of the mechanisms and pathways which lead to toxicities emphasizes the need for scientific collaboration at both the discovery level and in the development of effective interventions.

We are grateful to our collaborators and contributors and to the editorial staff at Springer for helping to bring this work to fruition.

Boston, MA, USA Stephen T. Sonis
Adelaide, Australia Dorothy M. Keefe

Contents

Part I
General Concepts Affecting Regimen-Related Toxicities

Epidemiology and Outcomes of Regimen-Related Toxicities

Linda S. Elting, Yu-Chia Chang, and Pratibha Parelkar

Abstract Treatment-related toxicities are common among patients with cancer; they lead to poor clinical outcomes, reduced functional status, higher consumption of healthcare resources, and excess costs. The toxicity, its risk and outcomes, varies depending on the regimen received. In this chapter, we describe the epidemiology of toxicities commonly observed among patients receiving current treatment regimens.

Based on the numerous descriptions of the epidemiology of a single treatment-related toxicity in the literature, one might assume that toxicities occur in isolation. However, in practice, it is the rare patient who experiences toxicities one at a time. Typically, regimen-related toxicities occur in concert, a result of common pathogenesis, similar timing, or causation (Fig. 1).

In some cases, toxicities are related only temporally. For example, both hypersensitivity reactions and nausea and vomiting occur immediately after initiation of some chemotherapy regimens, yet they are considered clinically unrelated. In contrast, neutropenia and anemia occur simultaneously because they share a common, pathobiologic cause, damage to bone marrow causing suppression of blood cell formation. Finally, neutropenia and infection occur together because they are linked causally; neutropenia increases the risk of infection.

As can be seen in Fig. 2, these relationships complicate the epidemiology of toxicities as well as the computation of the costs of such toxicities [1]. Thus, for ease of discussion, we group toxicities based on the temporal relationship between their onset and treatment administration. Causal relationships also are discussed. The pathobiology of toxicities is the topic of the remaining chapters in this volume.

L.S. Elting, Dr. P.H. (✉) • Y.-C. Chang, M.P.H. • P. Parelkar, M.P.H.
The Section of Health Services Research, Department of Biostatistics, Unit 1411,
The University of Texas M.D. Anderson Cancer Center, PO Box 301402,
Houston, TX 77230-1402, USA
e-mail: lelting@mdanderson.org

S.T. Sonis and D.M. Keefe (eds.), *Pathobiology of Cancer Regimen-Related Toxicities*,
DOI 10.1007/978-1-4614-5438-0_1, © Springer Science+Business Media New York 2013

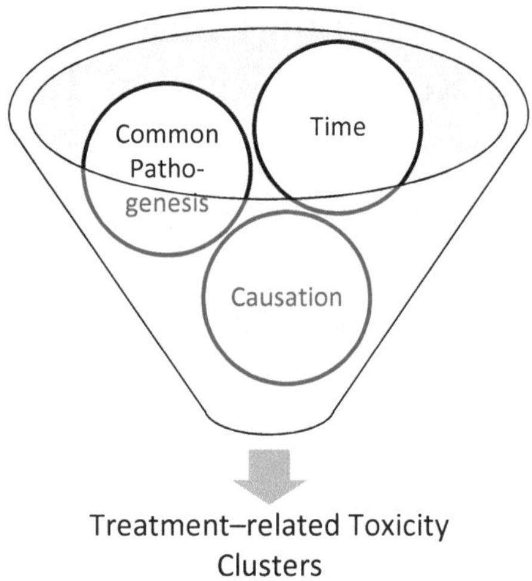

Fig. 1 Treatment-related toxicity clusters

Immediate Toxicities

In the absence of preventive medications or altered infusion schedules, hypersensitivity and infusion reactions occur commonly during administration of chemotherapy, although the incidence of serious reactions is low. Hypersensitivity reactions occur as a result of at least three mechanisms [2]. Some of these reactions (most notably those induced by platinum compounds) are IgE-mediated anaphylaxis and, thus, typically develop after multiple cycles of therapy. Non-IgE antibody-mediated reactions drive the development of anaphylactoid reactions to taxanes and to some monoclonal antibodies. These reactions typically develop during the first infusion, often during the first 10 min of the infusion. Finally, the cytokine release syndrome occurs as a result of massive release of cytokines, within an hour or two of infusion of monoclonal antibodies, particularly rituximab. With the exception of their timing, these reactions are clinically similar, being characterized by urticaria, flushing, rash, pruritus, hypotension, angioedema, dyspnea, and bronchospasm. An interesting distinguishing feature between antibody-mediated and cytokine release reactions is the absence of wheezing in the latter case [2].

The incidence of hypersensitivity reactions to platinum compounds is approximately 12–19% [3] while reactions occur among approximately 8–45% of paclitaxel recipients and 5–20% of docetaxel recipients [4]. Depending on the specific agent, reactions to monoclonal antibodies may be higher, occurring among 40% of trastuzumab recipients, as many as 77% of rituximab recipients, 16–19% of cetuximab recipients, but only 5% of panitumumab recipients [2, 5]. The incidence of serious hypersensitivity reactions is low (1–5%) for all agents with the possible exception of rituximab, which may approach 10%. Premedication with corticosteroids, diphenhydramine, H_2 antagonists, or acetaminophen as well as reduction of the rate of infusion has been effective in reducing the incidence of infusion reactions.

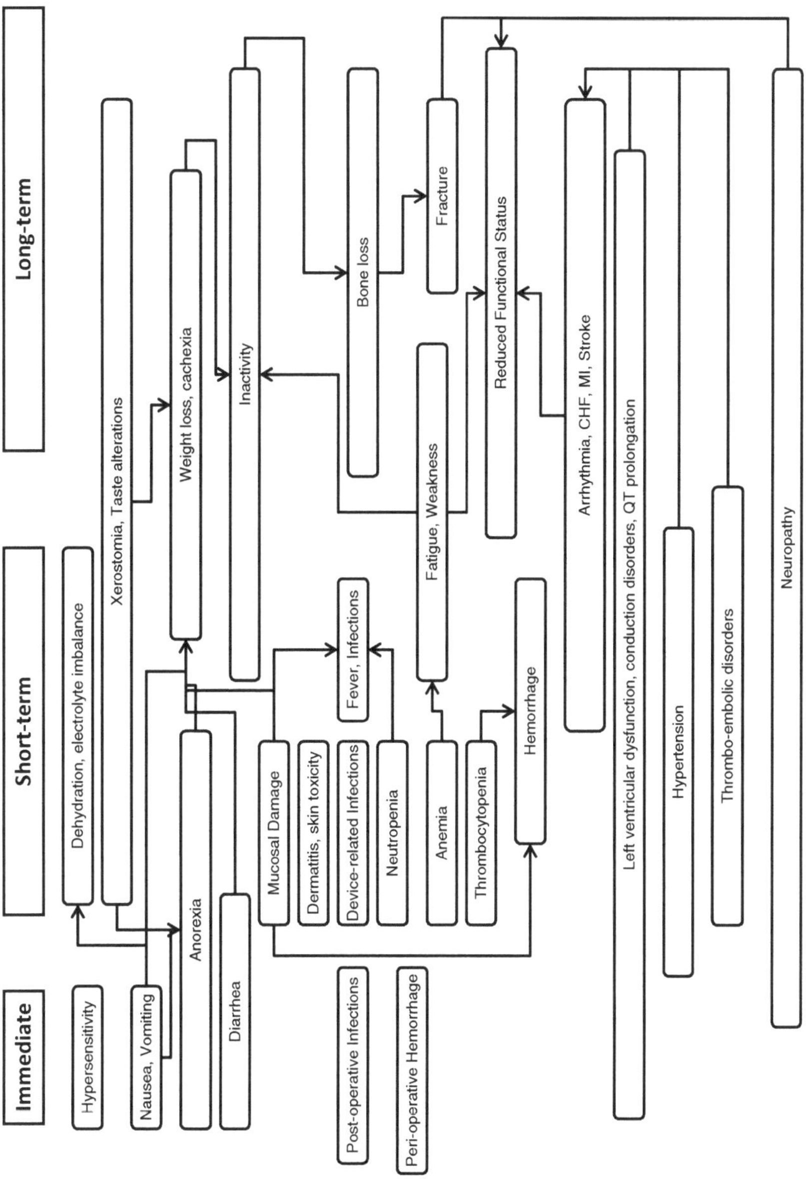

Fig. 2 Temporal and causal relationships among toxicities

Chemotherapy-induced nausea and vomiting may develop acutely, within a few minutes or hours of infusion, or the onset may be delayed 24 or more hours after the infusion. The incidence, duration, and severity of nausea and vomiting depend on the type, dose, and schedule of the chemotherapy agent as well patient factors such as age, sex, and history of alcohol use or motion sickness. Chemotherapy agents are classified by the expected incidence of vomiting and these classifications guide prophylaxis. Highly emetogenic agents (high doses of cisplatin, cyclophosphamide, doxorubicin) cause vomiting in 90% or more of patients. Moderately emetogenic agents (carboplatin, cytarabine, oxaliplatin, and lower doses of cisplatin, cyclophosphamide, and doxorubicin) cause vomiting in 30–90%, low emetic risk agents (docetaxel, paclitaxel, etoposide, gemcitabine) cause vomiting in 10–30%, and minimal emetic risk agents (decitabine, monoclonal antibodies) in fewer than 10% of patients [6–8]. Radiation induces nausea and vomiting in approximately 30–90% of patients when the upper abdomen is targeted and more than 90% of patients who receive total body irradiation prior to bone marrow transplantation [9]. Modern antiemetic prophylaxis is very effective and reduces the rate of vomiting by 70–75% although it is somewhat less effective in the case of delayed nausea and vomiting [9].

Other than their impact on quality of life, nausea and vomiting also may result in anorexia, weight loss, dehydration, and electrolyte imbalance. These outcomes sometimes are associated with unplanned clinic visits and occasionally with emergency department visits and hospitalization [10].

The neurotoxicity that is associated with chemotherapeutic agents is typically cumulative and dose dependent and thus develops after several cycles of chemotherapy. However, acute neurotoxicity, often developing immediately after an infusion, is observed in the majority (80–95%) of patients who receive oxaliplatin [11–13]. Typically, the neurotoxicity is grade 1 or 2 and in most cases, this is reversible within a week.

Short-Term Toxicities

We classify as "short-term" those toxicities that arise within 1–2 weeks of therapy and resolve by the end of therapy. Chemotherapy, radiation, and immunotherapies have profound effects on the entire gastrointestinal mucosa, from the oral cavity to the rectum. These effects are manifest as oral mucositis, esophagitis, gastritis, and diarrhea. In some cases this injury occurs as a direct result of injury to the tissue by radiation. However in many cases mucosal injury occurs systemically as a result of induction and amplification of cytokine pathways [14].

The risk of oral mucositis varies depending on the chemotherapy agents received as well as the concurrent use of radiation. Mucositis occurs in virtually all patients who receive radiation to the head and neck; among 80% of patients it is reported to be severe [15, 16]. Among patients who receive total body irradiation prior to transplantation, the risk exceeds 80% with more than 60% experiencing severe symptoms [14]. Among patients with solid tumors who receive myelosuppressive therapy, oral mucositis develops during approximately 30% of chemotherapy cycles [17]. Severe oral mucositis occurs in approximately 8–15% of patients who receive chemotherapy alone and 25–50% of patients who receive concurrent radiation therapy [14]. Highest rates are observed when the gastrointestinal tract is the target of radiation.

Diarrhea is a common toxicity of cancer therapy; in some situations it is both serious and dose-limiting. Diarrhea develops typically 2–14 days after chemotherapy. The incidence of diarrhea with 5-fluorouracil-based regimens ranges between 10 and 50% (lower incidence rates are observed with infusional rather than bolus administration) and rates with irinotecan approach or exceed 80%. The incidence of serious, grade 3–4 diarrhea with these agents is approximately 10–30% [18, 19] with the highest rates observed when chemotherapy is administered alongside radiation for anorectal cancers [20]. Reports from clinical trials of chemotherapy regimens suggest that standard dose chemotherapy regimens for cancers other than colorectal cancer result in diarrhea in 10–20% of patients; 5–8% of these are grade 3–4 [21]. However it is likely that diarrhea is underreported in such trials. As for the newer targeted therapies, diarrhea has been reported in 12–30% of trastuzumab recipients [22, 23], 20% of sunitinib recipients [24, 25], 30% of bortezomib recipients [24, 26], 33% of sorafenib recipients [24, 27], 45% of imatinib recipients [24, 28], 40–60% of lapatinib recipients [23, 24, 29, 30], 40–60% of gefitinib recipients [24, 31, 32], and 55–66% of erlotinib recipients [24, 33, 34].

Like nausea and vomiting, diarrhea may lead to weight loss, dehydration, and electrolyte imbalance. Estimates of the risk of such outcomes vary. However a study of consecutive cases of severe (grade 3–4) diarrhea requiring hospitalization among patients with colorectal cancer showed parenteral fluid support in 87%, a death rate of 14%, reduction and/or delay of the next chemotherapy cycle in 37%, and a change in regimen or discontinuation of chemotherapy in 52% [35]. The average cost of care of these patients exceeded $8,000 (2004 Canadian dollars).

Chemotherapeutic agents cause a variety of reversible toxicities involving the skin and hair. Alopecia occurs in 2–50% of patients depending on the drugs and combinations received. It is particularly common among patients who receive doxorubicin [36, 37]. Acral erythema, also known as hand–foot syndrome, occurs in 6–42% of chemotherapy recipients, overall [36, 38], and up to 50% of capecitabine recipients [37]. Approximately 30% of patients who receive sorafenib and 19% of those who receive sunitinib experience this syndrome [39–41]. Nail toxicities, in particular, paronychia, occur in approximately 14% of patients who receive cetuximab or erlotinib, 25% of those who receive panitumumab, and up to 40% of those who receive taxanes [37, 41, 42].

Skin rashes occasionally accompany administration of most chemotherapy agents. They occur in up to 39% of gemcitabine recipients, 6–21% of recipients of topoisomerase 1 inhibitors, and up to 100% of patients who receive pemetrexed [37]. However, skin rash is the hallmark of targeted therapy, particularly EGFR inhibitors. Since EGFR is expressed in healthy epithelial tissues such as skin, it comes as no surprise that skin toxicity is common among patients who receive antineoplastic agents that target that receptor. Among patients who receive EGFR inhibitors, skin rash is associated with response to therapy. The most severe skin rashes occur among patients with the highest response rates and the longest survival times [43]. Reversible and mostly mild–moderate skin rashes occur in up to 90% of imatinib, cetuximab, and panitumumab recipients, 60–75% of erlotinib recipients, 40–50% of lapatinib and sorafenib recipients, 19% of sunitinib recipients, and 8–15% of bortezomib recipients [33, 37, 39–41, 44–46].

Many chemotherapy agents as well as some radiation regimens suppress or damage bone marrow function, leading to reductions in critical blood cell components including neutropenia, anemia, and thrombocytopenia. Blood loss associated with surgery leads to similar reductions in blood cells. These regimen-related toxicities may exacerbate preexisting hereditary anemia and anemia and thrombocytopenia caused by bone marrow infiltration by tumor or chronic bleeding associated with tumor necrosis.

Neutropenia following chemotherapy is of primary concern owing to its associated risk of fever and infection. Febrile neutropenia is an ominous clinical sign, may lead to hospitalization and excessive costs [16, 47], and is a dose-limiting toxicity of therapy. The risk of febrile neutropenia varies depending on the chemotherapy agent, dose, and schedule. Regimens associated with a high (>20%) or intermediate (10–20%) risk of febrile neutropenia include most multidrug regimens based on doxorubicin (with or without cyclophosphamide), taxanes, dacarbazine, and ifosfamide [48]. Most dose-dense regimens also are associated with high risk. Patient characteristics such as age greater than 65 years and bone marrow involvement with tumor also increase the risk of febrile neutropenia.

Because its causes are multifactorial, anemia is quite common among patients who are receiving treatment for cancer, affecting 30–90% of patients, depending on cancer diagnosis and treatment. The platinum-based multidrug regimens used for treatment of ovarian, lung, and head and neck cancers cause anemia in 60–90% of recipients, although the majority of patients experience grade 1–2 anemia in this setting [49–52]. In contrast, dose-intense regimens result in profound anemia in virtually all patients. There is epidemiologic evidence that the risk of anemia increases with the number of cycles of chemotherapy, suggesting a cumulative effect over time [53].

Anemia leads to a number of symptoms that have negative impacts on quality of life, but most notable among these is fatigue. In fact, some reports indicate that fatigue troubles patients more than pain, nausea, and vomiting [54, 55]. Cancer-associated fatigue differs from fatigue in the general population in that it is persistent, even after rest, and profoundly reduces patients' functional capacity, physical well-being, and cognitive ability [54, 56–58]. Reductions in functional capacity may have a negative economic impact, requiring professional household or caregiving assistance and inability to work, often for many months during cancer treatment. Anemia-related fatigue may recur with each episode of anemia, and while it typically improves with resolution of anemia it may persist for months after the end of treatment.

Chemotherapy-associated thrombocytopenia (defined as a platelet count <50,000/μL) is a common complication of treatment for hematologic malignancies, occurring in approximately 75% of patients [59, 60]. It is less common among patients receiving therapy for solid tumors. Thrombocytopenia has been reported in approximately 11% of patients who receive platinum-based regimens, 5–6% of those who receive anthracycline-based regimens, 11% who receive gemcitabine-based regimens, 2% who receive taxane-based regimens, and 7% of all other solid tumor patients who receive chemotherapy [61].

Hemorrhage is the clinically significant outcome of thrombocytopenia. The risk of hemorrhage increases as the nadir of platelets decreases; the highest risk of bleeding is seen among patients with hematologic malignancies and a platelet count <10,000/μL. In this population, the risk of bleeding is approximately 15% although the risk of severe hemorrhage is <5% and fatal hemorrhages are very rare [62]. Patients with solid tumors display a similar pattern, with risk increasing as the platelet nadir decreases. However, in this population, a significant proportion of all serious hemorrhages occur at platelet counts higher than 50,000/μL, as a result of tumor necrosis [63]. Platelet transfusions may be used prophylactically to reduce the risk of severe hemorrhage as well as during episodes of bleeding. Patients who experience episodes of hemorrhage consume significantly more healthcare resources at a significantly higher cost than those without such episodes [63, 64].

Hemorrhage and its attendant outcomes and costs also occur as a direct result of agents that inhibit vascular endothelial growth factor (VEGF). Most notable among these is bevacizumab, which is associated with a 30–40% risk of all-grade hemorrhage and a 3–5% risk of grade 3–5 hemorrhage [65, 66]. Lower rates of all-grade (17%) and grade 3–5 (2%) hemorrhage are seen with sorafenib and sunitinib [67].

A number of reversible cardiac abnormalities are induced by chemotherapy agents. Reversible, left ventricular dysfunction (LVD) is observed rarely, in fewer than 1% of anthracycline recipients, immediately following the infusion [68]. Cyclophosphamide induces LVD approximately 1–10 days after infusion among 7–28% of patients. Although the acute LVD may be life-threatening in a proportion of these individuals, among those who survive therapy, ventricular function may return to baseline [69]. Taxanes induce LVD among 2–15% of recipients while reversible LVD is observed among 10–20% of sunitinib and sorafenib recipients [68]. The addition of trastuzumab doubles the risk of reduced left ventricular ejection fraction compared with doxorubicin-based regimens. This effect is not dose-dependent and with appropriate clinical management and discontinuation of the agent, this toxicity is reversible [68, 70].

Hypertension commonly accompanies a diagnosis of cancer and its incidence is increased by a number of agents [71]. Rarely, hypertensive crises, stroke, or reversible posterior leukoencephalopathy ensues, but in the majority of cases treatment-induced hypertension is managed successfully with medical intervention [68, 70, 71]. Hypertension is a hallmark of targeted therapies as it is caused by inhibition of the drugs' targets (i.e., VEGF) [72–74]. The incidence of clinically significant (grade 3 or 4) hypertension ranges from 7 to 47% during therapy with drugs that target VEGF (i.e., bevacizumab, sorafenib, sunitinib). Cisplatin also causes hypertension less frequently, perhaps as a result of renal toxicity [68, 70, 74].

Prolongation of the QT interval, occasionally leading to arrhythmias such as torsades de pointes, is observed in up to 60% of patients who receive arsenic trioxide [68]. In the absence of development of severe sequelae, QT intervals return to baseline following discontinuation of therapy. Medical management is typically adequate to ensure safe administration of arsenic trioxide. Prolongation of QT interval is also seen among patients treated with anthracyclines although clinically significant arrhythmias are usually reversible [74].

Long-Term Toxicities

We classify as long-term those toxicities that persist or develop after discontinuation of therapy.

Although reversible LVD occurs as described above, some agents are associated with chronic and nonreversible heart failure. Between 1.6% and 5% of anthracycline recipients develop chronic, progressive heart failure. The onset may be during therapy or more than a year following therapy. The incidence is dose related, approaching 45% at cumulative doses exceeding 550 mg/M2. This toxicity is not reversible and is potentially fatal. Trastuzumab doubles or triples the risk of heart failure to approximately 2–4% [68, 74], although rates as high as 27% were reported in early studies that enrolled patients with reduced baseline ejection fraction and included concomitant doxorubicin therapy [75]. Other targeted agents occasionally cause heart failure, with higher rates (8–10%) seen with sunitinib and sorafenib therapy and lower rates (1–4%) with bevacizumab, dasatinib, and imatinib therapy [68, 70, 74]. In contrast to other targeted agents, lapatinib is associated with a low risk of cardiac toxicity; some authors have hypothesized that its activation of AMPK protects cardiac tissue from damage induced by TNFa and other cytokines [72–74]. Heart failure is among the most costly conditions in the United States, resulting in incremental annual costs of approximately $2,000–5,500 [76, 77].

Asymptomatic ischemic changes on electrocardiogram occur in as many as 68% of patients who receive therapy with fluorouracil or capecitabine. Symptomatic ischemia occurs in 3–9% of patients. Cardiac symptoms and events typically occur within 5 days of treatment initiation, particularly among those who receive high doses of the drugs [71, 74]. Taxane administration is associated with ischemic symptoms or events in 1–5% of recipients. Increased incidence of thrombosis, including myocardial ischemia related to arterial thrombosis, is observed among patients receiving anti-angiogenesis agents. These agents (bevacizumab, erlotinib, sorafenib, sunitinib) double the rate of myocardial ischemia resulting in rates of 1–5% [71, 74].

The risk of venous thrombosis (VTE) also is increased with the administration of bevacizumab (RR = 1.33), thalidomide (RR = 2.6), and lenalidomide [78, 79]. The rate of VTE approaches 12% with bevacizumab [66, 78, 80] and 19% when thalidomide or lenalidomide is given in combination with other anti-neoplastic agents [79–81]. Thromboprophylaxis reduces the rate of VTE among patients who receive thalidomide; however the risk of hemorrhage is increased with such agents [78, 80].

Chemotherapy-induced neurotoxicity occurs commonly during therapy and persists for weeks to months after therapy in many cases. With the exception of the case of acute oxaliplatin toxicity mentioned above, the risk of neurotoxicity is dose- or cumulative dose-related [12]. It is important to note that other conditions that are associated with aging also increase the risk, most notably diabetes, infections, medications, immunologic and neurologic disorders, and metabolic or nutritional deficiencies [82]. Thus, as many as 4–5% of adult cancer patients may be experiencing neuropathies prior to receiving chemotherapy [13]. These patients are at higher risk for serious and prolonged chemotherapy-induced neuropathies.

The risk, clinical course, and reversibility of neuropathies vary depending on the chemotherapy agents used and the dose. The risk of neuropathies with platinum compounds, taxanes, and thalidomide exceeds 50% and approaches 100% in some series. The vinca alkaloids, lenalidomide, and bortezomib also lead to significant risk of neuropathies, typically in 30–50% of patients [12, 13].

In some individuals neuropathies are transient or mild. However, neuropathies also may be dose-limiting, debilitating, and persistent. Neuropathies induced by platinum compounds or thalidomide not only persist for many months, but also continue to progress for months after treatment is discontinued [13, 83]. Thalidomide-induced neurotoxicity may persist for more than a year, a pattern also observed occasionally with vincristine-induced neuropathies [13]. Because of the long-term persistence of symptoms, patients whose chemotherapy regimens are changed from one neurotoxic agent to another may experience severe neuropathies as a result of the combined toxicity of multiple agents.

Although fatigue is often related to short-term treatment induced anemia, it also occurs in the absence of anemia, presumably, in many cases, as a result of activation and amplification of cytokine pathways by cancer therapies. Fatigue is reported by the majority of patients who receive radiation therapy and by most patients who receive chemotherapy. Unlike anemia-related fatigue, this fatigue may persist long after treatment is completed, resulting in a correspondingly prolonged and profound negative impact on patients' and their families' economic circumstances [84].

Xerostomia is a subjective sensation of dry mouth that characterizes salivary hypofunction. In most cases, it is caused by irradiating the salivary glands. Considering all head and neck radiation regimens together, the prevalence of xerostomia exceeds 90% during therapy with slightly lower rates 1–2 years and >2 years post radiation (77 and 85%, respectively) [85]. Radiation-induced xerostomia is not typically reversible. In contrast, chemotherapy-associated xerostomia is uncommon and typically resolves after chemotherapy is discontinued [85, 86]. Xerostomia leads to increased risk of oral infections and dental caries, reduced oral functioning leading to impaired daily functioning, and reduced quality of life, which lead, in turn to increased use of healthcare services and increased cost [87–92].

Osteonecrosis complicating cancer treatment occurs as a result of radiation, bisphosphonate use, and, perhaps, chemotherapy. Osteoradionecrosis develops among 3–10% of patients who receive head and neck radiotherapy; it may affect the maxilla or the mandible and may develop as long as 6 months—years after therapy [93]. The risk of osteoradionecrosis is increased with higher dosages of radiation, poor dental status, and post-radiation extraction of diseased teeth [93]. The use of spacers during radiation therapy appears to reduce the risk of osteoradionecrosis, although precise estimates of the risk reduction are lacking. Osteonecrosis of the jaw develops among approximately 6–13% of patients who receive bisphosphonates [94]. Higher rates (as high as 24%) are seen when two bisphosphonates are combined and when dental extraction are done. There are recent reports of increased risk of osteonecrosis of the jaw among patients who receive anti-angiogenic agents such as bevacizumab [95, 96].

Graft-versus-host disease (GVHD) is the major therapy limiting toxicity associated with hematopoietic cell transplantation and the primary cause of morbidity and mortality following allogeneic transplant. GVDH results from the brisk immune response of donor T-cells to HLA proteins on host cells [97, 98]. Even in the case where there is a full match between patient and sibling donor HLA cells, GVHD occurs among 40% of recipients. One-antigen mismatch transplants result in GVHD in 80% of recipients or more [98, 99]. Acute GVHD occurs in the first 100 days following transplant. It is characterized by rash, nausea, anorexia, diarrhea, and hyperbilirubinemia [98]. This may progress to chronic GVHD, or chronic GVHD may arise de novo, affecting the skin, oral cavity, gastrointestinal tract, liver, lungs, heart, and marrow. Patients are at very high risk for secondary infections and GVHD is a potentially lethal condition; cumulative incidence of non-relapse mortality is approximately 30% at 5 years [98, 100].

Discussion

Two important themes arise from an inspection of the epidemiology and outcomes of common treatment-related toxicities. First, while each condition requires careful evaluation and management individually, toxicities rarely occur alone. We have described their clustering in time as well as their causal relationships. However, it is notable how frequently their etiologies are described as immune-mediated or resulting from up-regulation of cytokine pathways. These commonalities point to the promise of prevention and management strategies based on common pathobiology, but also to the necessity of assessing the combined risks and costs of multiple simultaneous toxicities.

Second, we note with concern the high incidence of clinically significant, costly, and increasingly long-term toxicities of our most promising agents, those targeted to specific receptors. The incidence of hemorrhage, neuropathies, cardiac and non-cardiac vascular conditions is particularly high. These toxicities are very costly and in many cases, they are not reversible. Thus, while these new agents increase cancer control and cure rates, they do so at the price of long-term and costly side effects. This observation has important implications for individual patients and their clinicians as well as for policy makers.

Complete information about the risk of individual toxicities as well as the combined risks of toxicities is critical to clinical decisions that balance both the benefits and harms of treatment. Knowing that a treatment confers a 5% risk of heart failure and a 15% risk of hypertension is not helpful. Does this mean a 20% risk of any side effect or is the overall risk only 15%? Are serious outcomes more common when multiple toxicities are present? Further enumeration of individual and combined risks must be a research priority.

Of equal importance is the burden of these toxicities on society, particularly on already strained healthcare services and limited healthcare dollars. For example, heart failure is among the most costly and resource intensive conditions among elderly patients, accounting for an incremental cost to Medicare of approximately $2,000–5,500 per patient per year, double the rate of hospitalizations and double the number of intensive care days [76, 77]. The incremental costs of hypertension are about half those of heart failure, but the condition is far more common following cancer therapy. If cancer survivors, who currently number more than 15,000,000 in the United States, develop hypertension and heart failure at higher rates or at earlier ages than the general population, the impact on the healthcare delivery and financing systems will be huge.

Modern treatment regimens are associated with increasingly higher cancer control and cure rates. Currently, the hope for long-term cancer control must be balanced against the risk for long-term toxicity. Research aimed at prevention and management of these toxicities must be a high priority.

References

1. Weiner MG, et al. Estimating the costs of chemotherapy-associated adverse event clusters. Health Serv Outcomes Res Method. 2007;7(1–2):1–21.
2. Maggi E, Vultaggio A, Matucci A. Acute infusion reactions induced by monoclonal antibody therapy. Exp Rev Clin Immunol. 2011;7(1):55–63.
3. Makrilia N, et al. Hypersensitivity reactions associated with platinum antineoplastic agents: a systematic review. Met Based Drugs 2010;2010.
4. Lee C, Gianos M, Klaustermeyer WB. Diagnosis and management of hypersensitivity reactions related to common cancer chemotherapy agents. Ann Allergy Asthma Immunol. 2009;102(3):179–87 [quiz 187–9, 222].
5. Lenz HJ. Management and preparedness for infusion and hypersensitivity reactions. Oncologist. 2007;12(5):601–9.
6. Hesketh PJ. Chemotherapy-induced nausea and vomiting. N Engl J Med. 2008;358(23): 2482–94.
7. Hesketh PJ, et al. Proposal for classifying the acute emetogenicity of cancer chemotherapy. J Clin Oncol. 1997;15(1):103–9.
8. Grunberg SM, et al. Evaluation of new antiemetic agents and definition of antineoplastic agent emetogenicity—state of the art. Support Care Cancer. 2011;19 Suppl 1:S43–7.
9. National Compresensive Cancer Network (NCCN). Clinical practice guidelines in oncology (NCCN guidelines): antiemesis. Version 1.2012; 2011.
10. Shih YC, Xu Y, Elting LS. Costs of uncontrolled chemotherapy-induced nausea and vomiting among working-age cancer patients receiving highly or moderately emetogenic chemotherapy. Cancer. 2007;110(3):678–85.
11. Cassidy J, Misset JL. Oxaliplatin-related side effects: characteristics and management. Semin Oncol. 2002;29(5 Suppl 15):11–20.
12. Park SB, et al. Mechanisms underlying chemotherapy-induced neurotoxicity and the potential for neuroprotective strategies. Curr Med Chem. 2008;15(29):3081–94.
13. Stubblefield MD, et al. NCCN task force report: management of neuropathy in cancer. J Natl Compr Canc Netw. 2009;7 Suppl 5:S1–26 [quiz S27–8].
14. Sonis ST, et al. Perspectives on cancer therapy-induced mucosal injury: pathogenesis, measurement, epidemiology, and consequences for patients. Cancer. 2004;100(9 Suppl): 1995–2025.

15. Elting LS, et al. Risk, outcomes, and costs of radiation-induced oral mucositis among patients with head-and-neck malignancies. Int J Radiat Oncol Biol Phys. 2007;68(4):1110–20.
16. Elting LS, et al. Patient-reported measurements of oral mucositis in head and neck cancer patients treated with radiotherapy with or without chemotherapy: demonstration of increased frequency, severity, resistance to palliation, and impact on quality of life. Cancer. 2008;113(10):2704–13.
17. Elting LS, et al. The burdens of cancer therapy. Clinical and economic outcomes of chemotherapy-induced mucositis. Cancer. 2003;98(7):1531–9.
18. Wadler S. Treatment guidelines for chemotherapy-induced diarrhea. Gastroenterol Endosc News 2004:29–33.
19. Maroun JA, et al. Prevention and management of chemotherapy-induced diarrhea in patients with colorectal cancer: a consensus statement by the Canadian Working Group on Chemotherapy-Induced Diarrhea. Curr Oncol. 2007;14(1):13–20.
20. Miller RC, et al. Acute treatment-related diarrhea during postoperative adjuvant therapy for high-risk rectal carcinoma. Int J Radiat Oncol Biol Phys. 1998;41(3):593–8.
21. Keefe DM, et al. Updated clinical practice guidelines for the prevention and treatment of mucositis. Cancer. 2007;109(5):820–31.
22. Al-Dasooqi N, et al. Trastuzumab induces gastrointestinal side effects in HER2-overexpressing breast cancer patients. Invest New Drugs. 2009;27(2):173–8.
23. Al-Dasooqi N, et al. HER2 targeted therapies for cancer and the gastrointestinal tract. Curr Drug Targets. 2009;10(6):537–42.
24. Loriot Y, et al. Drug insight: gastrointestinal and hepatic adverse effects of molecular-targeted agents in cancer therapy. Nat Clin Pract Oncol. 2008;5(5):268–78.
25. Motzer RJ, et al. Sunitinib in patients with metastatic renal cell carcinoma. JAMA. 2006;295(21):2516–24.
26. Fanucchi MP, et al. Randomized phase II study of bortezomib alone and bortezomib in combination with docetaxel in previously treated advanced non-small-cell lung cancer. J Clin Oncol. 2006;24(31):5025–33.
27. Escudier B, et al. Sorafenib for treatment of renal cell carcinoma: final efficacy and safety results of the phase III treatment approaches in renal cancer global evaluation trial. J Clin Oncol. 2009;27(20):3312–8.
28. Demetri GD, et al. Efficacy and safety of imatinib mesylate in advanced gastrointestinal stromal tumors. N Engl J Med. 2002;347(7):472–80.
29. Crown JP, et al. Pooled analysis of diarrhea events in patients with cancer treated with lapatinib. Breast Cancer Res Treat. 2008;112(2):317–25.
30. Geyer CE, et al. Lapatinib plus capecitabine for HER2-positive advanced breast cancer. N Engl J Med. 2006;355(26):2733–43.
31. Fukuoka M, et al. Multi-institutional randomized phase II trial of gefitinib for previously treated patients with advanced non-small-cell lung cancer (The IDEAL 1 Trial) [corrected]. J Clin Oncol. 2003;21(12):2237–46.
32. Herbst RS, et al. Gefitinib in combination with paclitaxel and carboplatin in advanced non-small-cell lung cancer: a phase III trial—INTACT 2. J Clin Oncol. 2004;22(5):785–94.
33. Shepherd FA, et al. Erlotinib in previously treated non-small-cell lung cancer. N Engl J Med. 2005;353(2):123–32.
34. Herbst RS, et al. TRIBUTE: a phase III trial of erlotinib hydrochloride (OSI-774) combined with carboplatin and paclitaxel chemotherapy in advanced non-small-cell lung cancer. J Clin Oncol. 2005;23(25):5892–9.
35. Dranitsaris G, Maroun J, Shah A. Estimating the cost of illness in colorectal cancer patients who were hospitalized for severe chemotherapy-induced diarrhea. Can J Gastroenterol. 2005;19(2):83–7.
36. Susser WS, Whitaker-Worth DL, Grant-Kels JM. Mucocutaneous reactions to chemotherapy. J Am Acad Dermatol. 1999;40(3):367–98 [quiz 399–400].
37. Payne AS, James WD, Weiss RB. Dermatologic toxicity of chemotherapeutic agents. Semin Oncol. 2006;33(1):86–97.

38. Baack BR, Burgdorf WH. Chemotherapy-induced acral erythema. J Am Acad Dermatol. 1991;24(3):457–61.
39. Chu D, et al. Risk of hand–foot skin reaction with sorafenib: a systematic review and meta-analysis. Acta Oncol. 2008;47(2):176–86.
40. Chu D, et al. Risk of hand–foot skin reaction with the multitargeted kinase inhibitor sunitinib in patients with renal cell and non-renal cell carcinoma: a meta-analysis. Clin Genitourin Cancer. 2009;7(1):11–9.
41. Balagula Y, Lacouture ME, Cotliar JA. Dermatologic toxicities of targeted anticancer therapies. J Support Oncol. 2010;8(4):149–61.
42. Roe E, et al. Description and management of cutaneous side effects during cetuximab or erlotinib treatments: a prospective study of 30 patients. J Am Acad Dermatol. 2006;55(3):429–37.
43. Burtness B, et al. NCCN Task Force Report: management of dermatologic and other toxicities associated with EGFR inhibition in patients with cancer. J Natl Compr Canc Netw. 2009;7 Suppl 1:S5–21 [quiz S22–4].
44. Jia Y, et al. Risk of skin rash associated with erlotinib in cancer patients: a meta-analysis. J Support Oncol. 2009;7(6):211–7.
45. Su X, et al. Risk of high-grade skin rash in cancer patients treated with cetuximab—an antibody against epidermal growth factor receptor: systemic review and meta-analysis. Oncology. 2009;77(2):124–33.
46. Lacouture ME, et al. Analysis of dermatologic events in patients with cancer treated with lapatinib. Breast Cancer Res Treat. 2009;114(3):485–93.
47. Lyman GH, Kuderer NM. Epidemiology of febrile neutropenia. Support Cancer Ther. 2003;1(1):23–35.
48. National Compresensive Cancer Network (NCCN). Clinical practice guidelines in oncology (NCCN guidelines): myeloid growth factors. Version 1.2011; 2011.
49. Groopman JE, Itri LM. Chemotherapy-induced anemia in adults: incidence and treatment. J Natl Cancer Inst. 1999;91(19):1616–34.
50. Knight K, Wade S, Balducci L. Prevalence and outcomes of anemia in cancer: a systematic review of the literature. Am J Med. 2004;116(Suppl 7A):11S–26.
51. Barrett-Lee PJ, et al. Independent risk factors for anemia in cancer patients receiving chemotherapy: results from the European Cancer Anaemia Survey. Oncology. 2006;70(1):34–48.
52. National Compresensive Cancer Network (NCCN). Clinical practice guidelines in oncology (NCCN guidelines): cancer- and chemotherapy-induced anemia. Version 2.2012; 2011.
53. Ludwig H, et al. The European Cancer Anaemia Survey (ECAS): a large, multinational, prospective survey defining the prevalence, incidence, and treatment of anaemia in cancer patients. Eur J Cancer. 2004;40(15):2293–306.
54. Curt GA, et al. Impact of cancer-related fatigue on the lives of patients: new findings from the Fatigue Coalition. Oncologist. 2000;5(5):353–60.
55. Vogelzang NJ, et al. Patient, caregiver, and oncologist perceptions of cancer-related fatigue: results of a tripart assessment survey. The Fatigue Coalition. Semin Hematol. 1997;34 (3 Suppl 2):4–12.
56. Cella D, et al. Fatigue in cancer patients compared with fatigue in the general United States population. Cancer. 2002;94(2):528–38.
57. Cella D, et al. The longitudinal relationship of hemoglobin, fatigue and quality of life in anemic cancer patients: results from five randomized clinical trials. Ann Oncol. 2004;15(6):979–86.
58. Hofman M, et al. Cancer-related fatigue: the scale of the problem. Oncologist. 2007;12 Suppl 1:4–10.
59. Kantarjian H, et al. The incidence and impact of thrombocytopenia in myelodysplastic syndromes. Cancer. 2007;109(9):1705–14.
60. Goldberg GL, et al. Clinical impact of chemotherapy-induced thrombocytopenia in patients with gynecologic cancer. J Clin Oncol. 1994;12(11):2317–20.

61. Wu Y, et al. Anemia and thrombocytopenia in patients undergoing chemotherapy for solid tumors: a descriptive study of a large outpatient oncology practice database, 2000–2007. Clin Ther. 2009;31(Pt 2):2416–32.
62. Schiffer CA, et al. Platelet transfusion for patients with cancer: clinical practice guidelines of the American Society of Clinical Oncology. J Clin Oncol. 2001;19(5):1519–38.
63. Elting LS, et al. Incidence, cost, and outcomes of bleeding and chemotherapy dose modification among solid tumor patients with chemotherapy-induced thrombocytopenia. J Clin Oncol. 2001;19(4):1137–46.
64. Havrilesky LJ, et al. Management of platinum-sensitive recurrent ovarian cancer: a cost-effectiveness analysis. Gynecol Oncol. 2007;107(2):211–8.
65. Hapani S, et al. Increased risk of serious hemorrhage with bevacizumab in cancer patients: a meta-analysis. Oncology. 2010;79(1–2):27–38.
66. Keefe D, et al. Noncardiac vascular toxicities of vascular endothelial growth factor inhibitors in advanced cancer: a review. Oncologist. 2011;16(4):432–44.
67. Je Y, Schutz FA, Choueiri TK. Risk of bleeding with vascular endothelial growth factor receptor tyrosine-kinase inhibitors sunitinib and sorafenib: a systematic review and meta-analysis of clinical trials. Lancet Oncol. 2009;10(10):967–74.
68. Curigliano G, et al. Cardiac toxicity from systemic cancer therapy: a comprehensive review. Prog Cardiovasc Dis. 2010;53(2):94–104.
69. Gottdiener JS, et al. Cardiotoxicity associated with high-dose cyclophosphamide therapy. Arch Intern Med. 1981;141(6):758–63.
70. Ederhy S, et al. Cardiac side effects of molecular targeted therapies: towards a better dialogue between oncologists and cardiologists. Crit Rev Oncol Hematol. 2011;80(3):369–79.
71. Yeh ET, Bickford CL. Cardiovascular complications of cancer therapy: incidence, pathogenesis, diagnosis, and management. J Am Coll Cardiol. 2009;53(24):2231–47.
72. Spector NL, et al. Activation of AMP-activated protein kinase by human EGF receptor 2/EGF receptor tyrosine kinase inhibitor protects cardiac cells. Proc Natl Acad Sci U S A. 2007;104(25):10607–12.
73. Azim H, Azim Jr HA, Escudier B. Trastuzumab versus lapatinib: the cardiac side of the story. Cancer Treat Rev. 2009;35(7):633–8.
74. Senkus E, Jassem J. Cardiovascular effects of systemic cancer treatment. Cancer Treat Rev. 2011;37(4):300–11.
75. Slamon DJ, et al. Use of chemotherapy plus a monoclonal antibody against HER2 for metastatic breast cancer that overexpresses HER2. N Engl J Med. 2001;344(11):783–92.
76. Liao L, et al. Long-term costs and resource use in elderly participants with congestive heart failure in the Cardiovascular Health Study. Am Heart J. 2007;153(2):245–52.
77. Blecker S, et al. Hospice care and resource utilization in Medicare beneficiaries with heart failure. Med Care. 2011;49(11):985–91.
78. Nalluri SR, et al. Risk of venous thromboembolism with the angiogenesis inhibitor bevacizumab in cancer patients: a meta-analysis. JAMA. 2008;300(19):2277–85.
79. El Accaoui RN, Shamseddeen WA, Taher AT. Thalidomide and thrombosis. A meta-analysis. Thromb Haemost. 2007;97(6):1031–6.
80. Zangari M, et al. Thrombotic events in patients with cancer receiving antiangiogenesis agents. J Clin Oncol. 2009;27(29):4865–73.
81. Bennett CL, et al. Thalidomide- and lenalidomide-associated thromboembolism among patients with cancer. JAMA. 2006;296(21):2558–60.
82. Hausheer FH, et al. Diagnosis, management, and evaluation of chemotherapy-induced peripheral neuropathy. Semin Oncol. 2006;33(1):15–49.
83. Siegal T, Haim N. Cisplatin-induced peripheral neuropathy. Frequent off-therapy deterioration, demyelinating syndromes, and muscle cramps. Cancer. 1990;66(6):1117–23.
84. Morrow GR, et al. Fatigue associated with cancer and its treatment. Support Care Cancer. 2002;10(5):389–98.

85. Jensen SB, et al. A systematic review of salivary gland hypofunction and xerostomia induced by cancer therapies: prevalence, severity and impact on quality of life. Support Care Cancer. 2010;18(8):1039–60.
86. Jensen SB, et al. Adjuvant chemotherapy in breast cancer patients induces temporary salivary gland hypofunction. Oral Oncol. 2008;44(2):162–73.
87. Connor NP, et al. Impact of conventional radiotherapy on health-related quality of life and critical functions of the head and neck. Int J Radiat Oncol Biol Phys. 2006;65(4):1051–62.
88. Epstein JB, et al. Quality of life and oral function following radiotherapy for head and neck cancer. Head Neck. 1999;21(1):1–11.
89. Harrison LB, et al. Detailed quality of life assessment in patients treated with primary radiotherapy for squamous cell cancer of the base of the tongue. Head Neck. 1997;19(3):169–75.
90. List MA, et al. Performance and quality of life outcome in patients completing concomitant chemoradiotherapy protocols for head and neck cancer. Qual Life Res. 1997;6(3):274–84.
91. List MA, et al. Quality of life and performance in advanced head and neck cancer patients on concomitant chemoradiotherapy: a prospective examination. J Clin Oncol. 1999;17(3): 1020–8.
92. Nijdam WM, et al. Longitudinal changes in quality of life and costs in long-term survivors of tumors of the oropharynx treated with brachytherapy or surgery. Brachytherapy. 2008;7(4):343–50.
93. Peterson DE, et al. Osteoradionecrosis in cancer patients: the evidence base for treatment-dependent frequency, current management strategies, and future studies. Support Care Cancer. 2010;18(8):1089–98.
94. Migliorati CA, et al. A systematic review of bisphosphonate osteonecrosis (BON) in cancer. Support Care Cancer. 2010;18(8):1099–106.
95. Almazrooa SA, Woo SB. Bisphosphonate and nonbisphosphonate-associated osteonecrosis of the jaw: a review. J Am Dent Assoc. 2009;140(7):864–75.
96. Estilo CL, et al. Osteonecrosis of the jaw related to bevacizumab. J Clin Oncol. 2008;26(24): 4037–8.
97. Ferrara JL. Advances in the clinical management of GVHD. Best Pract Res Clin Haematol. 2008;21(4):677–82.
98. Ferrara JL, et al. Graft-versus-host disease. Lancet. 2009;373(9674):1550–61.
99. Flowers ME, et al. Comparative analysis of risk factors for acute graft-versus-host disease and for chronic graft-versus-host disease according to National Institutes of Health consensus criteria. Blood. 2011;117(11):3214–9.
100. Arora M, et al. Chronic GVHD risk score: a Center for International Blood and Marrow Transplant Research analysis. Blood. 2011;117(24):6714–20.

The Biological Basis for Differences in Normal Tissue Response to Radiation Therapy and Strategies to Establish Predictive Assays for Individual Complication Risk

Christian Nicolaj Andreassen

Abstract Cancer patients vary considerably in their normal tissue response to radiation therapy. If this variation could be taken into account in the treatment planning phase, the therapeutic strategy could be individualized accordingly. Therefore, the ability to predict individual normal tissue radiosensitivity has been a long sought goal in radiobiology. Initially, interest was taken in predictive assays based on the cellular or subcellular response to irradiation in vitro. More recently, the attempts have focused on possible associations between various types of genetic sequence variation and the risk of adverse reactions to radiation therapy. Many of these studies have been hampered by certain methodological shortcomings and so far no clinically applicable predictive assay has been developed. Nevertheless, recent advances in high-throughput genotyping, bioinformatics, and the formation of collaborative research groups provide unprecedented opportunities to investigate the molecular and genetic factors underlying interindividual differences in normal tissue complication risk.

Introduction

Radiation therapy represents a very important treatment modality in the management of cancer, either alone or as part of combined modality therapy [1]. Compared to surgery, radiation therapy has the advantage of being potentially organ and function preserving. Nevertheless, radiation therapy may lead to severe acute or late side effects [2]. Apart from causing pain and distress during treatment, acute normal tissue reactions can result in treatment interruptions or may develop into lasting

---

C.N. Andreassen, M.D., Ph.D. (✉)
Department of Oncology, Aarhus University Hospital, Aarhus, Denmark

Department of Experimental Clinical Oncology, Aarhus University Hospital, Aarhus, Denmark
e-mail: nicolaj@oncology.dk

S.T. Sonis and D.M. Keefe (eds.), *Pathobiology of Cancer Regimen-Related Toxicities*,
DOI 10.1007/978-1-4614-5438-0_2,

damage. Late normal tissue reactions may cause chronic disability and compromise quality of life. Consequently, radiation-induced normal tissue reactions constitute a dose limiting factor in radiation therapy. Typically, radiation therapy regimens are designed to ensure that the risk of severe adverse effects does not exceed 5–10%. This basically means that a small fraction of radiosensitive patients limits the dose that can be given to the entire patient population, though the majority of patients could potentially tolerate a higher dose. Due to a relatively steep dose response relationship for tumor control, even a modest dose escalation would translate into a substantial increase in the chance of cure [1].

It has been estimated that more than 80% of the variability in normal tissue radiosensitivity can be attributed to patient-related factors rather than stochastic effects [3]. If this variability could be taken into account in the treatment planning phase, the therapeutic strategy could be individualized accordingly. Patients being relatively sensitive to the effects of ionizing radiation could (when possible) be offered a treatment strategy that does not include radiation therapy whereas the resistant patients could be dose escalated to some extent. This would lead to a substantial improvement in therapeutic index. Consequently, the ability to predict normal tissue complication risk prior to therapy has been a long sought goal among scientists and a predictive assay for normal tissue tolerance has occasionally been referred to as the "Holy Grail of radiobiology" [4].

Cellular and Subcellular In Vitro Assays

During the 1990s, radiation biology was heavily influenced by the so-called target cell hypothesis. According to this, tumor response as well the development of normal tissue toxicity was assumed to be the result of depletion of certain target cells. Studies conducted in the 1980s had shown that fibroblasts from patients with extreme reactions to radiotherapy exhibited a higher degree of in vitro radiosensitivity compared to fibroblasts from normal donors. Inspired by this, efforts were made to establish a predictive assay based on in vitro irradiation of cultured skin fibroblasts, typically measured as the survival fraction at 2 Gy (SF2). The underlying assumption was that the fibroblast should be regarded the primary target cell with respect to the development of fibrosis. Initially, five smaller trials reported significant associations between in vitro radiosensitivity of fibroblasts and fibrosis risk among patients with a normal tissue response being in the (Gaussian) normal range. Nevertheless, two larger confirmatory studies published in 1998 and 2000 did not demonstrate any significant predictive value of this assay. Retrospectively, it seems quite obvious that the early positive studies were quite heterogeneous in terms of methodology and therefore difficult to compare directly. In addition, certain questions can be raised concerning the statistical testing [5].

Interest has also been taken in the development of rapid assays measuring cellular or subcellular damage such as DNA lesions, chromosome aberrations, or apoptosis after irradiation of lymphocytes. A classical example is the so-called alkaline comet assay. A relatively large prospective study has reported a significant association between lymphocyte radiosensitivity and normal tissue toxicity after radiation therapy for cervical cancer [6]. Such approaches are still being explored and indications have been provided that a correlation exists between lymphocyte radiosensitivity and risk of adverse effects after radiation therapy for various types of cancer [2]. These observations are quite interesting considering that the lymphocyte can hardly be regarded as a target cell for typical normal tissue reactions. Thus, the correlation is consistent with the existence of a common host factor (presumably genetic) that influences different aspects of radio-responsiveness. Nevertheless, despite possible correlations between certain cellular/subcellular reactions and clinical radiosensitivity, it has not been possible to develop an in vitro assay that is suitable for routine clinical use. First of all, the predictive value in terms of sensitivity and specificity has turned out to be insufficient. Furthermore, some of the assays (particularly those based on cultured cells) are too time consuming to be practically feasible [1, 7].

Genetic Germline Alterations as Predictors of Normal Tissue Radiosensitivity

In the late 1990s increasing interest was taken in the hypothesis that clinical normal tissue radiosensitivity is under genetic control and that normal tissue complication risk could be predicted from genetic analysis. This concept received support from the observation that patients suffering from certain rare genetic syndromes such as ataxia telangiectasia, Blooms syndrome, Fanconi's anemia, and Nijmegen breakage syndrome seem to experience devastating normal tissue reactions if treated with radiation therapy [2]. All these syndromes are related to mutations in genes involved in detection of DNA damage and initiation of DNA repair. Nevertheless, these syndromes characterized by Mendelian inheritance are extremely rare and probably of little relevance when addressing the average cancer patient. However, it was hypothesized that heterozygous carriers of pathogenic (truncating) mutations in ATM and BRCA 1 and 2 could constitute a radiosensitive subpopulation. This assumption was supported by the observation that cells from heterozygous carriers of ATM mutations exhibited cellular radiosensitivity that was intermediate compared to ataxia telangiectasia patients and normal controls. Even though heterozygous carriers of these mutations are quite rare in general population (below 1%), they may be more frequent among breast cancer patients due to their cancer predisposing effect. Nevertheless, a number of relatively small studies did not provide any indications that such genetic alterations are overrepresented among patients with excessive normal tissue reactions nor that carriers of truncating ATM or BRCA mutations exhibit a higher normal tissue complication risk than the average patient. Similarly, no obvious association was found between clinical normal tissue radiosensitivity and mutations in other DNA repair genes such as RAD50, RAD21, NBN, or MRE11A. Nevertheless, the limited statistical power of these investigations should be taken into account in the interpretation of the results. A few studies have provided preliminary indications that patients with certain rare single base substitutions in the ATM gene may be more prone to adverse effects after radiation therapy [7].

Hypotheses Concerning the Genetic Background of Normal Tissue Radiosensitivity

At the turn of the millennium, increasing efforts were made to unravel the genetics underlying a variety of different biomedical phenotypes. This impetus in research activity was to a large extent fueled by rapid progress in genotyping technology and the sequencing of the human genome [8]. As a result of these advances, a more comprehensive hypothesis about the genetics of normal tissue radiosensitivity could be formulated. This hypothesis can be encompassed in the following three paragraphs:

1. *Normal tissue radiosensitivity should be regarded as a complex trait dependent on the combined influence of sequence alteration in several genes.*
   The very observation that normal tissue radiosensitivity is characterized by a continuous spectrum rather than falling into distinct categories is per se indicative of a polygenic inheritance. The biological response to irradiation is presumably very complex and involves a multitude of different pathways and genes [1, 7, 9]. Given the widespread occurrence of sequence alterations in the genome, numerous loci with potential impact on clinical radiosensitivity are likely to exist [7].
2. *Single nucleotide polymorphisms (SNPs) may make up a proportion of the genetic background underlying differences in clinical normal tissue radiosensitivity.*
   One of the important insights provided by the sequencing of the human genome was that SNPs are widespread throughout the genome. An SNP is defined as a single base substitution in which the least common allele has an abundance above 1% in the general population. It is estimated that around 11 million SNPs exist throughout the human genome and that SNPs account for the vast majority (approximately 90%) of naturally occurring interindividual sequence variation within human populations. SNPs in coding regions that causes an amino acid substitution may alter protein function whereas SNPs in regulatory regions may affect gene expression/protein secretion rate. Consequently, SNPs have the potential to affect various phenotypes including risk of normal tissue complications after radiation therapy. This assumption is the cornerstone of the so-called common disease—common variant hypothesis according to which common sequence alterations (such as SNPs) make up a large proportion of the inherited predisposition to various common conditions [8]. However, it also seems likely that rare sequence alterations may impact normal tissue radiosensitivity [7].
3. *Some genetic alterations are expressed selectively through certain types of normal tissue reactions, whereas others exhibit a "global" impact on radiosensitivity.*
   Clinical studies addressing normal tissue radiosensitivity have indicated that the risk of different types of adverse reactions after radiation therapy are not associated with each other when corrections are made for differences in treatment characteristics and length of follow-up. For instance, using such an analytical approach, no association was found between risk of acute skin erythema and fibrosis [10], nor between telangiectasia and fibrosis after radiotherapy for breast cancer [11]. This seems to indicate that some genetic alterations have to be expressed selectively through certain types of normal tissue reactions [7].

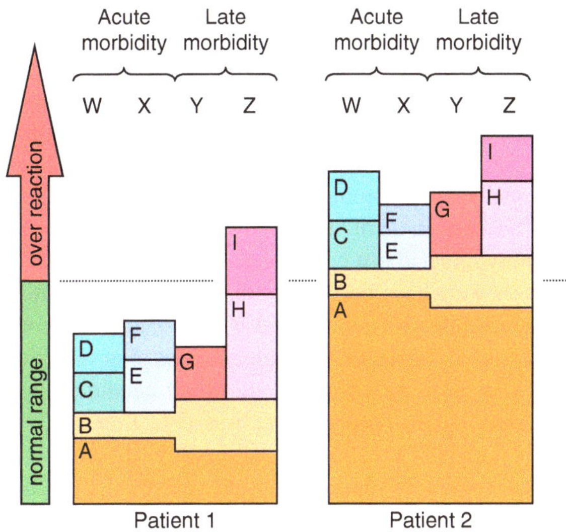

Fig. 1 A hypothetical and presumably very simplified model illustrating how different genetic alterations may influence clinical normal tissue radiosensitivity. The underlying assumption is that normal tissue complication risk is determined by the cumulative effect of a number of sequence variants (boxes A–I) that inflict a suboptimal function of gene products involved in the biological response to ionizing radiation. Some of these alterations (boxes C–I) are expressed selectively through certain types of normal tissue reactions, whereas others (boxes A and B) affect radiosensitivity in a generalized way. Patient 1 exhibits a selectively increased susceptibility to normal tissue reaction Z due to the variants H and I. In contrast patient 2 exhibits a severe "global" enhancement of radiosensitivity primarily due to a single sequence alteration (box A). Modified from Ref. [7] with permission

Nevertheless, patients suffering from one of the aforementioned "radiosensitive syndromes" seem to experience a general enhancement of clinical radiosensitivity apparently affecting both acute and late morbidity [2]. Based on this, it seems likely that some genetic alterations affect clinical radiosensitivity in a generalized manner whereas others exhibit a differential expression through different types of normal tissue reactions [7]. From a mechanistic perspective, these assumptions seem reasonable.

Over the last decade it has become increasingly clear that there is more to the development of radiation-induced normal tissue injury than just cellular death or survival (an insight which is at odds with the classical target cell theory). First, it was realized that the development of radiation-induced fibrosis was closely related to abnormal long-term tissue remodeling and dysregulation of certain cytokine cascades [1, 9]. In addition, it has now been recognized that also the development of acute reactions depend on the influence of certain inflammatory mediators [12]. Thus, the molecular pathways (and genes) underlying different types of normal tissue reactions probably differ substantially from each other, thereby giving way for a differential impact of certain genetic alterations [7]. These assumptions are graphically illustrated in Fig. 1.

SNPs and Normal Tissue Radiosensitivity

Within the last decade, the efforts to unravel the genetics of normal tissue radiosensitivity have primarily focused on SNPs. One reason for this is probably that such approach was in accordance with the abovementioned hypotheses including the quite influential "common disease—common variant hypothesis" [8]. Apart from that, the strategy certainly had an element of "looking for the ring under the light post." Rapid assays for SNP genotyping had just been commercially available and comprehensive public SNP databases were developed at the turn of the millennium.

More than 50 studies have been carried out to establish associations between selected SNPs and risk of normal tissue complications after radiation therapy [2, 4]. These studies have all been based on the so-called candidate gene approach [7] addressing genes involved in processes such as detection of DNA damage (i.e., ATM), DNA repair (i.e., XRCC1, XRCC3, and APEX) tissue remodeling (TGFB1 and TIMP) and scavenging of reactive oxygen species (i.e., SOD2 and GSTP1). More than 100 different genes have been investigated as part of this research. Remarkably, about 2/3 of these studies have reported significant associations between the assessed SNPs and various types of radiation-induced normal tissue reactions [4]. Unfortunately, this does not mean that a great deal of firm knowledge has been achieved. Generally, the findings have been hampered by conflicting results and lack of ability to replicate previous associations. For instance, of 13 studies investigating the impact of the XRCC1 codon 399 Arg/Gln SNP upon risk of various adverse reactions after radiation therapy, four have reported significant associations. Nevertheless, two of these studies demonstrated that the minority allele (Gln) was associated with enhanced complication risk whereas two studies found that the majority allele (Arg) was a risk factor [13]. Similar inconsistent findings have been obtained for many other SNPs [4]. Thus, none of the associations reported so far can by any reason be regarded as unambiguously proven.

This lack of conclusive evidence can presumably be ascribed to certain methodological problems related to most of the scientific work carried out until now. First of all, the studies have been very heterogeneous in terms of patient selection, tumor site, treatment regimens, and the assessed normal tissue endpoints [7]. Consequently the studies are difficult to compare directly. Furthermore, most studies have been relatively small with sample sizes between 25 and 778 subjects. Seventeen studies included <100 subjects and only nine studies had more than 400 patients. The median sample size of the studies was 144 [4]. This means that most studies have been severely underpowered to detect the impact of SNPs with only modest impact on normal tissue complication risk (as accounted for below, rather small effect sizes are probably what we should expect with SNPs). In that regard, it is thought-provoking that a study with 150 participants and a 1:2 distribution between high risk and low risk genotypes has a power of <30% to detect a 1.5-fold increase in complication risk from 20 to 30% ($\alpha=0.05$, two-tailed test). This lack of statistical power makes it particularly difficult to interpret negative studies. Furthermore, many of the studies investigated the impact of multiple SNPs and several different normal tissue reactions. In addition, subgroup analyses were occasionally conducted. Despite this, measures were rarely taken to counteract a "multiple testing problem." This results in a high risk of false positives by chance. For instance, in a study investigating the impact of seven SNPs upon two different normal tissue endpoints (corresponding to 14 independent comparisons) the risk of getting at least one positive finding by chance is above 50% assuming a 5% significance level in each comparison. In this context, it should be kept in mind that the human genome contains a very high number of SNPs and other sequence alterations. This presumably means that the prior probability that a given genetic variant is "truly associated" with the phenotype at interest may be very low. Given this assumption, the vast majority of positive associations reported are likely to be "false positives" using a typical 1% or 5% significance level. To put it simply, too many of the studies conducted so far have been designed in such a way that the probability of detecting the presence of a "true" association may be as low as 30% (in each comparison) whereas the risk of finding something that does not exist has been above 50% (in the entire study). This may very well explain why the results achieved so far have been rather inconsistent and conflicting [4].

TGFB1 SNPs and Risk of Late Toxicity: An Illustrative Case

TGFB1 is the gene of the versatile cytokine TGF-beta1. TGF-beta1 promotes the maturation of fibroblasts and stimulates the production of extracellular matrix proteins. TGF-beta1 is assumed to play a crucial role in the development of fibrosis and has been acclaimed "the master switch" of the processes that leads to radiation-induced fibrosis [7, 9]. Within the promoter region and starting sequence of the TGFB1 gene, a number of SNPs exist. Two of these SNPs, designated the position −509 T/C and codon 10 Leu/pro respectively, have gathered particular interest in the field of normal tissue radiobiology. A few smaller studies have previously reported that these polymorphisms may affect the risk of various fibrosis-related pathological conditions and influence the secretion rate of TGF-beta1 [7]. Consequently, these two SNPs represent obvious candidates for association studies addressing late normal tissue damage endpoints. Due to relatively strong linkage disequilibrium, the position −509 C and codon 10 pro (minority) alleles are often inherited together. Therefore, quite identical results are often achieved when the impact of these two SNPs are investigated [4].

Since 2003, a total of 15 studies have investigated the impact of one or both of these SNPs upon risk of various normal tissue reactions [4]. Most of these studies have been relatively small. Initially, relatively strong associations with risk of late toxicity were reported with regard to the position −509 C and/or codon 10 pro alleles. A combined analysis of data from two studies addressing changes in the breast after post-lumpectomy radiotherapy, demonstrated a threefold increase of toxicity risk in patients with the TGFB1 position −509 TT compared to those with the TC or CC genotype and an almost 15-fold increase when the TT genotype was

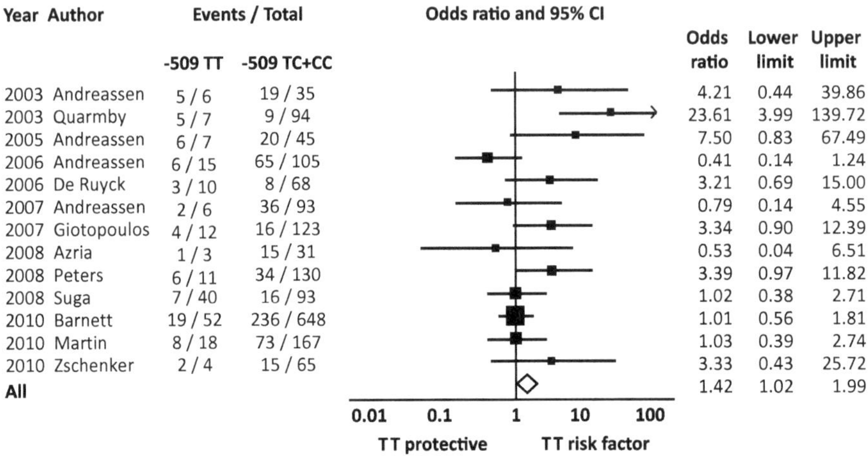

Fig. 2 Meta-analysis of studies addressing the impact of the TGFB1 position –509 C/T SNP upon risk of various late normal tissue reactions. The TT genotype was compared to the CT/CC genotype. The analysis indicates an enhanced risk of late toxicity in patients with the TT genotype (OR 1.42, 95% CI 1.02–1.99). Nevertheless, the distribution of the observations is highly suggestive of publication bias. Modified from Ref. [4] with permission

compared to the CC genotype [14]. Though not all studies have been positive, the TGFB1 position –509 and codon 10 SNPs have for almost a decade been regarded as some of the most promising SNP markers for prediction of radiation-induced late toxicity (particularly regarding endpoints that are assumed to have the development of fibrosis as a major underlying mechanism). Nevertheless, a more recent study with an accrual of almost 800 patients failed to demonstrate any association between the aforementioned SNPs and risk of various late reactions in the breast (including breast shrinkage and induration) after post-lumpectomy radiation therapy [15]. With the possible exception of a relatively short length of follow-up (2 years), the study was methodologically strong and by far the largest of its kind. Interestingly, the study was designed to provide 99% power to detect a threefold increase in toxicity risk and was fairly well powered to detect a 1.5-fold increase. This finding seriously questions the assumption that the assessed TGFB1 SNPs has a major impact on risk of radiation-induced fibrosis. A literature-based meta-analysis of studies addressing the impact of the position –509 SNP upon risk of late radiation-induced toxicity has recently been published [4]. The analysis summarized the influence of this SNP in 1,888 patients assessed for various late normal tissue effects after radiation therapy for different cancers. The meta-analysis demonstrated a modest enhancement of toxicity risk corresponding to an odds ratio of 1.42 with 95% confidence intervals not overlapping 1 (Fig. 2). This could of course be seen as an indication that the TGFB1 position –509 SNP does in fact inflict a smaller increment in late toxicity risk. Nevertheless, the distribution of the observations (small studies reporting high odds ratios, larger studies clustering around the line of equality and a relative absence of small studies reporting an inverse association) is highly suggestive of publication bias. Thus, despite substantial research efforts for almost a decade, it is in fact still difficult to conclude whether or not an association exists between the TGFB1 SNPs and risk of fibrosis-related late toxicity [4]. Consequently, the situation does in many ways resemble that when the predictive power of fibroblast survival after in vitro radiation was investigated in the 1990s [5].

Predictive Models Based on Multiple SNPs

In a number of studies, predictive models based on multiple SNPs have been established. Typically, the studies analyzed the association between normal tissue outcome and the number of so-called risk alleles harbored at the assessed polymorphic sites. Such an approach is indeed very appealing as it is completely in accordance with the assumption that normal tissue complication risk should be regarded as a complex trait dependent on the combined impact of sequence variation in several loci. Nevertheless, the models established so far should certainly be interpreted cautiously. First of all, many of the SNPs included in these models have yielded rather inconsistent results in studies investigating them individually [13]. Furthermore, certain methodological problems relate to these models. Often, the risk alleles (minority versus majority alleles) were not defined as risk alleles based on a biological hypothesis or based on findings in previous investigations. Instead, they were typically defined as risk alleles based on the plain observation that they were (occasionally non-significantly) associated with enhanced toxicity risk in the very study itself. This introduces some kind of circularity and the models therefore contain an inherent risk that coincidental statistical flukes are amplified into statistically significant results when the apparent high risk alleles are clustered together as part of the analysis. Thus, a significant finding in relation to the entire model cannot be used to indirectly verify nonsignificant associations for the individual SNPs. The importance of this potential problem is stressed by the fact that different alleles (majority versus minority alleles) were appointed as "risk alleles" in different studies. For instance, four independent studies have established models that had a number of SNPs in common (ATM codon 1853 Asp/Asn, TGFB1 codon 10 Leu/Pro, XRCC1 codon 399 Arg/Gln, and XRCC3 codon 241 Thr/Met). All four models seem to work convincingly with impressively low p-values. Nevertheless, a further analysis of the models reveals that the Arg allele in XRCC1 was appointed as the risk allele in two of the models whereas the Gln allele was appointed as the risk allele in two models. Similar inconsistencies were seen with regard to the other mentioned SNPs [4].

A few studies have utilized what could be referred as a "broad based candidate gene approach." These studies assessed several hundreds of SNPs and established predictive models based on the findings [2, 4]. In many of these investigations, the number of genotyped SNPs exceeded the number of participants which unavoidably leads to a "multiple test problem." Thus, independent validation of the results from such studies is of utmost importance.

Genome Wide Studies in Normal Tissue Radiobiology

The genome wide association study (GWAS) provides a radical alternative to the candidate gene approach. The GWAS makes use of the fact that SNPs cluster into haplotypes due to linkage disequilibrium. This means that the majority of the estimated 11 million SNPs that exist in the genome can be indirectly assessed by means of a microarray which genotypes around 250,000–1 million well-chosen "tagging SNPs." Thus, a GWAS provides the opportunity to conduct a hypothesis-free search for SNP associations without any need for a prior understanding of the biology underlying the phenotype of interest. Obviously, the GWAS has an inherent (severe) "multiple testing problem." In order to keep the risk of false positives at a reasonable level and still maintain statistical power, many GWASs have been based on very large patient cohorts occasionally approaching 50,000 subjects [8].

So far (October 2011), only one GWAS addressing normal tissue radiobiology has been published. This study investigated the risk of erectile dysfunction among 79 African-Americans treated with radiation therapy for prostate cancer [16]. The study utilized a microarray that genotyped around 900,000 SNPs. This investigation identified an SNP in the gene of the follicle-stimulating hormone receptor (FSHR) that was significantly associated with erective dysfunction with a *p* value of 5×10^{-8} corresponding to a Bonferoni corrected *p*-value of 0.023. Furthermore a predictive model was established based on the top-ranking four SNPs. These findings of course need independent validation. A few other relatively small GWASs addressing the risk of normal tissue toxicity after radiation therapy are expected to be published in the near future.

Lessons Learned in Other Scientific Fields

Within the last decade, substantial efforts have been made to unravel the genetics of various phenotypes that are assumed to have a complex polygenic background. Until 2006 these efforts were entirely based on a candidate gene approach. Generally, the studies carried out have been relatively unfruitful and despite thousands of publications only a very limited number of irrefutable associations have been established [8]. One major reason for this is probably that many of the studies have suffered from the same shortcomings (primarily lack of statistical power and multiple testing problems) as those carried out in the field of normal tissue radiobiology. Furthermore, these studies have only investigated very small proportion of the genetic variation that exists throughout the human genome.

Since 2006, more than 950 GWASs addressing various biomedical phenotypes have been carried out (updated lists are available at www.genome.gov/gwastudies) [17]. For at least two reasons, these studies represent an important breakthrough in the attempts to unravel the genetics of complex traits. Firstly, GWASs have dramatically increased the number of convincing SNP associations reported. Thus, they have been productive in a context where the candidate gene approach has generally yielded very limited success [8]. Secondly, GWASs shed important new light on the allelic architecture that may underlie most complex traits. The experiences gained in other scientific fields can be summarized as follows.

The Typical Impact of SNPs Is Presumably Relatively Small

It has been a common experience in most GWASs that the majority of SNPs that have been convincingly related to various biomedical traits only exhibited a modest impact on phenotype. Often, the effect size corresponded to an odds ratio around 1.2. Several of the GWASs were relatively well powered to detect associations with odds ratios above 1.5 but such associations were rarely reported [8, 17]. This probably implies that the typical impact of SNPs upon phenotype is rather small and emphasizes the need for well-powered studies.

The Candidate Gene Approach May Not Be of Much Use

Another striking finding in most GWASs is that many of the identified SNPs with an impact on phenotype were not located in any of the genes investigated as part of candidate gene studies or in genes involved in pathways assumed to be of major importance for the phenotype of interest. Often, the SNPs were located in noncoding sequences without any known function. This experience seriously questions the value of the candidate gene approach [4, 8].

There Is Probably (Much) More to Complex Trait Genetics Than SNPs

The genetic backgrounds of a number of different traits (i.e., cancer risk) have been quite extensively investigated using GWASs. Regardless of that, the identified SNPs affecting phenotype typically only accounted for a rather limited proportion of the expected heritable contribution to trait variance (often around only 5–25%). This can probably to some extent be attributed to methodological issues. Despite very large study cohorts, most GWASs provided limited statistical power to detect SNPs conferring odds ratios below 1.2. Thus, numerous "low impact" SNPs may easily have been missed. Furthermore, many of the utilized SNP genotyping platforms offered limited coverage for SNPs with population frequencies below 5–10%. Nevertheless, the abovementioned observation first of all implies that other types of sequence variants than SNPs (e.g., copy number variants, translocations, and inversions) probably have to constitute a substantial proportion of the genetic background underlying many complex traits. It is likely that rare genetic alterations (that are not

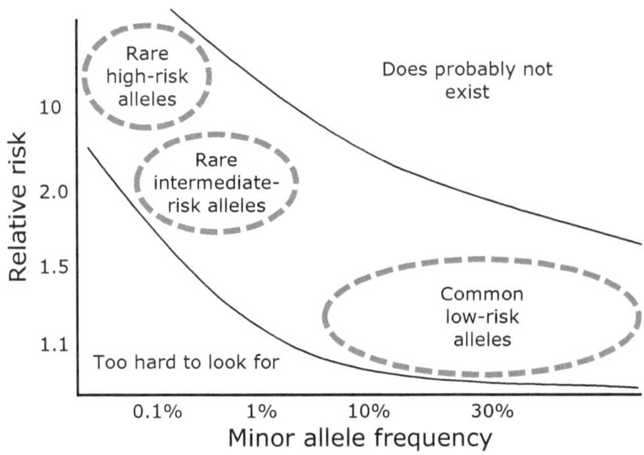

Fig. 3 A proposed model for the relationship between frequency and relative risk for genetic variants associated with normal tissue toxicity after radiotherapy. The model is inspired from studies addressing various phenotypes such as breast cancer risk. Modified from Ref. [13] with permission

captured by the current generation of GWASs) are of major importance with regard to complex trait genetics [17]. This recognition is somewhat at odds with the aforementioned rather influential "common disease—common variant hypothesis." To conclude, the current evidence is consistent with the assumption that polygenic/complex traits are dependent on a spectrum ranging from common low penetrance alleles to rare alterations with a more dramatic impact on phenotype (Fig. 3) [13].

How to Proceed?

The experiences gained from studies addressing the genetics of various complex traits should probably be taken into account in the future attempts to unravel the genetics if normal tissue complication risk after radiation therapy. As indicated above, a radical change in research strategy is probably needed if substantial progress should be made [4]. Statistical power and rigorous statistical testing are issues of immense importance. Under the assumption that each SNP only affects phenotype slightly (e.g., corresponding to an odds ratio around 1.2–1.5) future candidate gene studies should be based on thousands (rather than tens or hundreds) of subjects in order to provide precise risk estimates (Fig. 4). Furthermore, careful correction for multiple testing and validation of previous results are needed. Given the limited success of the candidate gene approach, GWASs are certainly warranted in normal tissue radiobiology. As mentioned previously, the GWASs currently planned are of limited sample size (a few hundred subjects). Thus, these are only powered to detect very strong SNP associations (when the mandatory correction is made for multiple

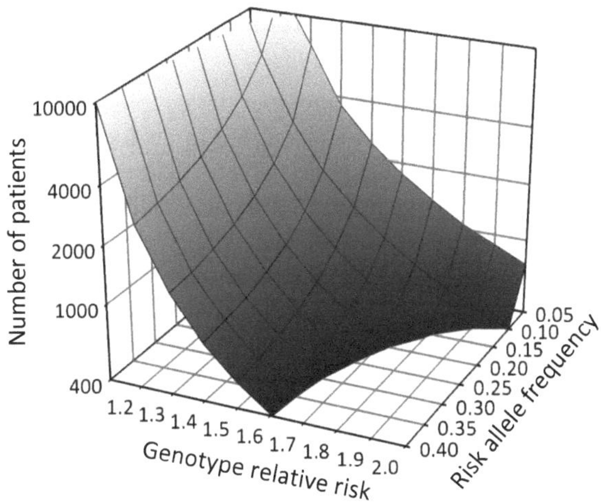

Fig. 4 An example of a power calculation showing the number of patients needed to obtain 80% power to detect genetic variants with risk allele frequencies ranging from 0.05 to 0.40 and relative risks from 1.1 to 2.0. Modified from Ref. [13] with permission

testing). Therefore, it seems quite likely that future GWASs will need to be expanded considerably with regard to sample size. In addition, novel catalogues of "low-frequency genetic variants" (i.e., those with minor allele frequencies around 0.5–5%) and high-density microarrays will enable the exploration of sequence alterations in the "sub-polymorphic" frequency range. However, such experiments also call for very large patient cohorts as the number of patients needed to obtain sufficient statistical power increases dramatically with decreasing minor allele frequencies [17] (Fig. 4). Under the assumption that rare alterations (arbitrarily defined as those with an abundance below 0.5%) constitute a substantial proportion of the genetic variation underlying differences in normal tissue radiosensitivity, SNP-based assays most likely need to be complemented by complete sequencing. Technologies that will enable cost-efficient high-throughput sequencing are currently emerging. Nevertheless, numerous obstacles (including immense statistical and financial challenges) still need to be overcome before association studies based on genome wide sequencing become feasible [8, 17].

To summarize, it seems increasingly clear that it represents a massive undertaking to pursue a comprehensive understanding of the genetics assumed to underlie differences in normal tissue radiosensitivity [13]. However, it seems equally clear that advances in bioinformatics, genotyping technology, and novel insights into population genetics provide unprecedented opportunities dissect the molecular and genetic basis of normal tissue radiosensitivity. Nevertheless, to fully exploit these new possibilities cooperation will be essential. One of the major challenges in that regard will be to establish sufficiently large cohorts of patients that are well categorized with regard to treatment characteristics and normal tissue outcome. This should encourage the formation of international networks and consortia such as the RAPPER, Radgenomics, Gene-pare, ESTRO GENEPI, and the International Radiogenomic Consortium [2, 4].

Should Alternative Strategies Be Considered?

As accounted for in the previous paragraphs, the genome has some inherent characteristics that makes it challenging to deal with (the number of variants to choose from is immense and many genetic determinants are likely to be either rare or of limited phenotypic impact). This may favor a return to research in predictive assays based on phenotype rather than genotype. Analysis of gene expression profiles has proven a useful tool in various settings. A few recent studies have established predictive models for late normal tissue reactions derived from the transcriptional response to in vitro irradiation of fibroblasts or lymphocytes [18]. Typically, the genes included in these classifiers were involved in processes such as apoptosis, cell cycle arrest, reactive radical scavenging, and tissue remodeling. These findings, of course, need independent validation but may indicate that studies addressing the (radiation-induced) transcriptome may represent a rewarding alternative to studies addressing genetic germline sequence. Furthermore, this kind of experiment may (in addition to GWASs) provide new insights into molecular radiobiology [19]. A possible spinoff from such insights could be the identification of pathways that could serve as targets for pharmacological intervention against radiation-induced normal tissue damage [9, 12].

Conclusion

Over the last decades various efforts have been made to establish a predictive test for normal tissue radiosensitivity. During the last decade, these attempts have primarily focused on the study of certain types of genetic germline alterations. Despite this, a predictive assay applicable for clinical use has not been established. Nevertheless dramatic advances in molecular biology provide unique opportunities to pursue a more complete understanding of the biology and genetics underlying differences in normal tissue radiosensitivity. Hopefully, this will translate into improved treatment regimens for cancer patients.

References

1. Barnett GC, West CM, Dunning AM, Elliott RM, Coles CE, Pharoah PD, Burnet NG. Normal tissue reactions to radiotherapy: towards tailoring treatment dose by genotype. Nat Rev Cancer. 2009;9(2):134–42.
2. West CM, Barnett GC. Genetics and genomics of radiotherapy toxicity: towards prediction. Genome Med. 2011;3(8):52.
3. Safwat A, Bentzen SM, Turesson I, Hendry JH. Deterministic rather than stochastic factors explain most of the variation in the expression of skin telangiectasia after radiotherapy. Int J Radiat Oncol Biol Phys. 2002;52(1):198–204.

4. Andreassen CN. Searching for genetic determinants of normal tissue radiosensitivity—are we on the right track? Radiother Oncol. 2010;97(1):1–8.
5. Bentzen SM. From cellular to high-throughput predictive assays in radiation oncology: challenges and opportunities. Semin Radiat Oncol. 2008;18(2):75–88.
6. Ozsahin M, Crompton NE, Gourgou S, Kramar A, Li L, Shi Y, Sozzi WJ, Zouhair A, Mirimanoff RO, Azria D. CD4 and CD8 T-lymphocyte apoptosis can predict radiation-induced late toxicity: a prospective study in 399 patients. Clin Cancer Res. 2005;11(20):7426–33.
7. Andreassen CN. Can risk of radiotherapy-induced normal tissue complications be predicted from genetic profiles? Acta Oncol. 2005;44(8):801–15.
8. Altshuler D, Daly MJ, Lander ES. Genetic mapping in human disease. Science. 2008; 322(5903):881–8.
9. Bentzen SM. Preventing or reducing late side effects of radiation therapy: radiobiology meets molecular pathology. Nat Rev Cancer. 2006;6(9):702–13.
10. Bentzen SM, Overgaard M. Relationship between early and late normal-tissue injury after postmastectomy radiotherapy. Radiother Oncol. 1991;20(3):159–65.
11. Bentzen SM, Overgaard M, Overgaard J. Clinical correlations between late normal tissue endpoints after radiotherapy: implications for predictive assays of radiosensitivity. Eur J Cancer. 1993;29A(10):1373–6.
12. Sonis ST. Oral mucositis. Anticancer Drugs. 2011;22(7):607–12.
13. Andreassen CN, Alsner J. Genetic variants and normal tissue toxicity after radiotherapy: a systematic review. Radiother Oncol. 2009;92(3):299–309.
14. Giotopoulos G, Symonds RP, Foweraker K, Griffin M, Peat I, Osman A, Plumb M. The late radiotherapy normal tissue injury phenotypes of telangiectasia, fibrosis and atrophy in breast cancer patients have distinct genotype-dependent causes. Br J Cancer. 2007;96(6):1001–7.
15. Barnett GC, Coles CE, Burnet NG, Pharoah PD, Wilkinson J, West CM, Elliott RM, Baynes C, Dunning AM. No association between SNPs regulating TGF-β1 secretion and late radiotherapy toxicity to the breast: results from the RAPPER study. Radiother Oncol. 2010;97(1):9–14.
16. Kerns SL, Ostrer H, Stock R, Li W, Moore J, Pearlman A, Campbell C, Shao Y, Stone N, Kusnetz L, Rosenstein BS. Genome-wide association study to identify single nucleotide polymorphisms (SNPs) associated with the development of erectile dysfunction in African-American men after radiotherapy for prostate cancer. Int J Radiat Oncol Biol Phys. 2010;78(5):1292–300.
17. Juran BD, Lazaridis KN. Genomics in the post-GWAS era. Semin Liver Dis. 2011;31(2): 215–22.
18. Mayer C, Popanda O, Greve B, Fritz E, Illig T, Eckardt-Schupp F, Gomolka M, Benner A, Schmezer P. A radiation-induced gene expression signature as a tool to predict acute radiotherapy-induced adverse side effects. Cancer Lett. 2011;302(1):20–8.
19. Sonis S, Haddad R, Posner M, Watkins B, Fey E, Morgan TV, Mookanamparambil L, Ramoni M. Gene expression changes in peripheral blood cells provide insight into the biological mechanisms associated with regimen-related toxicities in patients being treated for head and neck cancers. Oral Oncol. 2007;43(3):289–300.

The Ionizing Radiation-Induced Bystander Effect: Evidence, Mechanism, and Significance

Edouard I. Azzam, Sonia M. de Toledo, Andrew L. Harris, Vladimir Ivanov, Hongning Zhou, Sally A. Amundson, Howard B. Lieberman, and Tom K. Hei

Abstract It has long been considered that the important biological effects of ionizing radiation are a direct consequence of unrepaired or misrepaired DNA damage occurring in the irradiated cells. It was presumed that no effect would occur in cells in the population that receive no direct radiation exposure. However, in vitro evidence generated over the past two decades has indicated that non-targeted cells in irradiated cell cultures also experience significant biochemical and phenotypic changes that are often similar to those observed in the targeted cells. Further, non-targeted tissues in partial body-irradiated rodents also experienced stressful effects, including oxidative and oncogenic effects. This phenomenon, termed the "bystander response," has been postulated to impact both the estimation of health risks of exposure to low doses/low fluences of ionizing radiation and the induction of second primary cancers following radiotherapy. Several mechanisms involving secreted soluble factors, oxidative metabolism, gap-junction intercellular communication, and DNA repair, have been proposed to regulate radiation-induced bystander effects. The latter mechanisms are major mediators of the *system* responses to ionizing radiation exposure, and our knowledge of the biochemical and molecular events involved in these processes is reviewed in this chapter.

E.I. Azzam (✉) • S.M. de Toledo
Department of Radiology, New Jersey Medical School Cancer Center,
University of Medicine and Dentistry of New Jersey, Newark, NJ 07103, USA
e-mail: azzamei@umdnj.edu; detolesm@umdnj.edu

A.L. Harris
Department of Pharmacology and Physiology, New Jersey Medical School,
University of Medicine and Dentistry of New Jersey, Newark, NJ 07103, USA
e-mail: aharris@umdnj.edu

V. Ivanov • H. Zhou • S.A. Amundson • H.B. Lieberman • T.K. Hei
Center for Radiological Research, Department of Radiation Oncology,
College of Physicians and Surgeons, Columbia University Medical Center,
New York, NY, USA
e-mail: vni3@columbia.edu; hz63@columbia.edu; HBL1@columbia.edu; tkh1@columbia.edu

S.T. Sonis and D.M. Keefe (eds.), *Pathobiology of Cancer Regimen-Related Toxicities*,
DOI 10.1007/978-1-4614-5438-0_3, © Springer Science+Business Media New York 2013

Introduction

The absorption of ionizing radiation by living cells can directly disrupt atomic structures, producing chemical and biological changes. It can also act indirectly through radiolysis of water, thereby generating reactive chemical species that may damage nucleic acids, proteins, and lipids [1] (Fig. 1). Together, the direct and indirect effects of radiation initiate a series of biochemical and molecular signaling events that may repair the damage, or culminate in permanent physiological changes or cell death [2].

Interestingly, the early biochemical modifications, which occur during or shortly after radiation exposure, were thought to be responsible for most of the effects of ionizing radiation in mammalian cells. However, oxidative changes may continue to arise for days and months after the initial exposure, presumably because of inflammatory responses [3, 4] and continuous generation of reactive oxygen (ROS) and nitrogen (RNS) species [5]. Remarkably, these processes occur not only in the irradiated cells but also in their progeny [2, 6–9]. Furthermore, radiation-induced oxidative stress may spread from targeted cells to non-targeted bystander cells

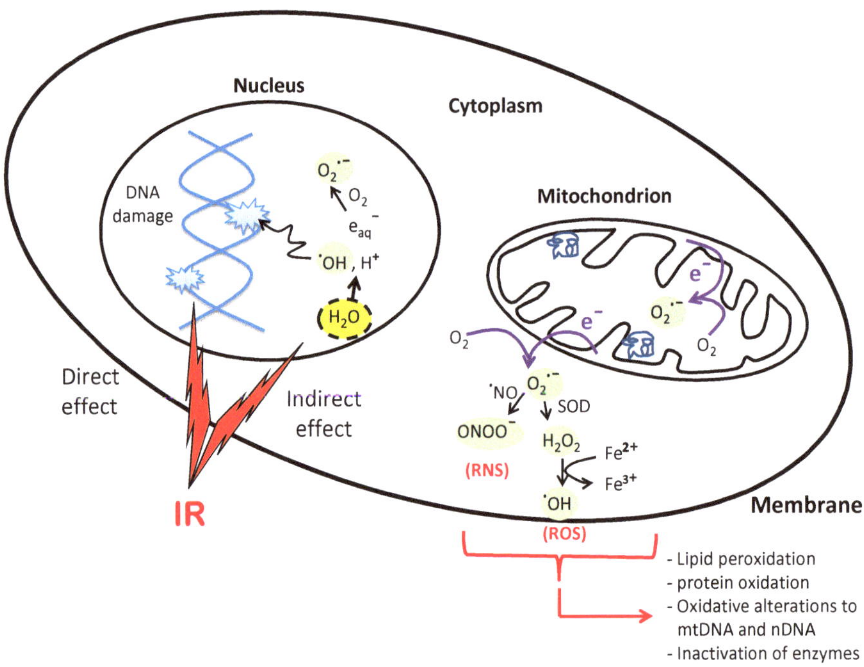

Fig. 1 The direct and indirect effects of ionizing radiation on cellular macromolecules. Absorption of ionizing radiation by living cells directly disrupts atomic structures, producing chemical and biological changes and indirectly through radiolysis of cellular water and generation of reactive chemical species by stimulation of oxidases and nitric oxide synthases. Ionizing radiation may also disrupt oxidative metabolism and other mitochondrial functions contributing to persistent alterations in lipids, proteins, nuclear DNA (nDNA), and mitochondrial DNA (mtDNA)

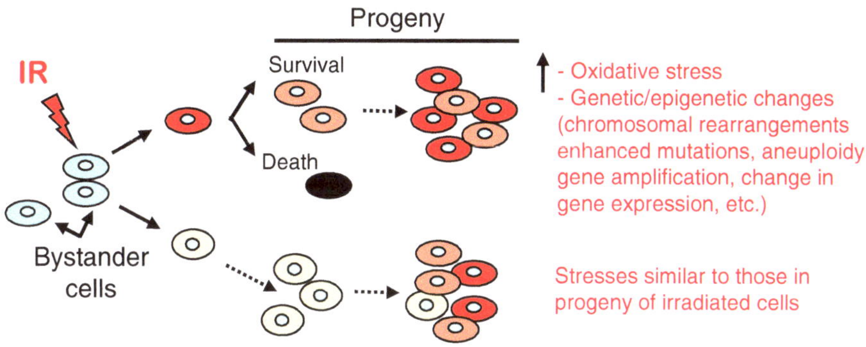

Fig. 2 Ionizing radiation (IR) induces targeted and non-targeted (bystander) effects. Communication of stress-inducing molecules from cells exposed to IR propagates stressful effects, including oxidative stress, as well as genetic and epigenetic changes, to the bystander cells and their progeny. The induced effects may be similar in nature to those observed in progeny of irradiated cells

through intercellular communication mechanisms (reviewed in [10–13]). The progeny of these bystander cells also experience perturbations in oxidative metabolism and exhibit a wide range of oxidative damages, including protein carbonylation, lipid peroxidation, and enhanced rates of spontaneous gene mutations as well as neoplastic transformation [14–16] (Fig. 2). The persistence of such stressful effects in progeny cells may have profound implications for long-term health risks, including the emergence of a second malignancy following radiotherapy treatments [17–20]. Understanding the mechanisms underlying non-targeted effects, together with those that mediate targeted effects, will be informative for counteracting adverse health effects caused by exposure to ionizing radiation, and may lead to formulation of countermeasures.

Ionizing Radiation Track Structure and the Nature of Induced Biological Effects

Strong evidence has shown that the magnitude and nature of radiation-induced bystander effects greatly depend on the biophysical properties of the impacting radiation. Thus, a review of bystander effects would be facilitated by a brief introduction to the different types of ionizing radiation and their energy deposition patterns.

Ionizing radiation is classified as either electromagnetic or particulate. Whereas X and γ rays belong to electromagnetic radiation, energetic electrons, protons, neutrons, α-particles, and heavy charged particles are different forms of particulate radiation [1]. Many of the damaging effects of radiation are due to the geometry of the physical energy deposition of the impacting radiation, referred to as

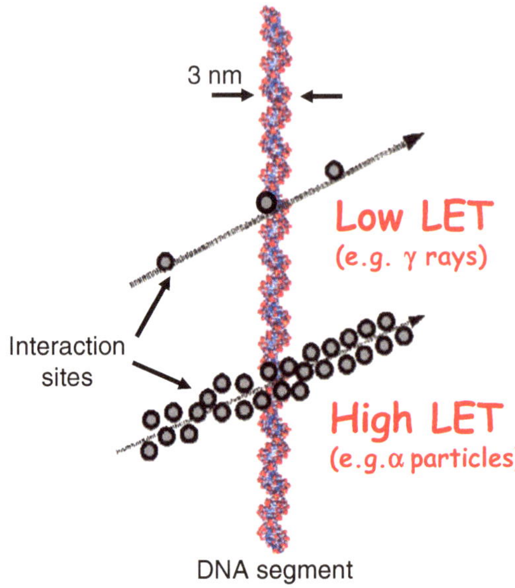

Fig. 3 Complexity of ionizing radiation-induced DNA damage. The complexity of DNA damage induced by ionizing radiation is highly dependent on the biophysical characteristics of the radiation, in particular its linear energy transfer (LET) properties

the track structure or linear energy transfer (LET) effects [21]. In irradiated cells, such energy deposition causes endogenous bursts of ROS that result mainly from water radiolysis in and around the radiation track as well as in the intercellular matrix. The differences in ionization patterns due to different types of radiation mainly arise from differences in charge-to-mass ratio of the impacting particles; for example, α-particles differ from electrons by a factor of ~8,000. Thus, whereas low-LET X and γ rays produce sparse ionization and excitation events along their trajectory, high LET α-particles or high charge (Z) and high energy (E) HZE particles produce a dense track of ionizations and excitations along the particle path [22]. The track structure determines the relative potency of different types of radiation in causing biological effects [23, 24] (Fig. 3). Following exposure to high LET radiations (e.g., α-particles, HZE particles), the yield of locally multiply damaged sites (LMDS) in DNA is greatly increased [6, 25, 26].

Whereas ~60 ROS per nanogram of tissue are generated within less than a microsecond from a hit caused by ^{137}Cs γ rays, ~2,000 ROS are generated from a 3.2 MeV α-particle traversal, which corresponds to an ROS concentration of ~19 nM in the nucleus [27]. Such a nuclear ROS concentration can obviously cause extensive oxidative injury and modify normal biochemical reactions [28, 29]. As a result, different signaling cascades responding to these stress conditions are triggered. For example, adaptive responses encompassing DNA repair and antioxidation reactions may be triggered following exposures to low doses of low LET radiations (X and γ rays) [30–32]. The protective mechanisms may overcompensate, resulting in stimulatory responses that enhance the well-being of the organism long after the exposure [33, 34]. In contrast, basal and induced signaling cascades do not seem to completely alleviate the complex damages induced by low fluences of high LET radiations (e.g., α- and HZE particles) [35–37] or high doses of X and γ rays [38]. Damaging effects endure and may spread to neighboring bystander cells [39] and persist in their progeny [15, 16] (Fig. 2). Thus, the track structure is crucial for dictating the size and precise location of the initial radiation-induced ROS bursts and their subsequent signaling or damaging effects [40–43]. The bursts of ROS, and also of RNS resulting from activation of nitric oxide synthases may affect directly or indirectly proteins/genes that participate in oxidative metabolism [44, 45]. The persistence of such perturbations in the normal oxidative metabolism is associated with chronic inflammatory responses [6, 46, 47], which impacts non-targeted effects (reviewed in [48, 49]).

Ionizing Radiation-Induced Bystander Responses

In a landmark study in 1992, Nagasawa and Little [39] presented evidence indicating that genetic changes occurred in a greater number of cells than expected when Chinese hamster ovary cell cultures were exposed to low fluences of α-particles that targeted only a small fraction of the cells. An enhanced frequency of sister chromatid exchanges (SCEs) in 20–40% of Chinese hamster ovary cells was observed in cultures exposed to fluences by which only 0.1–1% of the cells' nuclei were actually traversed by a particle track. These results indicated that the target for genetic damage by α-particles is much larger than the nucleus or in fact much larger than the cell itself. Those observations in α-particle-irradiated cell cultures were soon confirmed by other laboratories [50–55]. Similar effects were also detected in cocultures of bystander cells and cells exposed to external beams of low LET radiations [56–60] and high LET radiations besides α-particles [61, 62], thus highlighting their relevance to radiotherapy, diagnostic radiology, and health risks of environmental and occupational exposures (reviewed in [11–13, 63]). In related studies, it has also been shown that when cells are labeled with tritiated thymidine in a three-dimensional multicellular cluster model, a cytotoxic effect is transmitted to adjoining non-labeled cells present in the same cluster [56, 60].

Consistent with these studies, a cytotoxic bystander effect produced by tumor cells labeled with 5-[^{125}]iodo-2′-deoxyuridine was demonstrated in vitro [64] and in vivo [59]. Furthermore, media transfer experiments have shown that incubating nonirradiated cells with growth medium harvested from α-particle or γ-irradiated cell cultures can induce biological effects in the medium-recipient cells [65–67]. Together, the studies have shown that upregulation of stress-responsive genes and proteins, genetic and epigenetic changes, induction of cell cycle checkpoints and cell killing occur in both irradiated and neighboring bystander cells, and the effects occur in various cell types of human and rodent origin at different stages of growth (reviewed in [7, 11, 12, 63, 68–73]). Mechanistic studies have shown that direct and indirect modes of intercellular communication, oxidative metabolism, and DNA repair processes can mediate these effects [53, 74, 75]; however, the exact molecular steps involved have not been defined [76].

In addition to observations in tissue culture experiments, significant radiation-induced bystander effects were detected in human tissue models [77–79] and in animal experiments [7, 80–82]. Significantly, α-particle emitters concentrated in the liver of Chinese hamsters showed that all cells in the liver are at the same risk for the induction of chromosome damage, when only a small fraction of the total liver cell population is exposed to α-particles [83, 84]. With significance to cancer risk, non-targeted oncogenic radiation effects were observed in the cerebellum of radiosensitive mice, when only the rest of their body was X-irradiated [85]. Besides radiotherapy, where often only small areas of the body are irradiated, the occurrence of in vivo stressful non-targeted effects could have significant consequences during particular activities, such as mining or space travel, when often only parts of the human body are irradiated at any one time [17]. In the case of deep space travel, it has been estimated that an astronaut's body would be exposed, *daily*, to very low mean doses of densely ionizing radiations [86]. At the typical doses encountered [87, 88], only a very small fraction of cells in the human body would experience the large ionization events created along the tracks of such radiations [87, 88]. Moreover, the radiation traversals would be separated in both tissue location and time [61]. The possibility of increased risk of carcinogenesis caused by exposure to space radiation during prolonged space travel has been considered a limiting factor for human space exploration [89].

Whereas radiation-induced bystander effects have been extensively investigated in the past 20 years, interactions between irradiated and nonirradiated cells have been suggested from observations made decades earlier showing that blood plasma from individuals undergoing radiotherapy or from individuals who were accidentally irradiated has a clastogenic effect on normal nonexposed cells [90, 91].

Overall, a multitude of studies challenge the paradigm that radiation traversal through the nucleus of a cell is the only prerequisite for the production of genetic damage or a biological response. They indicate that cells in the vicinity of directly irradiated cells or those receiving media from irradiated cultures can respond to the radiation exposure. They denote that cell populations exposed to ionizing radiation respond as an integrated unit rather than separate individual cells that have been irradiated. They point to a critical role for intercellular communication in mediating bystander responses.

Bystander effects are not unique in demonstrating that biological changes can occur in cells that do not receive radiation directly. Evidence gathered over the last two decades from two different areas of study has also shown that genetic damage in cells need not be a direct consequence of direct nuclear irradiation. (1) Genomic instability experiments have shown that the progeny of cells that survive a radiation exposure harbor a spectrum of genetic lesions that are different in nature from the lesions that initially occurred in the irradiated parental cells [35, 38, 92, 93]. The endpoints studied have included malignant transformation [94, 95], chromosomal aberrations [35, 96], specific gene mutations [97], and cell survival [98–100]. Typically, this phenomenon has been studied by examining the occurrence of genetic effects in clonal populations derived from single cells surviving radiation exposure; though, chromosomal instability was also reported in the descendants of unirradiated surviving cells after α-particle-irradiation [101]. (2) Individual charged particles targeted to specific organelles in the cell using microbeam technology showed that gene mutations do occur following cytoplasmic irradiation [102]. Mutagenesis by cytoplasmic irradiation was induced even by a single α-particle traversal and reached a plateau after hits by four to eight particles, which elicited minimal cytotoxicity and induced a class of mutations that is entirely different from those produced by nuclear irradiation. The nature of the induced mutations consisted mainly of base damage and suggested the involvement of ROS.

Mechanisms of Radiation-Induced Bystander Effects

The role of transmissible factor(s) generated by irradiated cells: Support of a role for soluble transmissible factor(s) released by irradiated cells that in turn induces effects in nonirradiated cells came not only from medium transfer experiments but also from irradiation of subconfluent cultures with α-particles. Targeting the nuclei of a few cells in a subconfluent culture of mammalian cells with α-particles resulted in the induction of damage (micronucleus formation and apoptosis) in a greater fraction of cells than those that were irradiated [103]. In these studies, targeting the α-particles outside the cells failed to generate an effect, suggesting that the induced bystander effects were due to transmissible factor(s) released from the irradiated cells. Other studies showed that α-particle irradiation of cultured cells generated a factor(s) able to induce SCEs in bystander cells; the factor(s) survived freeze thawing and was heat labile [104]. In parallel experiments contrasting those where α-particles were targeted outside cells, the same group showed that α-particle irradiation of culture medium devoid of cells also caused the generation of SCE-inducing factor(s); such factors however were short-lived [104]. In both situations, the supernatant from irradiated cells or irradiated medium caused the induction of excessive SCEs in unirradiated cells to the same extent observed with direct α-particle-irradiated cell cultures. Interestingly, both the short-lived medium- and cell-derived SCE-inducing activities were inhibited by the antioxidant enzyme superoxide dismutase (SOD), suggesting that ROS are involved in the response.

The effects of α-particle-irradiated medium with or without cells on bystander responses were studied by a novel approach utilizing cells plated on either one or both sides of double-mylar dishes [105]. The distance between the mylar surfaces in such dishes was 9.5 mm. It was argued that since low energy α-particles can only travel a distance of about 50 μm when one side with or without cells was irradiated, cells on the other side would not receive any hits. Irradiation of Chinese hamster–human hybrid cells on one side of the dish with 1, 10, or 100 Gy resulted in cytotoxicity to the bystander cells when they were cocultured with the irradiated cells for 48 h. Using the same irradiation setup, Hu et al. [106] demonstrated that protein kinase C (PKC) epsilon is up-regulated in bystander fibroblasts. Blocking its expression with a small molecule inhibitor reduced the induction of micronuclei in bystanders by either γ rays or α-particles, supporting a role for PKC signaling in the bystander response.

It has been suggested that secreted TGF-β1 [107–110], IL-8 [111], or prostaglandins [112, 113] in the medium of α-particle-irradiated cultures may have a role in mediating bystander responses. As the signaling molecule(s) have proven difficult to isolate from the culture medium; however, an increasing number of studies have applied focused or whole-genome expression profiling techniques to help gain insight into potential signaling from the irradiated cells and the response in the bystander cells. Medium transfer experiments have been used to investigate bystander gene expression triggered by low LET γ- or X-ray exposures in both fibroblasts [114] and lymphoblasts [115, 116] or erythroleukemia cells [117]. Genes involved in oxidative phosphorylation and mitochondrial function and dysfunction were overrepresented among the genes responding in bystanders in experiments with fibroblasts [114] and lymphoblasts [118]. Interestingly, in the lymphoblasts, some mitochondrial genes responded the same in irradiated cells and bystanders, but a subset of these genes were up-regulated by direct irradiation and down-regulated in the bystanders. A similar result was reported in erythroleukemia cells, where 0.6% of the genes were found to change in the opposite direction in bystanders and irradiated cells [117].

Gene expression has also been profiled in fibroblast bystanders to high LET particle-irradiated cells, using medium transfer [118], carbon ion microbeam irradiation [119], and a two-layered Mylar strip dish [55]. The latter case employs custom culture dishes, consisting of two concentric rings with a thin Mylar bottom on the outer dish, and strips of thicker Mylar on the inner dish, allow α-particle bystander exposures in situ. The cells are grown as a contiguous monolayer, but only the cells growing directly on the thinner Mylar will be exposed to α-particles, while the adjacent bystanders growing on the thicker Mylar strips will be completely shielded. Using this approach, we have documented an attenuated response of genes regulated by TP53 in the bystander cells, and a robust response of NFκB-regulated genes [55]. The NFκB response was essentially identical in both magnitude and timing in the irradiated cells and bystanders. Further network analysis also implicated KDM5B and HDAC1 and 2 in gene regulation in bystander fibroblasts. These regulators were found to change at the protein level in both irradiated and bystander cells, suggesting a possible role for epigenetic regulation of bystander responses [56].

Network analysis also implicated AKT in early signal transduction, possibly through the GSK3B/CTNNB1 pathway. We found that AKT was phosphorylated in response to radiation by half an hour after exposure, when there was no change in bystanders. By 1–4 h after exposure, however, levels were similarly elevated in both the irradiated and bystander cells [120]. CTNNB1 was also dephosphorylated to similar levels between 4 and 8 h after exposure in both irradiated and bystander cells.

The same series of experiments also identified a number of genes coding for potential extracellular signaling molecules that were up-regulated in both irradiated and bystander fibroblasts. These included genes previously implicated in bystander response, such as IL8, IL1A, and IL1B, as well as new candidates, such as IL6,

IL33, LIF, and FGF2. We found that blocking the activity of IL33 by adding specific antibodies to the culture medium blocked activation of NFκB in both irradiated and bystander cells [113]. Regulation of IL33 expression in response to direct or bystander irradiation was also confirmed to be under the control of the AKT pathway.

In other experiments involving high LET radiations, a mitogenic bystander effect was also observed. Exposure of normal human lung fibroblasts (NHLF), maintained in culture, to a low mean dose of α-particles stimulated their proliferation; the response also occurred when unirradiated cells were treated with supernatants from α-particle or heavy ion-irradiated cells. TGF-β1 was implicated in mediating the observed effects in the α-particle experiments [121], and nitric oxide apparently contributed to the modest enhancement in cell proliferation and induction of micronuclei observed in bystander cells in the heavy ion study [122]. The occurrence of such mitogenic effects has been suggested to contribute to the hyperplastic responses in the conducting airways of the lower respiratory track that occur after inhalation exposure to radon and other environmental stresses [121]. The stimulatory growth response observed in these media transfer studies is contradictory to the observation that transient and permanent arrests in G_1 phase of the cell cycle are induced in normal human cell cultures exposed to mean doses as low as 1 cGy where bystander cells participate in the overall response of the exposed cell population [123]. It is possible that following an initial arrest in G_1, an enhancement in cell growth of the irradiated cells occurs. Further analyses of the kinetics of induction of molecules associated with cell growth or cell cycle arrest are needed in support of these studies.

In contrast to the above studies, conditioned medium harvested from keratinocytes exposed to γ rays (a low LET radiation) and added to recipient control keratinocytes or fibroblasts [65] resulted in a toxic effect in the recipient cells. The effect was dependent on the type and number of cells in the exposed cultures. Medium harvested from irradiated keratinocytes had a greater cytotoxic effect on fibroblast bystander cells than on keratinocytes, while medium from irradiated fibroblasts had no effect on either keratinocyte or fibroblast bystander cells [65]. The factor(s) leading to such bystander effects appeared to be released by the irradiated cells within the first few hours after exposure. It was suggested that the released factor(s) may be a protein as it was labile when heated but stable when frozen [65]. Suggesting a possible link between the bystander response and genomic instability effects, it was shown that medium harvested over several generations from cells surviving γ-irradiation is cytotoxic to nonirradiated bystander cells [124, 125]. Consistent with induction of apoptosis in the unirradiated cells, a rapid calcium flux, a subsequent loss of mitochondrial membrane potential and increases in ROS were observed in those cells [124].

Participation of gap-junction intercellular communication (GJIC): Evidence for the involvement of GJIC in propagation of bystander effects has been derived from studies with high and low LET radiations [53, 56, 75, 85, 126–131]. Gap junctions were shown to mediate the propagation of stressful effects not only between targeted and non-targeted cells (reviewed in [10, 70]), but also among the targeted cells [27, 132]. The intercellular channels that comprise gap junctions are formed by *connexin* proteins [133]. Manipulation (↓↑) of connexin expression/gap-junction gating by chemical agents, forced connexin expression by transfection, and connexin gene knockout studies provide substantial evidence for the participation of gap junctions in radiation-induced bystander effects [10, 73, 134]. This is supported by stabilization and up-regulation of connexin mRNA and protein by ionizing radiation [135]. Disruption of cholesterol rich areas of the plasma membrane, where gap-junction channels partition [136], attenuated propagation of radiation-induced effects to bystander cells [137].

The participation of gap junctions in stress-induced bystander effects is not unique to ionizing radiation; it was also described in high density cells exposed to chemotherapeutic agents. Toxicity of these compounds was enhanced by the presence of functional gap junctions between the target cells [138–141]. These effects bear a striking functional similarity to the ability of ganciclovir triphosphate generated by herpes simplex virus (HSV) thymidine kinase to pass through gap junctions and kill cells uninfected with HSV and therefore insensitive to ganciclovir [142–146]. Thus, many systems show that gap junctions enhance and spread the effects of toxic agents on target cells. It appears that a compound or "stress signal" that leads to toxic/clastogenic effects can pass through at least some types of gap-junction channels. This can result in stressful effects, including killing of cells adjacent to those exposed to the toxic treatment or a synergistic enhancement of toxicity between exposed cells (for example, as in the case of cells exposed to hyperthermic treatment [147]).

Gap junctions are dynamic structures that are critical for diverse physiological functions [145, 148–152]. By allowing direct intercellular transfer of cytoplasmic molecules, they provide a powerful pathway for direct molecular signaling between cells. Each of the ~20 forms of connexin [153] forms channels with distinct permeability properties. Though the properties of connexin channels differ, their pores are thought to allow permeation by molecules up to ~1,000 Da, well above the size of most second messengers [133]. Our ongoing studies focus on the effects of specific connexins in radiation-induced non-targeted effects. Connexin channels are highly selective among molecular permeants. The selectivity among cytoplasmic permeants is not simply on the basis of size or charge. Although connexin channels are permeable to second messengers [133], *different* connexins form channels with *different* selectivities for second messengers [154–156]. For example, ATP, ADP, AMP, glutamate, and glutathione are significantly more permeable through junctional Cx43 than Cx32 channels. On the other hand, adenosine and IP3 are more permeable through Cx32 than through Cx43 channels. Depending on their composition, connexin channels can discriminate between highly similar second messengers (e.g., cAMP and cGMP, and among inositol trisphosphates [157–161]). By understanding the effects of specific connexins in the nature of radiation-induced responses, countermeasures to the harmful effects of radiation may be formulated and strategies to enhance radiotherapy may be developed.

Gap junctions and tumor cells: Loss of GJIC is widely regarded to correlate with tumorigenic phenotypes, but there are exceptions. More importantly, it is now clear that connexins play distinct roles in specific stages of cancer progression. Specifically, increased levels of connexin expression and of GJIC are correlated with invasiveness, extravasation, and metastasis in a variety of cancer cells. It has also been noted that primary tumors that are initially GJIC impaired become GJIC competent at the metastatic stage [162, 163]. For tumor cells with reduced GJIC, development of drugs and methods that can recover or increase GJIC provide a new and potent way to enhance treatment. Several compounds, notably 4-phenylbutyrate, an inhibitor of histone deacetylases, have been shown to increase GJIC of otherwise GJIC-impaired tumor cells, which enhances toxic bystander effects as well as tends to restore growth control [164–166]. Thus, enhancement of GJIC by chemotherapeutic agents in tumor cells, coupled with radiotherapy and the associated transmission of toxic compounds between cells in the irradiated tumor, would offer a therapeutic gain. In contrast, transmission of toxic effects from irradiated to neighboring normal bystander cells would cause a health risk.

Oxidative metabolism mediates signaling events leading to radiation-induced bystander responses: Normal oxidative metabolism is a key endogenous generator of ROS and RNS [167], and homeostatic control of normal cellular growth pathways is tightly dependent on oxidants [168]. A disruption of the balance between oxidant production and antioxidant defense alters the homeostatic cellular redox environment, resulting in a state of oxidative stress that promotes many pathological conditions including degenerative diseases and cancer [169]. The endogenous targets of oxidants are diverse and include nucleic acids, proteins, and lipids.

An indication that ROS are involved in the induction of SCEs in bystander cells present in cell cultures exposed to very low fluences of α-particles was suggested when the bystander effect was inhibited by SOD, a superoxide radical scavenger [107]. Subsequent studies using more direct approaches have shown that low doses of α-particles initiate the intracellular production of ROS (superoxide anions and hydrogen peroxide) in human cells through involvement of the plasma bound NADPH-oxidase [104]. These studies suggested that the ROS response did not require direct nuclear or even cellular hits by α-particles [104]. In other studies, the antioxidant DMSO reduced the lethal effects imparted on bystander cells by ^{3}H-TdR-labeled cells present in the same 3D cell cluster [170]. Interestingly, maximum protection of the bystander cells was observed in the presence of both DMSO and lindane, an inhibitor of GJIC [170].

Oxidative metabolism has also been implicated in toxic bystander effects observed in media transfer experiments involving γ-radiation [171–173]. Treatment of the irradiated cultures with the antioxidants L-lactate and L-deprenyl [171–173] or with drugs that inhibit collapse of mitochondrial membrane potential prevented the cytotoxic effects from irradiated cell-conditioned medium [172]. In vivo experiments have also shown that inflammatory-type responses occur after exposure to ionizing radiation [80]. In those experiments, activation of macrophages and neutrophil infiltration were not direct effects of irradiation, but were a consequence of the recognition and clearance of radiation-induced apoptotic cells. The occurrence of such response has been suggested to provide a likely mechanism for the interactions between irradiated and nonirradiated hemopoietic cells both in vitro and in vivo [80]. Such interaction was also observed in out of field in vivo experiments examining the genetic effects of partial organ irradiation. Antioxidants and nitric oxide synthase inhibitors attenuated these effects [174], strongly supporting the role of ROS and RNS in mediating bystander effects [175, 176].

ROS scavengers inhibited the induction of mutations following cytoplasmic irradiation [102]. Further, induction of hypoxanthine guanine phosphoribosyltransferase (*HPRT*) mutations in CHO bystander cells from cultures exposed to low fluences of α-particles was consistent with the involvement of oxidative metabolism in the effect. Whereas the mutations induced in cells directly irradiated through the nucleus with α-particles were primarily partial and total gene deletions, over 90% of those arising in bystander cells were point mutations [177]. Interestingly, point mutations were mainly generated following cytoplasmic irradiations [102].

Further evidence that oxidative metabolism is up-regulated in bystander cells was generated from gene expression studies in human diploid fibroblast cultures exposed to very low fluences of α-particles [74]. Whereas $p21^{Waf1}$ expression examined in situ in cell cultures exposed to low mean doses of α-particles occurred in clusters of adjacent cells that far exceed the fraction of cells that were irradiated, active and not boiled SOD inhibited the effect [74]. Enzyme activity analyses indicated that the exogenously added SOD enzymatic activity becomes significantly associated with the cells. Whether this association is limited to the plasma membrane or is internalized by the cells remains to be tested.

Radiation-induced ROS are known to cause damage to various cellular components (reviewed in [5, 178]) and produce double-strand breaks in addition to base damage and single-strand breaks (reviewed in [179–181]). Alpha-particle-induced metabolic ROS production was also shown to activate signaling pathways mediated by p53, MAPK, and PI3K-AKT-GSK3beta in bystander cells [74, 120]. Active SOD and catalase enzymes were capable of suppressing these effects and also inhibited the activation in bystander cells of redox-sensitive transcription factors (e.g., NFκB, AP-1, and ATF2) [74]. Similar to GJIC inhibitors, antioxidant enzymes significantly reduced the excess formation of micronuclei in bystander cells [74]. Of interest is the finding that ROS-activated kinase(s) (e.g., member(s) of the MAPK superfamily) have a role in activation of gap-junction proteins [182]. Binding sites for the redox-sensitive AP-1 and NFκB transcription factors, which are activated by low fluences of α-particles, exist in the connexin43 gene promoter region [183].

In addition to membrane bound oxidases [74, 107], involvement of cell membranes in bystander responses was highlighted by the complete suppression of SCEs and *HPRT* mutation induction in CHO cells exposed to low fluences of α-particles in the presence of Filipin, a drug that disrupts lipid rafts [137]. It is of interest to note that gap junctions have been reported to partition in lipid rafts [136]. Further, critical molecules that participate in inflammatory responses, such as prostaglandin-endoperoxide synthase 2 (*PTGS2*), also known as cyclooxygenase-2 (COX-2), also localize in cholesterol-rich domains of plasma membrane [184]. Using a signal transduction pathway-specific SuperArray, we compared differentially expressed genes among nonirradiated control NHLF and bystander cells that were in coculture with α-particle-irradiated NHLF [112]. Among the 96 genes represented on the platform, one gene, COX-2, was found to be consistently up-regulated by more than threefold, while the RNA level of insulin growth factor binding protein-3 (IGFBP3) was found to be consistently lower by more than sevenfold in multiple analyses of independent bystander samples. The expression of the COX-2 protein in the nonirradiated bystander cells was further confirmed by western blotting. Addition of the COX-2 inhibitor NS-398 (50 μM) suppressed COX-2 activity in NHLF cells and finally, after 24 h, reduced the COX-2 protein level in bystander cells to a non-detectable level [112]. These results indicated that expression of COX-2 is associated with the bystander effect. Subsequent experiments showed that a non-cytotoxic and non-mutagenic dose of NS-398 inhibited the propagation of signaling events leading to bystander mutagenesis at the HPRT locus in NHLF cell cultures exposed to α-particles. Therefore, it is attractive to speculate that several mechanisms act in concert to promote the bystander effect.

Involvement of Rad9 in signaling radiation-induced bystander responses: The Rad9 protein was shown to participate in the bystander response to radiation exposure since mouse embryonic stem (ES) cells null for the corresponding gene demonstrate enhanced bystander micronuclei formation and apoptosis relative to wild-type *Rad9* controls [185]. These results suggest that Rad9 might normally suppress the bystander signal, and when the protein is not functioning properly a stronger and perhaps longer lasting persistent signal is produced postirradiation. Rad9 is a multifunctional protein with a variety of activities that promote repair of DNA damage, including roles in several DNA repair pathways and cell cycle checkpoints [186]. If DNA damage induced by radiation is the initiating event for a bystander response, then it is reasonable to speculate that in the absence of Rad9, damage is not properly repaired and thus an enhanced bystander response will be in effect. Consistent with this model is the previous finding that Chinese hamster ovary cells bearing an xrs-5 mutation, which reduces the ability to repair DNA double-strand breaks, are sensitized relative to wild-type cells to the formation of chromosome aberrations caused by bystander effects induced by a low fluence of α-particles [187]. However, since Rad9 has many functions in addition to DNA repair and cell cycle checkpoint control, such as the ability to transactivate transcription of specific downstream target genes and regulate apoptosis, the contribution of these other activities to the bystander response is unknown and thus should also be determined.

Conclusions

In vitro and *in vivo* observations have provided strong evidence indicating that molecular events leading to various biological effects, including genetic damage, can be transmitted from irradiated to nonirradiated cells. The phenomenon occurs in

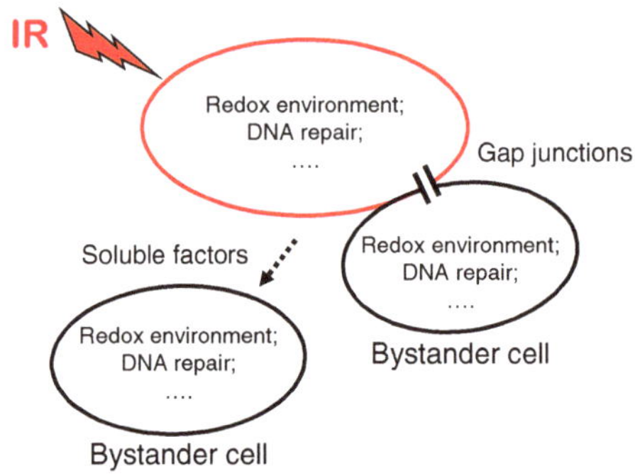

Fig. 4 Mechanisms underlying ionizing radiation-induced bystander effects. Signaling molecules are propagated among irradiated and bystander cells through direct intercellular communication via gap junctions or through diffusible secretion in the surrounding environment. The expression of propagation of bystander effects is highly dependent upon the phenotype of both the irradiated and bystander cells

a variety of cell types of human and rodent origin and involves GJIC, oxidative metabolism, secreted diffusible factors, and DNA repair (Fig. 4). Preliminary evidence points to a crucial role of membrane originating effects where gap junctions and critical enzymes such as COX-2 are located. Further, the expression of connexin proteins has been reported to be modulated by the cellular redox environment and by redox-sensitive soluble factors released by cells exposed to ionizing radiation (e.g., TNF-α, IL-1β) [188, 189]. However, it is also possible that specific bystander effects are regulated by some mechanism(s) and not by others [105]; this may depend on cell type, cellular growth state, type of radiation, and the biological endpoint being measured.

While direct approaches were used to investigate the role of GJIC in the radiation-induced bystander effect, this remains to be adopted in studies involving oxidative metabolism. In particular, genomic approaches, by which antioxidant enzymes or ROS generating enzymes are overexpressed or underexpressed in cells, coupled with coculture experiments will help identify whether cellular redox environment contributes to the bystander effect at the level of the irradiated and/or the bystander cells.

Current evidence indicates that genetic damage occurs in bystander cells; however, very few studies have examined the reparability of such damage. The fact that cells pre-exposed to an adapting γ-ray dose are less susceptible to genetic damage induced by bystander mechanisms following exposure to low fluences of α-particles [190], and the indication that greater bystander damage occurs in repair-deficient cells [191] is in support of the concept that under certain conditions, bystander damage is amenable to repair or is preventable. Studies with rad9 wild type and mutant cells support this concept.

The radiation-induced lesion(s) that signal expression of the bystander response is under intense investigation [192, 193]: DNA strand breaks, oxidative damage, and other targets have been proposed. Studies that focus on identifying such lesion(s) should enhance our understanding of the mechanisms underlying the bystander effect.

Significance

Radiation protection and radiotherapy are two areas where radiation-induced bystander effects could have significant ramifications.

Radiation protection: The occurrence of a bystander effect in cell populations exposed to low fluences of high LET radiation such as α-particles could have a significant impact on the estimation of risks after such exposure [194, 195]. It suggests that cell populations or tissues respond as a whole to radiation exposure and the response is not restricted to that of the individual traversed cells but involves the non-traversed cells also. This would imply that the modeling of dose response relationships at low mean doses, based on the number of cells hit or even on the type of DNA damage they receive, may not be a valid approach. These studies are directly relevant to public health issues where humans are exposed to low fluences of high LET particles. For example, it has been estimated that 10–14% of lung cancer cases are linked to radon gas in the environment and its α-particle-emitting decay products [196]. These estimates were derived by extrapolation from data for high dose exposures to low doses assuming a linear, no threshold dose response. At exposures similar to those from indoor radon, most cells in the bronchial epithelium would not be traversed by an irradiating particle at all and most of the irradiated cells would be traversed by a single particle only [197]. A cell traversed by one α-particle receives a substantial dose of radiation (~0.1–0.5 Gy) and thus would be prone to the deleterious effects of radiation. The studies reviewed here indicate that non-traversed bystander cells exhibit similar genetic alterations, including DNA damage, and hence could contribute to the risk from such exposure. Thus, bystander effect studies, along with other approaches (e.g., epidemiological and toxicological) should contribute to the establishment of not only adequate environmental radiation protection guidelines but also occupational radiation protection standards. An increasing number of workers are currently deployed to decommission and clean nuclear installations, isolate nuclear material, or man space stations. These workers can be accidentally exposed to ionizing radiation including low fluences of α-particles and high energy heavy ions.

Radiotherapeutic gain: The induction of cytotoxic effects in bystander cells adjoining irradiated cells has the potential to enhance radiotherapy and help achieve tumor eradication. Evidence for the existence of radiation-induced bystander effects in vivo and elucidation of underlying mechanisms will not only help optimize the contribution of bystander effects to radiotherapy but also provide an explanation to various reports of cytotoxic effects observed in solid tumors located at distant sites from those targeted by radiation. Such abscopal effects, reported as early as 1952 (reviewed in [69]), led to the regression of a variety of tumors (e.g., [198, 199]). Studies have suggested that ionizing radiation induces the release of cytokines into the circulation which in turn mediate a systemic antitumor effect [199]. Such an effect may involve enhancement of immune activity [200]. Interestingly, in vivo mouse experiments have shown that the Trp53 protein is a key mediator of the radiation-induced abscopal effect [201]. p53 was previously shown to have a role in the secretion of stress-induced growth inhibitors [202]. The secretion of factors capable of inhibitory abscopal/bystander effects when p53 wild-type tumors are irradiated would potentiate the effect of radiation in eradicating tumors.

Bystander effects are thought to have a role in fractionated radiotherapy [203]. Growth medium harvested from cultured cells receiving fractionated irradiation resulted in greater cytotoxic effects when added to bystander nonirradiated cells than growth medium harvested from cultures receiving a single dose of irradiation. This cell killing effect of conditioned medium from irradiated cultures is contrasted with the significant split dose recovery observed in cultures directly exposed to fractionated irradiation [203]. It was argued that if bystander factors were produced in vivo, they may reduce the sparing effect observed in dose fractionation regimens. However, the existence of such factors is likely to be patient, tissue, and life-style specific [204].

A direct role for GJIC has been implicated in antitumor suicide gene therapy protocols by which apoptotic and/or toxic metabolites are transferred through gap-junction channels from affected to bystander cells [205]. Our emerging studies support the role of GJIC in enhancing cell killing when all cells in the population are irradiated [27, 132]. A limitation to enhancing radiotherapeutic gain by such cohort effects is due to the fact that functional GJIC is generally compromised in tumor cells. Specific chemical treatments have been proposed to increase the GJIC capacity of tumor cells and cAMP, retinoic acids, carotenoids, glucocorticoids, and flavenoids have been shown to have such an effect (reviewed in [205]). Up-regulation of gap-junction communication in tumor cells could contribute to the propagation of cell death signals (e.g., calcium ions [206]) generated by cells in tumors that uptake α-particle emitters. These radionuclides are being investigated in the treatment of cancer [207–209].

Emergence of second primary cancers following radiotherapy: Since genomic instability is considered a predisposition factor for carcinogenesis, it has been postulated that radiation-induced non-targeted/bystander effects may promote secondary cancer induction in radiotherapy patients [210]. From animal studies with X-rays, there is evidence that irradiation of part of the lung in mice can induce a non-targeted response in the nonirradiated part of the lung through the induction of inflammatory cytokines [211]. Furthermore, our recent evidence (unpublished) and that of others [212] indicates that irradiation of the lower abdomen of mice with X-rays results in the induction of inflammatory response as well as mutations and COX-2 induction [213] in out of field lung tissues. Using the radiosensitive Patched-$1^{+/-}$ (Ptch$1^{+/-}$) mouse model system that has a defect in radiation-induced activation of the ATR-Chk1 checkpoint signaling pathway, Mancuso et al. reported induction of medulloblastoma in the nonirradiated brain tissues after partial irradiation of the lower half of the animal with a 3 Gy dose of X-rays [85]. Further, based on human serum analyses, there is clear evidence that plasma clastogenic factors are present many years after radiation exposure from the Japanese atomic bomb survivors, Chernobyl liquidators, and from radiotherapy patients [91, 214, 215]. Recently, the frequency distribution of second primary tumor sites in relation to previous irradiation volumes was estimated in a cohort of 115 pediatric patients who developed such cancers [216]. It was estimated that ~22% of secondarily derived tumors arise from a distance of at least 5 cm from the irradiated site and ~6% arise from a distance that is >10 cm away. A peak second primary tumor frequency of ~31% was identified in volumes receiving less than 2.5 Gy and a total of 10–15% of these tumors are estimated to arise in tissues receiving less than 0.5 Gy. Although these findings are suggestive, nonetheless, the data highlight the potential of second tumor development outside the treatment field and at much lower dose level.

In summary, bystander effect studies have led to a paradigm shift in our understanding of the target theory. They are enhancing our general understanding of intercellular communication under stress conditions. The outcome may contribute to both radiation protection and radiotherapy.

Acknowledgment Grant CA049062 from the National Institutes of Health supported research in the authors' laboratories.

References

1. Hall EJ, Giaccia AJ. Radiobiology for the radiologist. 6th ed. Philadelphia, PA: Lippincott Williams & Wilkins; 2006.
2. Spitz DR, Azzam EI, Li JJ, Gius D. Metabolic oxidation/reduction reactions and cellular responses to ionizing radiation: a unifying concept in stress response biology. Cancer Metastasis Rev. 2004;23:311–22.
3. Hayashi T, Kusunoki Y, Hakoda M, Morishita Y, Kubo Y, Maki M, et al. Radiation dose-dependent increases in inflammatory response markers in A-bomb survivors. Int J Radiat Biol. 2003;79:129–36.
4. Neriishi K, Nakashima E, Delongchamp RR. Persistent subclinical inflammation among A-bomb survivors. Int J Radiat Biol. 2001;77:475–82.
5. Petkau A. Role of superoxide dismutase in modification of radiation injury. Br J Cancer Suppl. 1987;8:87–95.
6. Kryston TB, Georgiev AB, Pissis P, Georgakilas AG. Role of oxidative stress and DNA damage in human carcinogenesis. Mutat Res. 2011;711:193–201.
7. Tamminga J, Kovalchuk O. Role of DNA damage and epigenetic DNA methylation changes in radiation-induced genomic instability and bystander effects in germline in vivo. Curr Mol Pharmacol. 2011;4:115–25.
8. Limoli CL, Giedzinski E, Morgan WF, Swarts SG, Jones GD, Hyun W. Persistent oxidative stress in chromosomally unstable cells. Cancer Res. 2003;63:3107–11.
9. Limoli CL, Hartmann A, Shephard L, Yang C, Boothman DA, Bartholomew J, et al. Apoptosis, reproductive failure, and oxidative stress in Chinese hamster ovary cells with compromised genomic integrity. Cancer Res. 1998;58:3712–8.
10. Azzam EI, de Toledo SM, Little JB. Oxidative metabolism, gap junctions and the ionizing radiation-induced bystander effect. Oncogene. 2003;22:7050–7.

11. Hei TK, Zhou H, Chai Y, Ponnaiya B, Ivanov VN. Radiation induced non-targeted response: mechanism and potential clinical implications. Curr Mol Pharmacol. 2011;4:96–105.
12. Mothersill C, Seymour CB. Radiation-induced bystander effects—implications for cancer. Nat Rev Cancer. 2004;4:158–64.
13. Prise KM, O'Sullivan JM. Radiation-induced bystander signalling in cancer therapy. Nat Rev Cancer. 2009;9:351–60.
14. Buonanno M, de Toledo SM, Azzam EI. Increased frequency of spontaneous neoplastic transformation in progeny of bystander cells from cultures exposed to densely-ionizing radiation. PloS One. 2011;6:art. no. e21540.
15. Buonanno M, De Toledo SM, Pain D, Azzam EI. Long-term consequences of radiation-induced bystander effects depend on radiation quality and dose and correlate with oxidative stress. Radiat Res. 2011;175:405–15.
16. Ponnaiya B, Suzuki M, Tsuruoka C, Uchihori Y, Wei Y, Hei TK. Detection of chromosomal instability in bystander cells after Si490-ion irradiation. Radiat Res. 2011;176:280–90.
17. Cucinotta FA, Chappell LJ. Non-targeted effects and the dose response for heavy ion tumor induction. Mutat Res. 2010;687:49–53.
18. Hall EJ, Henry S. Kaplan Distinguished Scientist Award 2003. The crooked shall be made straight; dose-response relationships for carcinogenesis. Int J Radiat Biol. 2004;80:327–37.
19. Tubiana M. Can we reduce the incidence of second primary malignancies occurring after radiotherapy? A critical review. Radiother Oncol. 2009;91:4–15; discussion 1–3.
20. National Research Council. Health risks from exposure to low levels of ionizing radiation. Washington, DC: National Research Council of the National Academies; 2005.
21. Roth RA, Sharma SC, Katz R. Systematic evaluation of cellular radiosensitivity parameters. Phys Med Biol. 1976;21:491–503.
22. Goodhead DT. Spatial and temporal distribution of energy. Health Phys. 1988;55:231–40.
23. Ottolenghi A, Merzagora M, Paretzke HG. DNA complex lesions induced by protons and alpha-particles: track structure characteristics determining linear energy transfer and particle type dependence. Radiat Environ Biophys. 1997;36:97–103.
24. Goodhead DT. Energy deposition stochastics and track structure: what about the target? Radiat Prot Dosimetry. 2006;122:3–15.
25. Ward JF. Biochemistry of DNA lesions. Radiat Res Suppl. 1985;8:S103–11.
26. Georgakilas AG. From chemistry of DNA damage to repair and biological significance. Comprehending the future. Mutat Res. 2011;711:1–2.
27. Autsavapromporn N, de Toledo SM, Little JB, Jay-Gerin J-P, Harris AL, Azzam EI. The role of gap junction communication and oxidative stress in the propagation of toxic effects among high-dose α-particle-irradiated human cells. Radiat Res. 2011;175:347–57.
28. Sutherland BM, Bennett PV, Sidorkina O, Laval J. Clustered DNA damages induced in isolated DNA and in human cells by low doses of ionizing radiation. Proc Natl Acad Sci USA. 2000;97:103–8.
29. Nikjoo H, Goodhead DT, Charlton DE, Paretzke HG. Energy deposition in small cylindrical targets by monoenergetic electrons. Int J Radiat Biol. 1991;60:739–56.
30. de Toledo SM, Asaad N, Venkatachalam P, Li L, Howell RW, Spitz DR, et al. Adaptive responses to low-dose/low-dose-rate γ rays in normal human fibroblasts: the role of growth architecture and oxidative metabolism. Radiat Res. 2006;166:849–57.
31. Feinendegen LE, Pollycove M, Neumann RD. Whole-body responses to low-level radiation exposure: new concepts in mammalian radiobiology. Exp Hematol. 2007;35:37–46.
32. Wolff S. The adaptive response in radiobiology: evolving insights and implications. Environ Health Perspect. 1998;106 Suppl 1:277–83.
33. Nomura T, Li XH, Ogata H, Sakai K, Kondo T, Takano Y, et al. Suppressive effects of continuous low-dose-rate gamma irradiation on diabetic nephropathy in Type II diabetes mellitus model mice. Radiat Res. 2011;176:356–65.
34. Azzam EI. Exposure to low level environmental agents: the induction of hormesis. Mutat Res. 2011;726:89–90.

35. Kadhim MA, Macdonald DA, Goodhead DT, Lorimore SA, Marsden SJ, Wright EG. Transmission of chromosomal instability after plutonium α-particle irradiation [see comments]. Nature. 1992;355:738–40.
36. Kronenberg A, Gauny S, Criddle K, Vannais D, Ueno A, Kraemer S, et al. Heavy ion mutagenesis: linear energy transfer effects and genetic linkage. Radiat Environ Biophys. 1995;34:73–8.
37. Kronenberg A, Little JB. Molecular characterization of thymidine kinase mutants of human cells induced by densely ionizing radiation. Mutat Res. 1989;211:215–24.
38. Grosovsky AJ, Parks KK, Giver CR, Nelson SL. Clonal analysis of delayed karyotypic abnormalities and gene mutations in radiation-induced genetic instability. Mol Cell Biol. 1996;16:6252–62.
39. Nagasawa H, Little JB. Induction of sister chromatid exchanges by extremely low doses of α-particles. Cancer Res. 1992;52:6394–6.
40. Cucinotta FA, Katz R, Wilson JW. Radial distribution of electron spectra from high-energy ions. Radiat Environ Biophys. 1998;37:259–65.
41. Goodhead DT, Thacker J, Cox R. Effects of radiations of different qualities on cells: molecular mechanisms of damage and repair (Weiss Lecture). Int J Radiat Biol. 1993;63:543–56.
42. Muroya Y, Plante I, Azzam EI, Meesungnoen J, Katsumura Y, Jay-Gerin J-P. High-LET ion radiolysis of water: visualization of the formation and evolution of ion tracks and relevance to the radiation-induced bystander effect. Radiat Res. 2006;165:485–91.
43. Plante I, Ponomarev A, Cucinotta FA. 3D visualisation of the stochastic patterns of the radial dose in nano-volumes by a Monte Carlo simulation of HZE ion track structure. Radiat Prot Dosimetry. 2011;143:156–61.
44. Mikkelsen RB, Wardman P. Biological chemistry of reactive oxygen and nitrogen and radiation-induced signal transduction mechanisms. Oncogene. 2003;22:5734–54.
45. Wu C, Parrott AM, Fu C, Liu T, Marino SM, Gladyshev VN, et al. Thioredoxin 1-mediated post-translational modifications: reduction, transnitrosylation, denitrosylation, and related proteomics methodologies. Antioxid Redox Signal. 2011;15:2565–604.
46. Formenti SC, Demaria S. Local control by radiotherapy: is that all there is? Breast Cancer Res. 2008;10:215.
47. Robbins ME, Zhao W. Chronic oxidative stress and radiation-induced late normal tissue injury: a review. Int J Radiat Biol. 2004;80:251–9.
48. Dedon PC, Tannenbaum SR. Reactive nitrogen species in the chemical biology of inflammation. Arch Biochem Biophys. 2004;423:12–22.
49. Cadet J, Douki T, Ravanat JL. Measurement of oxidatively generated base damage in cellular DNA. Mutat Res. 2011;711:3–12.
50. Deshpande A, Goodwin EH, Bailey SM, Marrone BL, Lehnert BE. Alpha-particle-induced sister chromatid exchange in normal human lung fibroblasts—evidence for an extranuclear target. Radiat Res. 1996;145:260–7.
51. Belyakov OV, Malcolmson AM, Folkard M, Prise KM, Michael BD. Direct evidence for a bystander effect of ionizing radiation in primary human fibroblasts. Br J Cancer. 2001;84:674–9.
52. Zhou H, Randers-Pehrson G, Waldren CA, Vannais D, Hall EJ, Hei TK. Induction of a bystander mutagenic effect of alpha particles in mammalian cells. Proc Natl Acad Sci USA. 2000;97:2099–104.
53. Azzam EI, de Toledo SM, Little JB. Direct evidence for the participation of gap-junction mediated intercellular communication in the transmission of damage signals from alpha-particle irradiated to non-irradiated cells. Proc Natl Acad Sci USA. 2001;98:473–8.
54. Sawant SG, Randers-Pehrson G, Geard CR, Brenner DJ, Hall EJ. The bystander effect in radiation oncogenesis: I. Transformation in C3H 10T1/2 cells in vitro can be initiated in the unirradiated neighbors of irradiated cells. Radiat Res. 2001;155:397–401.
55. Ghandhi SA, Yaghoubian B, Amundson SA. Global gene expression analyses of bystander and alpha particle irradiated normal human lung fibroblasts: synchronous and differential responses. BMC Med Genomics. 2008;1:63.

56. Bishayee A, Rao DV, Howell RW. Evidence for pronounced bystander effects caused by nonuniform distributions of radioactivity using a novel three-dimensional tissue culture model. Radiat Res. 1999;152:88–97.
57. Gerashchenko BI, Howell RW. Cell proximity is a prerequisite for the proliferative response of bystander cells co-cultured with cells irradiated with gamma-rays. Cytometry. 2003;56A:71–80.
58. Mothersill C, Seymour C. Survival of human epithelial cells irradiated with cobalt 60 as microcolonies or single cells. Int J Radiat Biol. 1997;72:597–606.
59. Xue LY, Butler NJ, Makrigiorgos GM, Adelstein SJ, Kassis AI. Bystander effect produced by radiolabeled tumor cells in vivo. Proc Natl Acad Sci USA. 2002;99:13765–70.
60. Persaud R, Zhou H, Baker SE, Hei TK, Hall EJ. Assessment of low linear energy transfer radiation-induced bystander mutagenesis in a three-dimensional culture model. Cancer Res. 2005;65:9876–82.
61. Held KD. Effects of low fluences of radiations found in space on cellular systems. Int J Radiat Biol. 2009;85:379–90.
62. Matsumoto H, Hayashi S, Hatashita M, Shioura H, Ohtsubo T, Kitai R, et al. Induction of radioresistance to accelerated carbon-ion beams in recipient cells by nitric oxide excreted from irradiated donor cells of human glioblastoma. Int J Radiat Biol. 2000;76:1649–57.
63. Azzam EI, Little JB. The radiation-induced bystander effect: evidence and significance. Hum Exp Toxicol. 2004;23:61–5.
64. Akudugu JM, Azzam EI, Howell RW. Induction of lethal bystander effects in human breast cancer cell cultures by DNA-incorporated iodine-125 depends on phenotype. Int J Radiat Biol. 2012 April 10; Epub ahead of print.
65. Mothersill C, Seymour C. Medium from irradiated human epithelial cells but not human fibroblasts reduces the clonogenic survival of unirradiated cells. Int J Radiat Biol. 1997;71:421–7.
66. Mothersill C, Seymour C. Radiation-induced bystander effects, carcinogenesis and models. Oncogene. 2003;22:7028–33.
67. Mothersill C, Seymour CB. Cell-cell contact during gamma irradiation is not required to induce a bystander effect in normal human keratinocytes: evidence for release during irradiation of a signal controlling survival into the medium. Radiat Res. 1998;149:252–62.
68. Morgan WF. Non-targeted and delayed effects of exposure to ionizing radiation: II. Radiation-induced genomic instability and bystander effects *in vivo*, clastogenic factors and transgenerational effects. Radiat Res. 2003;159:581–96.
69. Mothersill C, Seymour C. Radiation-induced bystander effects: past history and future directions. Radiat Res. 2001;155:759–67.
70. Hamada N, Maeda M, Otsuka K, Tomita M. Signaling pathways underpinning the manifestations of ionizing radiation-induced bystander effects. Curr Mol Pharmacol. 2011;4:79–95.
71. Blyth BJ, Sykes PJ. Radiation-induced bystander effects: what are they, and how relevant are they to human radiation exposures? Radiat Res. 2011;176:139–57.
72. Hei TK, Zhou H, Ivanov VN, Hong M, Lieberman HB, Brenner DJ, et al. Mechanism of radiation-induced bystander effects: a unifying model. J Pharm Pharmacol. 2008;60:943–50.
73. Little JB. Genomic instability and bystander effects: a historical perspective. Oncogene. 2003;22:6978–87.
74. Azzam EI, de Toledo SM, Spitz DR, Little JB. Oxidative metabolism modulates signal transduction and micronucleus formation in bystander cells from alpha-particle-irradiated normal human fibroblast cultures. Cancer Res. 2002;62:5436–42.
75. Zhou HN, Suzuki M, Randers-Pehrson R, Chen G, Trosko J, Vannais D, et al. Radiation risk at low doses may be greater than we thought. Proc Natl Acad Sci USA. 2001;98:14410–5.
76. Ghandhi SA, Sinha A, Markatou M, Amundson SA. Time-series clustering of gene expression in irradiated and bystander fibroblasts: an application of FBPA clustering. BMC Genomics. 2011;12:2.
77. Belyakov OV, Mitchell SA, Parikh D, Randers-Pehrson G, Marino SA, Amundson SA, et al. Biological effects in unirradiated human tissue induced by radiation damage up to 1 mm away. Proc Natl Acad Sci USA. 2005;102:14203–8.

78. Sedelnikova OA, Nakamura A, Kovalchuk O, Koturbash I, Mitchell SA, Marino SA, et al. DNA double-strand breaks form in bystander cells after microbeam irradiation of three-dimensional human tissue models. Cancer Res. 2007;67:4295–302.
79. Kovalchuk O, Zemp FJ, Filkowski J, Altamirano A, Dickey JS, Jenkins-Baker G, et al. MicroRNAome changes in bystander three-dimensional human tissue models suggest priming of apoptotic pathways. Carcinogenesis. 2010;31:1882–8.
80. Lorimore SA, Coates PJ, Scobie GE, Milne G, Wright EG. Inflammatory-type responses after exposure to ionizing radiation in vivo: a mechanism for radiation-induced bystander effects? Oncogene. 2001;20:7085–95.
81. Koturbash I, Kutanzi K, Hendrickson K, Rodriguez-Juarez R, Kogosov D, Kovalchuk O. Radiation-induced bystander effects in vivo are sex specific. Mutat Res. 2008;642:28–36.
82. Jain MR, Li M, Chen W, Liu T, De Toledo SM, Pandey BN, et al. In vivo space radiation-induced non-targeted responses: late effects on molecular signaling in mitochondria. Curr Mol Pharmacol. 2011;4:106–14.
83. Brooks AL, Retherford JC, McClellan RO. Effect of 239PuO2 particle number and size on the frequency and distribution of chromosome aberrations in the liver of the Chinese hamster. Radiat Res. 1974;59:693–709.
84. Brooks AL, Benjamin SA, Hahn FF, Brownstein DG, Griffith WC, McClellan RO. The induction of liver tumors by 239Pu citrate or 239PuO2 particles in the Chinese hamster. Radiat Res. 1983;96:135–51.
85. Mancuso M, Pasquali E, Leonardi S, Tanori M, Rebessi S, Di Majo V, et al. Oncogenic bystander radiation effects in patched heterozygous mouse cerebellum. Proc Natl Acad Sci USA. 2008;105:12445–50.
86. Blakely EA. Biological effects of cosmic radiation: deterministic and stochastic. Health Phys. 2000;79:495–506.
87. Charlton DE, Sephton R. A relationship between microdosimetric spectra and cell survival for high-LET irradiation. Int J Radiat Biol. 1991;59:447–57.
88. Cucinotta F, Nikjoo H, Goodhead DT. The effect of delta rays on the number of particle traversals per cell in laboratory and space exposures. Radiat Res. 1998;150:115–9.
89. National Research Council. Managing space radiation risk in the new era of space exploration. Washington, DC: The National Academies Press; 2008.
90. Hollowell Jr JG, Littlefield LG. Chromosome damage induced by plasma of x-rayed patients: an indirect effect of x-ray. Proc Soc Exp Biol Med. 1968;129:240–4.
91. Goh K, Sumner H. Breaks in normal human chromosomes: are they induced by a transferable substance in the plasma of persons exposed to total-body irradiation? Radiat Res. 1968;35:171–81.
92. Morgan WF, Day JP, Kaplan MI, McGhee EM, Limoli CL. Genomic instability induced by ionizing radiation. Radiat Res. 1996;146:247–58.
93. Kadhim MA, Lorimore SA, Hepburn MD, Goodhead DT, Buckle VJ, Wright EG. Alpha-particle-induced chromosomal instability in human bone marrow cells. Lancet. 1994;344:987–8.
94. Kennedy AR, Fox M, Murphy G, Little JB. Relationship between X-ray exposure and malignant transformation in C3H 10T1/2 cells. Proc Natl Acad Sci. 1980;77:7262–6.
95. Kennedy AR, Little JB. Evidence that a second event in X-ray induced oncogenic transformation in vitro occurs during cellular proliferation. Radiat Res. 1984;99:228–48.
96. Sabatier L, Dutrillaux B, Martin MB. Chromosomal instability. Nature. 1992;357:548.
97. Chang WP, Little JB. Persistently elevated frequency of spontaneous mutations in progeny of CHO clones surviving X-irradiation: association with delayed reproductive death phenotype. Mutat Res. 1992;270:191–9.
98. Seymour CB, Mothersill C, Alper T. High yields of lethal mutations in somatic mammalian cells that survive ionizing radiation. Int J Radiat Biol Relat Stud Phys Chem Med. 1986;50:167–79.
99. Chang WP, Little JB. Delayed reproductive death as a dominant phenotype in cell clones surviving X-irradiation. Carcinogenesis. 1992;13:923–8.

100. Belyakov OV, Prise KM, Trott KR, Michael BD. Delayed lethality, apoptosis and micronucleus formation in human fibroblasts irradiated with X-rays or alpha-particles. Int J Radiat Biol. 1999;75:985–93.
101. Lorimore SA, Kadhim MA, Pocock DA, Papworth D, Stevens DL, Goodhead DT, et al. Chromosomal instability in the descendants of unirradiated surviving cells after alpha-particle irradiation. Proc Natl Acad Sci USA. 1998;95:5730–3.
102. Wu LJ, Randers-Pehrson GR, Xu A, Waldren CA, Geard CR, Yu ZL, et al. Targeted cytoplasmic irradiation with alpha particles induces mutations in mammalian cells. Proc Natl Acad Sci USA. 1999;96:4959–64.
103. Prise KM, Belyakov OV, Folkard M, Micheal BD. Studies on bystander effects in human fibroblasts using a charged particle microbeam. Int J Radiat Biol. 1998;74:793–8.
104. Lehnert BE, Goodwin EH. Extracellular factor(s) following exposure to alpha particles can cause sister chromatid exchanges in normal human cells. Cancer Res. 1997;57:2164–71.
105. Zhou H, Suzuki M, Geard CR, Hei TK. Effects of irradiated medium with or without cells on bystander cell responses. Mutat Res. 2002;499:135–41.
106. Hu B, Shen B, Su Y, Geard CR, Balajee AS. Protein kinase C epsilon is involved in ionizing radiation induced bystander response in human cells. Int J Biochem Cell Biol. 2009;41:2413–21.
107. Narayanan PK, Goodwin EH, Lehnert BE. Alpha particles initiate biological production of superoxide anions and hydrogen peroxide in human cells. Cancer Res. 1997;57:3963–71.
108. Abdelrazzak AB, Stevens DL, Bauer G, O'Neill P, Hill MA. The role of radiation quality in the stimulation of intercellular induction of apoptosis in transformed cells at very low doses. Radiat Res. 2011;176:346–55.
109. Dickey JS, Baird BJ, Redon CE, Sokolov MV, Sedelnikova OA, Bonner WM. Intercellular communication of cellular stress monitored by gamma-H2AX induction. Carcinogenesis. 2009;30:1686–95.
110. Shao C, Folkard M, Prise KM. Role of TGF-beta1 and nitric oxide in the bystander response of irradiated glioma cells. Oncogene. 2008;27:434–40.
111. Narayanan PK, LaRue KEA, Goodwin EH, Lehnert BE. Alpha particles induce the production of interleukin-8 by human cells. Radiat Res. 1999;152:57–63.
112. Zhou H, Ivanov VN, Gillespie J, Geard CR, Amundson SA, Brenner DJ, et al. Mechanism of radiation-induced bystander effect: role of the cyclooxygenase-2 signaling pathway. Proc Natl Acad Sci USA. 2005;102:14641–6.
113. Ivanov VN, Zhou H, Ghandhi SA, Karasic TB, Yaghoubian B, Amundson SA, et al. Radiation-induced bystander signaling pathways in human fibroblasts: a role for interleukin-33 in the signal transmission. Cell Signal. 2010;22:1076–87.
114. Kalanxhi E, Dahle J. Genome-wide microarray analysis of human fibroblasts in response to gamma radiation and the radiation-induced bystander effect. Radiat Res. 2012;177:35–43.
115. Asur R, Balasubramaniam M, Marples B, Thomas RA, Tucker JD. Bystander effects induced by chemicals and ionizing radiation: evaluation of changes in gene expression of downstream MAPK targets. Mutagenesis. 2010;25:271–9.
116. Chaudhry MA, Omaruddin RA. Mitochondrial gene expression in directly irradiated and nonirradiated bystander cells. Cancer Biother Radiopharm. 2011;26(5):657–63.
117. Herok R, Konopacka M, Polanska J, Swierniak A, Rogolinski J, Jaksik R, et al. Bystander effects induced by medium from irradiated cells: similar transcriptome responses in irradiated and bystander K562 cells. Int J Radiat Oncol Biol Phys. 2010;77:244–52.
118. Chaudhry MA. Bystander effect: biological endpoints and microarray analysis. Mutat Res. 2006;597:98–112.
119. Iwakawa M, Hamada N, Imadome K, Funayama T, Sakashita T, Kobayashi Y, et al. Expression profiles are different in carbon ion-irradiated normal human fibroblasts and their bystander cells. Mutat Res. 2008;642:57–67.
120. Ghandhi SA, Ming L, Ivanov VN, Hei TK, Amundson SA. Regulation of early signaling and gene expression in the alpha-particle and bystander response of IMR-90 human fibroblasts. BMC Med Genomics. 2010;3:31.

121. Iyer R, Lehnert BE. Factors underlying the cell growth-related bystander responses to alpha particles. Cancer Res. 2000;60:1290–8.
122. Shao C, Furusawa Y, Aoki M, Matsumoto H, Ando K. Nitric oxide-mediated bystander effect induced by heavy-ions in human salivary gland tumour cells. Int J Radiat Biol. 2002;78:837–44.
123. Azzam EI, de Toledo SM, Waker AJ, Little JB. High and low fluences of alpha-particles induce a G1 checkpoint in human diploid fibroblasts. Cancer Res. 2000;60:2623–31.
124. Lyng FM, Seymour CB, Mothersill C. Initiation of apoptosis in cells exposed to medium from the progeny of irradiated cells: a possible mechanism for bystander-induced genomic instability? Radiat Res. 2002;157:365–70.
125. Nagar S, Smith LE, Morgan WF. Characterization of a novel epigenetic effect of ionizing radiation: the death-inducing effect. Cancer Res. 2003;63:324–8.
126. Azzam EI, de Toledo SM, Gooding T, Little JB. Intercellular communication is involved in the bystander regulation of gene expression in human cells exposed to very low fluences of alpha particles. Radiat Res. 1998;150:497–504.
127. Shao C, Furusawa Y, Kobayashi Y, Funayama T, Wada S. Bystander effect induced by counted high-LET particles in confluent human fibroblasts: a mechanistic study. FASEB J. 2003;17:1422–7.
128. Harada K, Nonaka T, Hamada N, Sakurai H, Hasegawa M, Funayama T, et al. Heavy-ion-induced bystander killing of human lung cancer cells: role of gap junctional intercellular communication. Cancer Sci. 2009;100:684–8.
129. Suzuki M, Tsuruoka C. Heavy charged particles produce a bystander effect via cell-cell junctions. Biol Sci Space. 2004;18:241–6.
130. Vance MM, Wiley LM. Gap junction intercellular communication mediates the competitive cell proliferation disadvantage of irradiated mouse preimplantation embryos in aggregation chimeras. Radiat Res. 1999;152:544–51.
131. Mitchell SA, Randers-Pehrson G, Brenner DJ, Hall EJ. The bystander response in C3H 10T1/2 cells: the influence of cell-to-cell contact. Radiat Res. 2004;161:397–401.
132. Autsavapromporn N, de Toledo SM, Buonanno M, Jay-Gerin JP, Harris AL, Azzam EI. Intercellular communication amplifies stressful effects in high-charge, high-energy (HZE) particle-irradiated human cells. J Radiat Res (Tokyo). 2011;52:408–14.
133. Harris AL. Emerging issues of connexin channels: biophysics fills the gap. Q Rev Biophys. 2001;34:325–472.
134. Hall EJ, Hei TK. Genomic instability and bystander effects induced by high-LET radiation. Oncogene. 2003;22:7034–42.
135. Azzam EI, de Toledo SM, Little JB. Expression of CONNEXIN43 is highly sensitive to ionizing radiation and environmental stresses. Cancer Res. 2003;63:7128–35.
136. Schubert AL, Schubert W, Spray DC, Lisanti MP. Connexin family members target to lipid raft domains and interact with caveolin-1. Biochemistry. 2002;41:5754–64.
137. Nagasawa H, Cremesti A, Kolesnick R, Fuks Z, Little JB. Involvement of membrane signaling in the bystander effect in irradiated cells. Cancer Res. 2002;62:2531–4.
138. Jensen R, Glazer PM. Cell-interdependent cisplatin killing by Ku/DNA-dependent protein kinase signaling transduced through gap junctions. Proc Natl Acad Sci USA. 2004;101:6134–9.
139. Kalvelyte A, Imbrasaite A, Bukauskiene A, Verselis VK, Bukauskas FF. Connexins and apoptotic transformation. Biochem Pharmacol. 2003;66:1661–72.
140. Tanaka M, Grossman HB. Connexin 26 gene therapy of human bladder cancer: induction of growth suppression, apoptosis, and synergy with Cisplatin. Hum Gene Ther. 2001;12:2225–36.
141. Hong X, Wang Q, Yang Y, Zheng S, Tong X, Zhang S, et al. Gap junctions propagate opposite effects in normal and tumor testicular cells in response to cisplatin. Cancer Lett. 2012;317(2):165–71.
142. Fick J, Barker II FD, Dazin P, Westphale EM, Beyer E, Israel MA. The extent of hepatocellular communication mediated by gap junctions is predictive of bystander tumor cytotoxicity in vitro. Proc Natl Acad Sci USA. 1995;92:11071–5.

143. Freeman SM, Abboud CN, Whartenby KA, Packman CH, Koeplin DS, Moolten FL, et al. The "bystander effect": tumor regression when a fraction of the tumor mass is genetically modified. Cancer Res. 1993;53:5274–83.
144. Kuriyama S, Nakatani T, Masui K, Sakamoto T, Tominaga K, Yoshikawa M, et al. Bystander effect caused by suicide gene expression indicates the feasibility of gene therapy for hepatocellular carcinoma. Hepatology. 1995;22:1838–46.
145. Mesnil M, Piccoli C, Tiraby G, Willecke K, Yamasaki H. Bystander killing of cancer cells by herpes simplex virus thymidine kinase gene is mediated by connexins. Proc Natl Acad Sci USA. 1996;93:1831–5.
146. van Dillen IJ, Mulder NH, Vaalburg W, de Vries EF, Hospers GA. Influence of the bystander effect on HSV-tk/GCV gene therapy. A review. Curr Gene Ther. 2002;2:307–22.
147. Hamada N, Kodama S, Suzuki K, Watanabe M. Gap junctional intercellular communication and cellular response to heat stress. Carcinogenesis. 2003;24:1723–8.
148. Loewenstein WR, Rose B. The cell-cell channel in the control of growth. Semin Cell Biol. 1992;3:59–79.
149. Trosko JE, Chang CC, Madhukar BV, Klaunig JE. Chemical, oncogene and growth factor inhibition gap junctional intercellular communication: an integrative hypothesis of carcinogenesis. Pathobiology. 1990;58:265–78.
150. Yamasaki H, Mesnil M, Omori Y, Mironov N, Krutovskikh V. Intercellular communication and carcinogenesis. Mutat Res. 1995;333:181–8.
151. Mehta PP, Bertram JS, Loewenstein WR. Growth inhibition of transformed cells correlates with their junctional communication with normal cells. Cell. 1986;44:187–96.
152. Yamasaki H, Naus CC. Role of connexin genes in growth control. Carcinogenesis. 1996;17:1199–213.
153. Willecke K, Eiberger J, Degen J, Eckardt D, Romualdi A, Guldenagel M, et al. Structural and functional diversity of connexin genes in the mouse and human genome. Biol Chem. 2002;383:725–37.
154. Bedner P, Niessen H, Odermatt B, Kretz M, Willecke K, Harz H. Selective permeability of different connexin channels to the second messenger cyclic AMP. J Biol Chem. 2005;281:6673–81.
155. Goldberg GS, Moreno AP, Lampe PD. Gap junctions between cells expressing connexin 43 or 32 show inverse permselectivity to adenosine and ATP. J Biol Chem. 2002;277:36725–30.
156. Niessen H, Harz H, Bedner P, Kramer K, Willecke K. Selective permeability of different connexin channels to the second messenger inositol 1,4,5-trisphosphate. J Cell Sci. 2000;113 (Pt 8):1365–72.
157. Bevans CG, Kordel M, Rhee SK, Harris AL. Isoform composition of connexin channels determines selectivity among second messengers and uncharged molecules. J Biol Chem. 1998;273:2808–16.
158. Elsayed W, Harris AL. Selective permeability of connexin channels among inositol phosphates. Biophys J. 2004;86:583a.
159. Elsayed WA, Koreen IV, Liu YL, Harris AL. Inositol phosphates are selectively permeable through heteromeric connexin channels. Mol Biol Cell. 2004;15:183a.
160. Locke D, Stein T, Davies C, Morris J, Harris AL, Evans WH, et al. Altered permeability and modulatory character of connexin channels during mammary gland development. Exp Cell Res. 2004;298:643–60.
161. Ayad WA, Locke D, Koreen IV, Harris AL. Heteromeric, but not homomeric, connexin channels are selectively permeable to inositol phosphates. J Biol Chem. 2006;281:16727–39.
162. Cronier L, Crespin S, Strale PO, Defamie N, Mesnil M. Gap junctions and cancer: new functions for an old story. Antioxid Redox Signal. 2009;11:323–38.
163. Mesnil M, Crespin S, Avanzo JL, Zaidan-Dagli ML. Defective gap junctional intercellular communication in the carcinogenic process. Biochim Biophys Acta. 2005;1719:125–45.
164. Ammerpohl O, Trauzold A, Schniewind B, Griep U, Pilarsky C, Grutzmann R, et al. Complementary effects of HDAC inhibitor 4-PB on gap junction communication and cellular export mechanisms support restoration of chemosensitivity of PDAC cells. Br J Cancer. 2007;96:73–81.
165. Chen W, Yan C, Hou J, Pu J, Ouyang J, Wen D. ATRA enhances bystander effect of suicide gene therapy in the treatment of prostate cancer. Urol Oncol. 2008;26:397–405.

166. Xing Y, Lu G, Xiao Y, Zeng F, Zhang Q, Xiong P, et al. [Bystander effect mediated by herpes simplex virus-thymidine kinase/ganciclovir approach on prostatic cancer cells and its regulation]. Zhonghua Yi Xue Za Zhi. 2002;82:1484–7.
167. Droge W. Free radicals in the physiological control of cell function. Physiol Rev. 2002;82:47–95.
168. Burdon RH. Control of cell proliferation by reactive oxygen species. Biochem Soc Trans. 1996;24:1028–32.
169. Finkel T, Holbrook NJ. Oxidants, oxidative stress and the biology of ageing. Nature. 2000;408:239–47.
170. Bishayee A, Hill HZ, Stein D, Rao DV, Howell RW. Free-radical initiated and gap junction-mediated bystander effect due to nonuniform distribution of incorporated radioactivity in a three-dimensional tissue culture model. Radiat Res. 2001;155:335–44.
171. Lyng FM, Seymour CB, Mothersill C. Production of a signal by irradiated cells which leads to a response in unirradiated cells characteristic of initiation of apoptosis. Br J Cancer. 2000;83:1223–30.
172. Lyng FM, Seymour CB, Mothersill C. Oxidative stress in cells exposed to low levels of ionizing radiation. Biochem Soc Trans. 2001;29:350–3.
173. Mothersill C, Stamato TD, Perez ML, Cummins R, Mooney R, Seymour C. Involvement of energy metabolism in the production of 'bystander effects' by radiation. Br J Cancer. 2000;82:1740–6.
174. Khan MA, Hill RP, Van Dyk J. Partial volume rat lung irradiation: an evaluation of early DNA damage. Int J Radiat Oncol Biol Phys. 1998;40:467–76.
175. Matsumoto H, Hayashi S, Hatashita M, Ohnishi K, Shioura H, Ohtsubo T, et al. Induction of radioresistance by a nitric oxide-mediated bystander effect. Radiat Res. 2001;155:387–96.
176. Shao C, Stewart V, Folkard M, Michael BD, Prise KM. Nitric oxide-mediated signaling in the bystander response of individually targeted glioma cells. Cancer Res. 2003;63:8437–42.
177. Huo L, Nagasawa H, Little JB. HPRT mutants induced in bystander cells by very low fluences of alpha particles result primarily from point mutations. Radiat Res. 2001;156:521–5.
178. Halliwell B. Free radicals, proteins and DNA: oxidative damage versus redox regulation. Biochem Soc Trans. 1996;24:1023–7.
179. Ames BN, Shigenaga MK, Hagen TM. Oxidants, antioxidants, and the degenerative diseases of aging. Proc Natl Acad Sci USA. 1993;90:7915–22.
180. Emerit I. Reactive oxygen species, chromosome mutation, and cancer: possible role of clastogenic factors in carcinogenesis. Free Radic Biol Med. 1994;16:99–109.
181. Cerutti P, Ghosh R, Oya Y, Amstad P. The role of the cellular antioxidant defense in oxidant carcinogenesis. Environ Health Perspect. 1994;102 Suppl 10:123–9.
182. Lampe PD, Lau AF. Regulation of gap junctions by phosphorylation of connexins. Arch Biochem Biophys. 2000;384:205–15.
183. Echetebu CO, Ali M, Izban MG, MacKay L, Garfield RE. Localization of regulatory protein binding sites in the proximal region of human myometrial connexin 43 gene. Mol Hum Reprod. 1999;5:757–66.
184. Ackerman WE, Robinson JM, Kniss DA. Despite transcriptional and functional coordination, cyclooxygenase-2 and microsomal prostaglandin E synthase-1 largely reside in distinct lipid microdomains in WISH epithelial cells. J Histochem Cytochem. 2005;53:1391–401.
185. Zhu A, Zhou H, Leloup C, Marino SA, Geard CR, Hei TK, et al. Differential impact of mouse Rad9 deletion on ionizing radiation-induced bystander effects. Radiat Res. 2005; 164:655–61.
186. Lieberman HB. Rad9, an evolutionarily conserved gene with multiple functions for preserving genomic integrity. J Cell Biochem. 2006;97:690–7.
187. Nagasawa H, Little JB. Bystander effect for chromosomal aberrations induced in wild-type and repair deficient CHO cells by low fluences of alpha particles. Mutat Res. 2002;508:121–9.
188. Weill D, Gay F, Tovey MG, Chouaib S. Induction of tumor necrosis factor alpha expression in human T lymphocytes following ionizing gamma irradiation. J Interferon Cytokine Res. 1996;16:395–402.
189. Hosoi Y, Miyachi H, Matsumoto Y, Enomoto A, Nakagawa K, Suzuki N, et al. Induction of interleukin-1beta and interleukin-6 mRNA by low doses of ionizing radiation in macrophages. Int J Cancer. 2001;96:270–6.

190. Sawant SG, Randers-Pehrson G, Metting NF, Hall EJ. Adaptive response and the bystander effect induced by radiation in C3H 10T(1/2) cells in culture. Radiat Res. 2001;156:177–80.
191. Little JB, Nagasawa H, Li GC, Chen DJ. Involvement of the nonhomologous end joining DNA repair pathway in the bystander effect for chromosomal aberrations. Radiat Res. 2003;159:262–7.
192. Ward JF. The radiation-induced lesions which trigger the bystander effect. Mutat Res. 2002;499:151–4.
193. Jay-Gerin JP, Meesungnoen J, Banville P, Mankhetkorn S. Comment on "The radiation-induced lesions which trigger the bystander effect" by J.F. Ward [Mutat. Res. 499 (2002) 151–154]. Mutat Res. 2003;525:125–7.
194. Brenner DJ, Little JB, Sachs RK. The bystander effect in radiation oncogenesis: II. A quantitative model. Radiat Res. 2001;155:402–8.
195. Little MP, Wakeford R. The bystander effect in C3H 10T cells and radon-induced lung cancer. Radiat Res. 2001;156:695–9.
196. NAS. Health effects of exposure to radon (BEIR VI). Washington, DC: National Academy Press; 1998.
197. Little JB. What are the risks of low-level exposure to alpha radiation from radon? Proc Natl Acad Sci USA. 1997;94:5996–7.
198. Rees GJ, Ross CM. Abscopal regression following radiotherapy for adenocarcinoma. Br J Radiol. 1983;56:63–6.
199. Ohba K, Omagari K, Nakamura T, Ikuno N, Saeki S, Matsuo I, et al. Abscopal regression of hepatocellular carcinoma after radiotherapy for bone metastasis. Gut. 1998;43:575–7.
200. Uchida A, Mizutani Y, Nagamuta M, Ikenaga M. Effects of X-ray irradiation on natural killer (NK) cell system. I. Elevation of sensitivity of tumor cells and lytic function of NK cells. Immunopharmacol Immunotoxicol. 1989;11:507–19.
201. Camphausen K, Moses MA, Menard C, Sproull M, Beecken WD, Folkman J, et al. Radiation abscopal antitumor effect is mediated through p53. Cancer Res. 2003;63:1990–3.
202. Komarova EA, Diatchenko L, Rokhlin OW, Hill JE, Wang ZJ, Krivokrysenko VI, et al. Stress-induced secretion of growth inhibitors: a novel tumor suppressor function of p53. Oncogene. 1998;17:1089–96.
203. Mothersill C, Seymour CB. Bystander and delayed effects after fractionated radiation exposure. Radiat Res. 2002;158:626–33.
204. Mothersill C, O'Malley K, Seymour CB. Characterisation of a bystander effect induced in human tissue explant cultures by low let radiation. Radiat Prot Dosimetry. 2002;99:163–7.
205. Mesnil M, Yamasaki H. Bystander effect in herpes simplex virus-thymidine kinase/ganciclovir cancer gene therapy: role of gap-junctional intercellular communication. Cancer Res. 2000;60:3989–99.
206. Krutovskikh VA, Piccoli C, Yamasaki H. Gap junction intercellular communication propagates cell death in cancerous cells. Oncogene. 2002;21:1989–99.
207. Kennel SJ, Mirzadeh S. Vascular targeted radioimmunotherapy with ^{213}Bi—an alpha-particle emitter. Nucl Med Biol. 1998;25:241–6.
208. Huneke RB, Pippin CG, Squire RA, Brechbiel MW, Gansow OA, Strand M. Effective α-particle-mediated radioimmunotherapy of murine leukemia. Cancer Res. 1992;52:5818–20.
209. Fisher DR. Alpha-particle emitters in medicine. In: Adelstein SJ, Kassis AI, Burt RW, editors. Dosimetry of administered radionuclides. Washington, DC: American College of Nuclear Physicians; 1990. p. 194–214.
210. Sigurdson AJ, Jones IM. Second cancers after radiotherapy: any evidence for radiation-induced genomic instability? Oncogene. 2003;22:7018–27.
211. Khan MA, Van Dyk J, Yeung IW, Hill RP. Partial volume rat lung irradiation; assessment of early DNA damage in different lung regions and effect of radical scavengers. Radiother Oncol. 2003;66:95–102.

212. Van der Meeren A, Monti P, Vandamme M, Squiban C, Wysocki J, Griffiths N. Abdominal radiation exposure elicits inflammatory responses and abscopal effects in the lungs of mice. Radiat Res. 2005;163:144–52.
213. Leonard JM, Ye H, Wetmore C, Karnitz LM. Sonic Hedgehog signaling impairs ionizing radiation-induced checkpoint activation and induces genomic instability. J Cell Biol. 2008;183:385–91.
214. Emerit I, Quastel M, Goldsmith J, Merkin L, Levy A, Cernjavski L, et al. Clastogenic factors in the plasma of children exposed at Chernobyl. Mutat Res. 1997;373:47–54.
215. Pant GS, Kamada N. Chromosome aberrations in normal leukocytes induced by the plasma of exposed individuals. Hiroshima J Med Sci. 1977;26:149–54.
216. Diallo I, Haddy N, Adjadj E, Samand A, Quiniou E, Chavaudra J, et al. Frequency distribution of second solid cancer locations in relation to the irradiated volume among 115 patients treated for childhood cancer. Int J Radiat Oncol Biol Phys. 2009;74:876–83.

The Role of Genes on the Metabolism of Chemotherapeutic Agents and Their Impact on Toxicity

Young S. Yi and Gil Alterovitz

Abstract The efficacy of chemotherapy depends on several different patient factors, including age, gender, ethnicity, and genetics. A patient's genetic composition strongly influences how chemotherapeutic agents are absorbed, distributed, metabolized, and excreted by the body. Consequently, variations in genes that are responsible for these activities play a significant role in determining the prognostic outcome for patients. Studying and identifying these associated genes have been main foci of chemotherapy research, and much progress has been made in understanding the metabolic relationship between genes and chemotherapeutic agents. This understanding can be used to personalize chemotherapy regimens and ensure that treatment benefits outweigh treatment risks.

Introduction

Although chemotherapy saves many lives, it comes with a new set of problems, including drug-resistance and chemo-toxicity [1]. Chemo-toxicity in particular can lead to adverse effects that negatively impact the overall course of treatment and in some cases even death (Table 1) [2, 3]. A recent study has shown that more than half of all patients undergoing chemotherapy experience at least one potential drug interaction episode, with one-third of the patients showing either life-threatening symptoms or consequences leading to permanent physical sequelae [4]. For some chemotherapy agents, more than half of their use has been associated with various cardiac problems, including left ventricular dysfunction, cardiac ischemia, and QTc prolongation [5]. The severity and types of clinical presentations of chemo-toxicity are heterogeneous

Y.S. Yi
Harvard School of Dental Medicine, Boston, MA, USA

G. Alterovitz (✉)
Harvard Medical School, Boston, MA, USA
e-mail: ga@alum.mit.edu

S.T. Sonis and D.M. Keefe (eds.), *Pathobiology of Cancer Regimen-Related Toxicities*, DOI 10.1007/978-1-4614-5438-0_4, © Springer Science+Business Media New York 2013

Table 1 Chemotherapy agents and commonly reported adverse signs and symptoms

Chemotherapy agents	Adverse signs and symptoms
Thiopurine [18]	Leukopenia, hematopoietic toxicity
5-FU [9]	Leukopenia, anemia, diarrhea, mucositis, nausea, vomiting
Irinotecan [1, 27]	Diarrhea, leukopenia
Platinum agents [30]	Neuropathy, sensory alteration, renal tubular damage, nausea, vomiting, ototoxicity, thrombocytopenia
EGFR inhibitors [39]	Hypertension, arterial thromboembolism, bowel perforation, reversible posterior leukoencephalopathy syndrome, cardiomyopathy, renal damage, hemorrhage

among different populations and may seem completely unpredictable at first glance [1]. Symptoms commonly present as combinations of nausea, vomiting, mucositis, fatigue, anxiety, hematologic toxicity, and pain [3]. Many studies have identified gender, age, and ethnicity to be major risk factors for developing certain symptoms of chemo-toxicity [1, 6–9] but at this time do not fully explain the varying degrees and types of these symptoms following chemotherapeutic intervention [2]. Furthermore, studies have shown that the patient stratification originated from these major risk factors are attributed to genetic differences existing in various patient groups [9].

Genetic polymorphisms are genetic variants that are observed in at least 1% of individuals in a given population [10]. These may be changes in single nucleotides, in the number of repeated nucleotide sequences, or in the number of tandem repeats [11]. Genetic polymorphisms may also affect gene expression via changes in DNA sequence around actual genes, a phenomenon known as epigenetic effects [11]. Some of these polymorphisms have been identified to be responsible for making patients extremely susceptible to toxicity of specific chemotherapy regimens [12]. Naturally, most of these genetic alterations are located in genes involved with processing of various chemotherapeutic agents. These sites include genes associated with absorption, digestion, distribution, and excretion of drugs, in addition to chemokine expression which influences recruitment of other cellular factors [2, 6, 13]. Differences between some of these genetic polymorphisms have been shown to increase or decrease enzymatic activity by up to 200-fold [7].

Additionally, polymorphisms that are associated with general detoxification have been examined closely. For example, breast cancer patients with polymorphisms that lead to an increased level of superoxide dismutase-2, an enzyme crucial in antioxidant activity of cells, exhibit a lesser degree of drug toxicity but also a higher level of cancer recurrence. These beneficial and harmful effects can both be attributed to superoxide dismutase-2's role in drug metabolism [14].

Identification of genetic variants is essential not only for assessment of toxicity risk but also for prediction of chemotherapy efficacy. For instance, cytarabine is a nucleoside analogue commonly used for treatment of acute myeloid leukemia [15]. The success rate of patients treated with this drug has been shown to be highly variable due to differences in metabolic enzyme activity [15]. This observation is also true for other nucleoside analogues, including gemcitabine, decitabine, and clofarabine [15]. In particular, gemcitabine, an antimetabolite drug for solid tumors, causes varying degrees of toxicities depending on an individual's allele for cytidine deaminase (CDA), a hepatic enzyme that eliminates the agent [16]. Naturally, individuals with decreased CDA activity are more likely to experience toxicity, and functional testing of this enzyme may potentially be a relatively simple and inexpensive tool for stratifying patient risk [16].

Therefore, systematically examining both systemic and drug-specific metabolic pathways can reveal unidentified genetic polymorphisms that may play an important role in chemotherapeutic agent efficacy and predicting drug toxicity [17]. Consequently, it is essential to understand the mechanism of action and pharmacokinetics for each chemotherapeutic agent. While this process may seem daunting at first, it should be noted that many of these agents share common mechanisms of action and metabolic pathways and can thus be classified into and addressed as broader drug categories [1, 2].

Mechanisms: Thiopurines

Thiopurines are a class of drugs that is used for acute lymphocytic leukemia, acute myeloid leukemia, rheumatic disease, and dermatologic disorders [1, 13]. They are prodrugs that must be converted to thioguanine nucleotides to become active and able to exert cytotoxic effects on target cells [1]. Multiple enzymes crucial in this activation cascade have been identified. For instance, hypoxanthine phosphoribosyltransferase (HPRT) is an activator, and xanthine oxidase (XO) and thiopurine methyltransferase (TPMT) are inhibitors of the prodrug conversion pathway [1]. Therefore, genetic polymorphisms in these genes that result in changes in the enzyme level can significantly affect how a patient will respond to thiopurines. For instance, one study has shown that individuals with debilitating polymorphisms in one or both copies of XO or TPMT genes have significantly lower levels of these enzymes, making the circulating level of activated thiopurines higher compared to normal individuals (Fig. 1). As a consequence, these individuals are more likely to exhibit moderate to severe hematopoietic toxicity. This same study further suggests that individuals with such mutations should be treated with 7–10% of a standard dose to reach similar therapeutic efficacy as those without these mutations on a standard dose [18].

Screening all patients for all potential polymorphisms in the thiopurines activation pathway may not be as cumbersome as it may first seem. Epidemiological studies have shown that three alleles account for more than 95% of significant polymorphisms, making relatively simple PCR-based assays a feasible screening method [1, 13]. One of these three alleles, TPMP-2, is particularly important as it has been implicated in reducing enzyme activity by more than 100-fold [19].

One caveat is that several small studies have failed to show any significant difference between individuals with wild-type and polymorphic gene expressions, trivializing previous claims that these polymorphisms may be a direct cause of thiopurine toxicity. Therefore, further research may be needed before testing becomes standard practice.

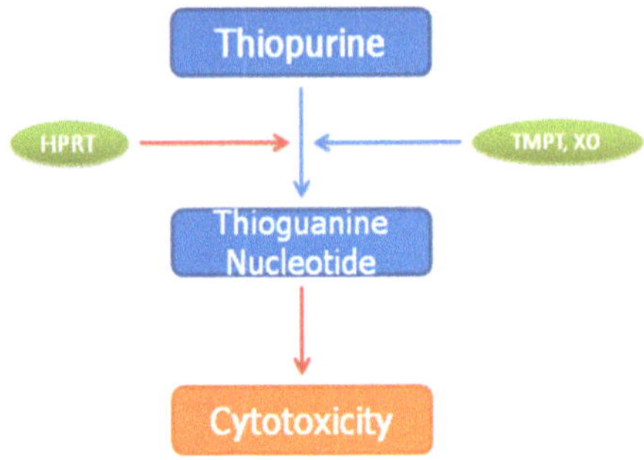

Fig. 1 Mechanism of Thiopurine agent. Thiopurine agents must be first converted to thioguanine nucleotides in order to cause cytotoxic effects to target cells. TMPT and XO increase this conversion rate whereas HPRT decreases it

Mechanisms: 5-Flurouracil

Drugs in the 5-flurouracil (5-FU) class are used for various types of solid tumors, including colorectal and breast cancers [1, 20]. These drugs must be activated to 5-fluoro-2-deoxyuridine monophosphate (5-FdUMP), which in turn inhibits thymidylate synthase (TS), an essential enzyme in DNA synthesis [1]. Therefore, the overall effect of 5-flurouracil on DNA synthesis is to halt tumor cell DNA replication. A liver enzyme, dihydropyrimidine dehydrogenase (DPD), is responsible for natural inactivation of approximately 85% of 5-FU, and, therefore, polymorphisms resulting in a significant decrease in DPD's activity increase the circulating level of 5-FU (Fig. 2). Patients with these genetic polymorphisms are more likely to experience gastrointestinal, hematopoietic, and neurological toxicities; in extreme cases, 5-FU toxicity has led to death [1, 21]. Another enzyme playing a crucial role in metabolism of 5-FU is methylene tetrahydrofolate reductase (MTHFR), an enzyme that produces a folate cofactor which inhibits TS [22]. A prospective study of 683 patients undergoing FU monotherapy concluded that various genotypes of TS, DPD, and MTHFR differentially affect the likelihood of patients' developing 5-FU toxicity [23]. The authors of this study also observed that males seemed more likely to experience toxicity than females, suggesting that patient gender may be an effect modifier with regard to the association between genetic polymorphisms and the severity of toxicity. Because the original publication did not specify gender distribution within its experimental and control groups, others questioned the validity of such a conclusion; however, this challenge was successfully rebutted by the authors, who showed that the patient gender was equally distributed within each chemotherapy regimen group [24].

However, the controversy regarding whether polymorphisms of the TS gene may influence 5-FU-induced toxicity does not end there. Another prospective study of 349 patients failed to show a significant correlation between more severe FU-induced toxicity and 20 common polymorphisms of the TS gene [25]. In addition, the fact that these polymorphisms are rare, yielding low positive predictive value of any polymorphism identification has further hampered transition to standard TS gene

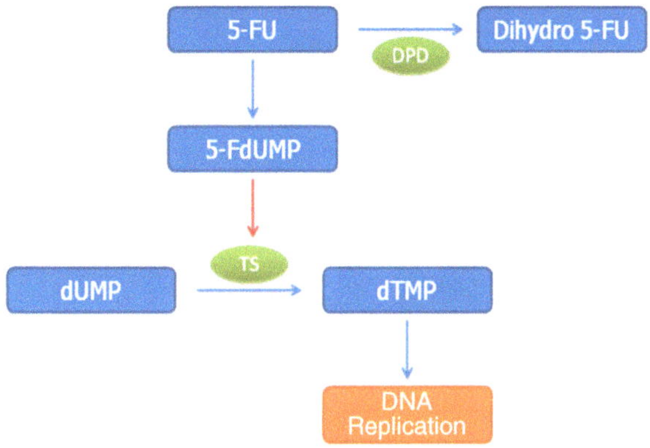

Fig. 2 Mechanism of 5-FU agent. 5-FU is first converted to its active form, 5-FdUMP, which inhibits TS activity. Inhibited TS activity decreases conversion of dUMP to dTMP, which eventually decreases DNA replication rate. 5-FU drug is normally inactivated by a liver enzyme, DPD, to dyhdro 5-FU

screening in clinical practice [1]. While the causal relationship between TS polymorphisms and FU-induced toxicity seems promising and logical, more data are needed for 5-FU toxicity risk stratification [26].

Mechanisms: Irinotecan

Irinotecan is a class of drugs used for solid tumors, including advanced colorectal cancer and lung cancer [1, 20, 27]. Carboxylesterase hydrolyzes these drugs into SN38, a topoisomerase I inhibitor [1, 6, 20]. Hepatic UDP-glucuronosyl-transferase (UGT) 1A1 inactivates this active metabolite via glucuronidation, and genetic polymorphisms of this gene lead to significantly increased levels of the topoisomerase I inhibitor. Patients with such genetic alterations exhibit severe diarrhea, neutropenia, and leucopenia upon prolonged treatment with irinotecan drugs [1, 6, 27]. In addition, these polymorphisms can lead to other clinical presentations, depending on other normal functions carried out by UGT1A1. Normally, UGT1A1, one of nine functional proteins in UGT1A enzyme family, glucurodinates xenobiotic compounds and endogenous steroids and conjugates bilirubin (Fig. 3). Therefore, a lack of normal activity of this enzyme may lead to an increased amount of unconjugated bilirubin, resulting in jaundice, a clinical manifestation that may easily be detected and could therefore serve as a marker of chemo-toxicity. However, it has been suggested that the degree of reduction in enzyme activity must be greater for jaundice symptoms to occur compared to what is needed for chemo-toxicity [28].

Much like other chemotherapies discussed earlier, UGT1A1 enzyme activity and its association with chemotherapy toxicity remains uncertain [6, 25]. No significant difference in irinotecan toxicity has been shown between advanced

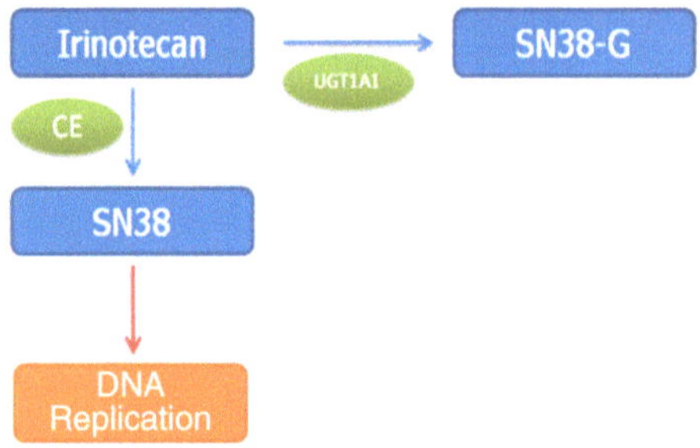

Fig. 3 Mechanism of Irinotecan agent. Irinotecan is first converted to SN38, a topoisomerase I inhibitor, by CE. The topoisomerase I inhibitor then inhibits DNA replication of rapidly growing tumor cells. UGT1A1 normally converts irinotecan to its inactive form, SN38-G

lung cancer patients with wild-type vs. polymorphic genotypes undergoing low-dose irinotecan therapy [29]. Furthermore, the effects of polymorphisms in other functional protein groups of UGT1A, namely UGT1A7 and UGT1A9, are controversial [1].

The US Food and Drug Administration (FDA) recognizes individuals with UGT1A1 polymorphisms as potentially high-risk for irinotecan toxicity and recommends that homozygotic individuals with UGT1A1 polymorphisms be treated at least one level lower than the standard irinotecan starting dose, in particular when the chemotherapy regimen is comprised of at least one other chemotherapeutic agent [6].

In addition to the polymorphisms involved with topoisomerase activity of the irinotecan drugs, the ones that alter protein binding rate of SN38 and extracellular transport efficiency have also been shown to cause interindividual variations in overall metabolic activity of the chemotherapeutic agent [6]. Therefore, future studies should focus not only on the polymorphisms of UGT1A but also on other polymorphisms that may theoretically affect irinotecan function and metabolism.

Mechanisms: Platinum Agents

Cisplatin, a platinum agent, is the most commonly used chemotherapeutic agent in the United States [30]. All three generations of platinum agents are mainly used for treatment of solid cancers [30] and are considered part of the standard chemotherapy regimen for non-small cell lung cancer (NSCLC) [31]. Additionally, randomized clinical trains have demonstrated that cisplatin's efficacy in treating adjuvant and advanced gastric cancer is promising [32].

Platinum agents cause structural changes in DNA, which eventually inhibits cellular replication [1]. Polymorphisms in genes crucial to DNA repair machinery are thought to account for up to 95% of platinum agent toxicity [33]. These genes include PIK3, PTEN, AKT, and genes involved in the mTOR pathway, and studies have demonstrated their influence on clinical outcomes of treatment of NSCLC patients with platinum-based chemotherapy agents [34].

One of the most common and well-established polymorphisms relevant to platinum agents is in glutathione S-transferase (GST) genes [33]. The GSTs naturally limit the circulating active platinum agents and are frequently overexpressed in tumor cells [33]. This overexpression compromises the efficacy of platinum agents against these growing cells, which clearly demonstrates the importance of these enzymatic activities in cancer treatment. A deficiency in XPD activity, a DNA excision repair gene, has been shown to increase the effect of platinum agents [1, 33]. Also, polymorphic XRCC1, a single-strand repair gene, and GSH-dependent enzymes, transporters of platinum agents, put patients at higher risk for neurotoxicity, renal toxicity, irreversible ototoxicity, and hematologic toxicity [1, 31].

Toxicity of platinum agents, even without considering these complex genetic factors, is a serious matter. Oxaliplatin, a third-generation platinum agent, has been associated with up to 70% prevalence of sensory neuropathy, and the FDA has officially recognized this as a serious adverse effect that needs to be taken into consideration prior to prescription [30]. Even with all of these overwhelming data on platinum agent toxicity, further studies are needed before an effective genetic screening tool can be developed and put to use. This is due to several studies that showed clinical results contradicting predictions made based on previous studies. For instance, one study on cisplatin-based chemotherapy found no association between genetic polymorphisms and adverse toxicity in NSCLC patients [35]. Another similar study in advanced NSCLC patients also did not find any correlation between chemo-toxicity and XPD or ERCC1 polymorphisms [36]. The same study, however, found that patients with Lys27Lys polymorphism in their CDA responded more favorably to cisplatin treatment, suggesting that CDA enzyme activity is of interest in determining efficacy of and risk for adverse effects due to platinum agents. Patient ethnicity may also be an important factor, as patients of Asian ethnicity are thought to have an "enriched phenotype". In other words, these patients who are undergoing platinum-based chemotherapy exhibit more frequent and severe toxicity symptoms in comparison to patients of other ethnic groups [37]. Therefore, studying the genetic differences between Asians and other ethnic groups may help in identifying novel genetic determinants that govern differences in platinum chemo-toxicity susceptibility.

Mechanisms: Endothelial Growth Factor Receptor Inhibitors

Two major types of Endothelial Growth Factor Receptor (EGFR) inhibitors are tyrosine kinase inhibitors (TKI) and monoclonal antibodies [38, 39]. Examples of TKIs include erlotinib and gefitinib [8, 38], which prevent autophosphorylation of a tyrosine kinase domain and subsequently block downstream activation of MAPK, PI3K/AKT, and STAT signaling pathways. Cetuximab is a type of monoclonal antibody that binds to an extracellular domain of EGFRs and prevents ligand-receptor binding, resulting in the same outcome [8, 38]. These pathways, when inhibited, cause decreased proliferation and growth of affected cells, via mechanisms such as blocking angiogenesis of rapidly growing tumor cells, which are essential in development and progression of various solid tumors [38–40]. Clinical efficacy of EGFRs has already been shown against lung, colorectal, breast, and kidney cancers [39].

These drugs are particularly used for treatment of NSCLC and renal cell carcinoma, and several genetic predictors for therapeutic outcomes and adverse effects of the agent have been identified [8, 38, 39]. A higher EGFR copy-number within the cancer cells increases a patient's response rate to TKIs, resulting in a better response rates to therapy [38]. On the other hand, ATP-binding cassette transporter G2 (ABCG2) actively transports intracellular TKIs out to the extracellular space. As a consequence, genetic variants that decrease the rate of transport lead to accumulation of the agent intracellularly, therefore increasing the likelihood of EGFR inhibitor toxicity [38]. Further complicating the matter of studying the relationship between these polymorphisms and the drugs is the complexity of these signaling pathways and multitude of transducers involved [8].

Chapter Conclusion

These five classes of chemotherapeutic agents are by no means comprehensive. Genetic polymorphisms in genes essential for the metabolism of doxorubicin and cyclophosphamide, both of which are used in breast cancer patients, affect how patients respond to these therapies [41]. Rituximab is anti-CD20 monoclonal antibody that can be used alone or as part of a comprehensive chemotherapy regimen. Standard therapy in treating diffuse large B-cell lymphoma includes use of rituximab along with cyclophosphamide, hydroxydaunorubicin, oncovin, and prednisone (CHOP) [20, 42, 43]. Molecular cascades involved with activation, detoxification, and extracellular efflux have already been identified as potential genetic polymorphisms sites that could affect adverse and therapeutic effects of the chemotherapy regimen [42]. ABCG2 is a type of active transmembrane transporters that binds doxorubicin, in addition to mitoxantrone, topotecan, irinotecan, flavopiridol, and methotrexate [43]. Genetic changes in expression or function of ABCG2 have been linked to various toxicity symptoms, including diarrhea in patients undergoing rituximab-CHOP therapy [43].

Societal and individual economic consequences of screening for various genetic polymorphisms have been examined by several scientific investigators. While data exist only for limited types of cancer and chemotherapy, many studies show a favorable financial outcome for genetic screening prior to administration of anticancer treatment [44]. With modern technology, searching the genome of a patient for polymorphisms is also relatively a simple task. The true difficulty lies in identifying genetic polymorphisms that are significantly associated with a patient's susceptibility to toxicity caused by a particular chemotherapeutic agent. Furthermore, distinguishing whether an exhibited symptom is from cancer, chemotherapy, or other associated diseases remains as a challenge [3]. Due to the complexity of most chemotherapeutic regimens and their biochemical mechanisms, blindly studying all possible polymorphisms for such association would be extremely inefficient.

Therefore, future systematic studies should utilize already-established polymorphisms and metabolic pathways [44].

Once relevant polymorphisms are identified for each class of chemotherapeutic agents, their presence in the genome of cancer patients can be evaluated. This, in turn, will help to risk-stratify patients for specific chemotherapeutic interventions. Mathematical computations of the risk can accurately predict the likelihood of a patient for developing chemo-toxicity, which will help personalize a patient's cancer treatment. Tailoring of chemotherapeutic interventions has already been shown to be of significant clinical benefit in the treatment of several cancers. For example, pretreatment genotyping using peripheral blood samples of colorectal cancer patients is already being used to personalize their treatment [45]. Colorectal cancer is notoriously known for its resistance to chemotherapy and several genetic markers have already been linked to accurate prediction of treatment effectiveness, highlighting a clinical advantage of genetic screening in treating cancer patients [46].

Large clinical trials are necessary to develop and fine-tune mathematical models for computing the toxicity susceptibility. Yet, they can be expensive and time-consuming. Therefore, studying animal models may be a helpful alternative before such trials are implemented [44, 47]. Knock-out models can mimic the enzymatic effects of certain polymorphisms, and these animal models can be treated with chemotherapeutic drugs [44]. On the other hand, peripheral blood samples, which are easier to collect than a cohort of patients for a prospective study, can be used for retrospective polymorphism analysis in patients who experience chemo-toxicity [48].

One final caveat is that, even when potential drug-associated risks are identified, there are seldom alternative treatment options to achieve the same therapeutic effect. Then, the question becomes where to draw the line for balancing the harmful chemo-toxicity and effective treatment. Therefore, in order to optimize the survival of cancer patients and their quality of life during chemotherapy, future studies should focus not only on finding genetic polymorphisms but also on developing alternative chemotherapeutic agents.

References

1. Watters JW, McLeod HL. Cancer pharmacogenomics: current and future applications. Biochim Biophys Acta. 2003;1603(2):99–111.
2. Efferth T, Volm M. Pharmacogenetics for individualized cancer chemotherapy. Pharmacol Ther. 2005;107(2):155–76.
3. Aprile G, et al. Application of distance matrices to define associations between acute toxicities in colorectal cancer patients receiving chemotherapy. Cancer. 2008;112(2):284–92.
4. van Leeuwen RW, et al. Potential drug interactions in cancer therapy: a prevalence study using an advanced screening method. Ann Oncol. 2011;22(10):2334–41.
5. Curigliano G, et al. Cardiac toxicity from systemic cancer therapy: a comprehensive review. Prog Cardiovasc Dis. 2010;53(2):94–104.
6. Hasegawa Y, et al. Pharmacogenetic approach for cancer treatment-tailored medicine in practice. Ann N Y Acad Sci. 2006;1086:223–32.

7. Nebert DW, McKinnon RA, Puga A. Human drug-metabolizing enzyme polymorphisms: effects on risk of toxicity and cancer. DNA Cell Biol. 1996;15(4):273–80.
8. Giovannetti E, et al. Association of polymorphisms in AKT1 and EGFR with clinical outcome and toxicity in non-small cell lung cancer patients treated with gefitinib. Mol Cancer Ther. 2010;9(3):581–93.
9. O'Donnell PH, Dolan ME. Cancer pharmacoethnicity: ethnic differences in susceptibility to the effects of chemotherapy. Clin Cancer Res. 2009;15(15):4806–14.
10. Coate L, et al. Germline genetic variation, cancer outcome, and pharmacogenetics. J Clin Oncol. 2010;28(26):4029–37.
11. Huang RS, Dolan ME. Approaches to the discovery of pharmacogenomic markers in oncology: 2000-2010-2020. Pharmacogenomics. 2010;11(4):471–4.
12. Bozina N, Bradamante V, Lovric M. Genetic polymorphism of metabolic enzymes P450 (CYP) as a susceptibility factor for drug response, toxicity, and cancer risk. Arh Hig Rada Toksikol. 2009;60(2):217–42.
13. Robert J, et al. Predicting drug response and toxicity based on gene polymorphisms. Crit Rev Oncol Hematol. 2005;54(3):171–96.
14. Yao S, et al. Manganese superoxide dismutase polymorphism, treatment-related toxicity and disease-free survival in SWOG 8897 clinical trial for breast cancer. Breast Cancer Res Treat. 2010;124(2):433–9.
15. Lamba JK. Genetic factors influencing cytarabine therapy. Pharmacogenomics. 2009;10(10):1657–74.
16. Ciccolini J, et al. Cytidine deaminase residual activity in serum is a predictive marker of early severe toxicities in adults after gemcitabine-based chemotherapies. J Clin Oncol. 2010;28(1):160–5.
17. Gasparini G, et al. Is tailored therapy feasible in oncology? Crit Rev Oncol Hematol. 2006;57(1):79–101.
18. McLeod HL, et al. Genetic polymorphism of thiopurine methyltransferase and its clinical relevance for childhood acute lymphoblastic leukemia. Leukemia. 2000;14(4):567–72.
19. Krynetski EY, et al. A single point mutation leading to loss of catalytic activity in human thiopurine S-methyltransferase. Proc Natl Acad Sci U S A. 1995;92(4):949–53.
20. Lee SY, McLeod HL. Pharmacogenetic tests in cancer chemotherapy: what physicians should know for clinical application. J Pathol. 2011;223(1):15–27.
21. Fidlerova J, et al. Contribution of dihydropyrimidinase gene alterations to the development of serious toxicity in fluoropyrimidine-treated cancer patients. Cancer Chemother Pharmacol. 2010;65(4):661–9.
22. Robien K, Boynton A, Ulrich CM. Pharmacogenetics of folate-related drug targets in cancer treatment. Pharmacogenomics. 2005;6(7):673–89.
23. Schwab M, et al. Role of genetic and nongenetic factors for fluorouracil treatment-related severe toxicity: a prospective clinical trial by the German 5-FU Toxicity Study Group. J Clin Oncol. 2008;26(13):2131–8.
24. Ezzeldin HH, Diasio RB. Predicting fluorouracil toxicity: can we finally do it? J Clin Oncol. 2008;26(13):2080–2.
25. Boige V, et al. Pharmacogenetic assessment of toxicity and outcome in patients with metastatic colorectal cancer treated with LV5FU2, FOLFOX, and FOLFIRI: FFCD 2000–05. J Clin Oncol. 2010;28(15):2556–64.
26. Lecomte T, et al. Thymidylate synthase gene polymorphism predicts toxicity in colorectal cancer patients receiving 5-fluorouracil-based chemotherapy. Clin Cancer Res. 2004;10(17):5880–8.
27. Marsh S, Hoskins JM. Irinotecan pharmacogenomics. Pharmacogenomics. 2010;11(7):1003–10.
28. Strassburg CP, et al. Family 1 uridine-5'-diphosphate glucuronosyltransferases (UGT1A): from Gilbert's syndrome to genetic organization and variability. Arch Toxicol. 2008;82(7):415–33.
29. Sugiyama T, et al. The UGT1A1*28 genotype and the toxicity of low-dose irinotecan in patients with advanced lung cancer. Oncol Res. 2010;18(7):337–42.
30. McWhinney SR, Goldberg RM, McLeod HL. Platinum neurotoxicity pharmacogenetics. Mol Cancer Ther. 2009;8(1):10–6.

31. Isla D, et al. Single nucleotide polymorphisms and outcome in docetaxel-cisplatin-treated advanced non-small-cell lung cancer. Ann Oncol. 2004;15(8):1194–203.
32. Shimoyama S. Pharmacogenetics of fluoropyrimidine and cisplatin. A future application to gastric cancer treatment. J Gastroenterol Hepatol. 2009;24(6):970–81.
33. Khrunin AV, et al. Genetic polymorphisms and the efficacy and toxicity of cisplatin-based chemotherapy in ovarian cancer patients. Pharmacogenomics J. 2010;10(1):54–61.
34. Pu X, et al. PI3K/PTEN/AKT/mTOR pathway genetic variation predicts toxicity and distant progression in lung cancer patients receiving platinum-based chemotherapy. Lung Cancer. 2011;71(1):82–8.
35. Chen S, et al. Association of MDR1 and ERCC1 polymorphisms with response and toxicity to cisplatin-based chemotherapy in non-small-cell lung cancer patients. Int J Hyg Environ Health. 2010;213(2):140–5.
36. Tibaldi C, et al. Correlation of CDA, ERCC1, and XPD polymorphisms with response and survival in gemcitabine/cisplatin-treated advanced non-small cell lung cancer patients. Clin Cancer Res. 2008;14(6):1797–803.
37. O'Donnell PH, et al. Population differences in platinum toxicity as a means to identify novel genetic susceptibility variants. Pharmacogenet Genomics. 2010;20(5):327–37.
38. Heist RS, Christiani D. EGFR-targeted therapies in lung cancer: predictors of response and toxicity. Pharmacogenomics. 2009;10(1):59–68.
39. Vaziri SA, et al. Vascular endothelial growth factor polymorphisms: role in response and toxicity of tyrosine kinase inhibitors. Curr Oncol Rep. 2010;12(2):102–8.
40. Hamadeh HK, et al. Application of genomics for identification of systemic toxicity triggers associated with VEGF-R inhibitors. Chem Res Toxicol. 2010;23(6):1025–33.
41. Bray J, et al. Influence of pharmacogenetics on response and toxicity in breast cancer patients treated with doxorubicin and cyclophosphamide. Br J Cancer. 2010;102(6):1003–9.
42. Rossi D, et al. Analysis of the host pharmacogenetic background for prediction of outcome and toxicity in diffuse large B-cell lymphoma treated with R-CHOP21. Leukemia. 2009;23(6):1118–26.
43. Kim IS, et al. ABCG2 Q141K polymorphism is associated with chemotherapy-induced diarrhea in patients with diffuse large B-cell lymphoma who received frontline rituximab plus cyclophosphamide/doxorubicin/vincristine/prednisone chemotherapy. Cancer Sci. 2008;99(12):2496–501.
44. Abraham J, et al. Pharmacogenetics of cancer chemotherapy. Biochim Biophys Acta. 2006;1766(2):168–83.
45. Ross JS, Ginsburg GS. Integrating diagnostics and therapeutics: revolutionizing drug discovery and patient care. Drug Discov Today. 2002;7(16):859–64.
46. Bandres E, et al. Pharmacogenomics in colorectal cancer: the first step for individualized-therapy. World J Gastroenterol. 2007;13(44):5888–901.
47. Roberts RB, Arteaga CL, Threadgill DW. Modeling the cancer patient with genetically engineered mice: prediction of toxicity from molecule-targeted therapies. Cancer Cell. 2004;5(2):115–20.
48. Sonis S, et al. Gene expression changes in peripheral blood cells provide insight into the biological mechanisms associated with regimen-related toxicities in patients being treated for head and neck cancers. Oral Oncol. 2007;43(3):289–300.

Animal Models of Regimen-Related Toxicities

Gregory Lyng, Richard Logan, and Maria Mancini

Abstract Radiation and chemotherapy are commonly used in the treatment of human cancer. Whilst often effective, they are known to cause a range of acute and chronic toxicities, including hematological complications (such as anemia) and non-hematological complications (such as mucositis). These toxicities can have profound effects on patients' quality of life as well as detrimentally affect treatment outcomes. Understanding these toxicities is therefore highly important. This chapter describes how animal models have been used to increase the understanding of antineoplastic therapy toxicity and how their use has led to improvements in clinical management of toxicity, including drug development.

Introduction

Radiation and chemotherapy are commonly used with great success in the treatment of multiple types of human cancer. While these therapies are extremely beneficial for the treatment of tumors, they are known to induce a wide range of acute and chronic toxicities. Not only are these side effects associated with poor health outcomes, but they limit patients' ability to tolerate treatment and incur significant increases in resources and cost. Universally, these toxicities impair patients' quality of life (QoL) and recovery in both the short and the long term.

Common complications associated with radiation and chemotherapy include hematological disorders such as anemia, thrombocytopenia, and neutropenia; however cancer patients are also at risk for a wide range of non-hematological

G. Lyng, Ph.D. • M. Mancini, Ph.D.
Biomodels, LLC, Watertown, MA, USA

R. Logan, B.D.S., Ph.D. (✉)
University of Adelaide, Adelaide, Australia
e-mail: richard.logan@adelaide.edu.au

S.T. Sonis and D.M. Keefe (eds.), *Pathobiology of Cancer Regimen-Related Toxicities*, DOI 10.1007/978-1-4614-5438-0_5, © Springer Science+Business Media New York 2013 toxicities. These may occur during or soon after cancer treatment (acute toxicities), or develop well after the completion of treatment (≥100 days, late toxicities) and become chronic and linger years after the patient's cancer is controlled. The overall incidence of some form of cancer treatment-related toxicity is almost 100%. Toxicities include those that are tissue-specific such as mucosal injury to parts of the gastrointestinal tract (mucositis and proctitis), cutaneous damage (dermatitis), salivary gland dysfunction, and venous thrombosis. Alternatively, patients may develop more systemic forms of toxicity that result in conditions such as fatigue, depression, cognitive impairment, and cachexia.

The clinical ramifications of regimen-related toxicities are diverse and significant. Some, such as oral mucositis, are associated with pain of such intensity that the patient suffers from an inability to eat and requires opioid analgesics (Sonis et al. 2004). Other toxicities, like enteritis, cause diarrhea and put patients at risk for bacteremia or sepsis. Almost all toxicities negatively impact patients' quality of life and ability to tolerate treatment, and will often necessitate less than optimal dosing regimens or early termination of treatment (Elting et al. 2008). Toxicities are also associated with higher mortality risks and higher healthcare costs. Patients with acute toxicities may require unplanned office and emergency room visits, or hospital admissions for fluid support, pain control, or infection management. The economic impact of this extra use of resources is substantial. For example, the incremental cost of oral mucositis among patients being treated for cancers of the lung or head and neck is in excess of $17,000 (Nonzee et al. 2008). Toxicities that linger in patients who have completed treatment such as fatigue and depression impair their ability to work and otherwise function normally.

Understanding chemotherapy- and radiation-induced toxicities is of high importance due to the direct impact on patients' symptoms and quality of life. This, combined with the possibility of having to limit an individuals' treatment regimen due to a decreased tolerance of the therapies employed, provides a basis for intense study. The discovery that many toxicities seem to cluster suggests shared pathoetiologies. Previous studies defining the biology of toxicities have stimulated the quest for appropriate and effective pharmacologic and biologic interventions. In recent years, the use of animal models that mimic toxicities seen in humans has broadly expanded and can reliably serve as predictive platforms for drug development. This chapter will focus on some examples of the current animal models of toxicities relating to injury in various tissues as a result of treatment with chemotherapy and radiation.

Animal Models of Oral Mucositis

Oral mucositis is one of the most prevalent toxicities associated with regimen-related therapies and is therefore among the most commonly studied. The maintenance of oral mucosal integrity has been closely linked to successful patient outcomes following radiation and/or chemotherapy treatment, as disruption of the mucosal barrier greatly increases the risk of systemic infection and results in painful and potentially dose-limiting lesions. While there have been a number of published reports suggesting novel therapies to prevent, accelerate the healing, or provide relief from the pain associated with oral mucositis, there has been limited clinical success and therefore a tremendous clinical need for effective therapies remains (Sonis 2007). With the importance placed upon finding new clinical therapies, there comes a need for powerful and predictive animal models to ensure that research efforts provide clinically meaningful data. As with any disease indication, the ultimate value of animal models of oral mucositis is defined by the model's ability to mimic the condition as it occurs in patients. For models of oral mucositis, this means that the changes to the oral mucosa should be clinically/visually identifiable and follow in its course, appearance, resolution, and dose response to stomatotoxic therapy. Additionally, the pathogenesis of mucositis should be similar to that of the human condition at the molecular, cellular, and tissue levels (Sonis 2004). Finally, the models need to respond to clinically effective therapies in a manner similar to that observed in human patients.

Current Models

Three main species are currently used for studies of oral mucositis: mice, rats, and hamsters. Since there is not one model/species that works perfectly for every indication, there are certain advantages and disadvantages to each that will be discussed. Murine models have been used to study both radiation- and chemotherapy-induced mucositis. The wide variety of available syngeneic, transgenic, and knock-out animal models, combined with the accessibility of immune reagents and gene chips, makes the mouse a good choice for answering specific questions associated with the pathogenesis of mucosal injury. In general, the endpoints used to assess mucositis in mice have relied heavily on histological outcomes, since clinical changes tend to be subtle and focus on erythema, rather than ulceration as a primary endpoint. A number of studies focusing on the epithelial biology of oral radiation have been performed using murine lip, snout, or tongue models. Xu et al. (1984) demonstrated that acute reactions of the lip mucosa, represented by focal desquamation as a result of single and fractionated radiation schedules, could be reliably scored. Alternatively, Kilic et al. (2007) developed a model where the ventral surface of the mouse tongue is isolated through a hole in an aluminum block. In this model, the body of the mouse is protected from the majority of the radiation whereas the exposed tongue receives an acute dose of radiation. While the endpoints in this model are primarily histological, we have modified this protocol in our lab utilizing lead shielding on the body of the mouse and isolating the tongue. By extending the tongue through a small gap in the shield and using sutures to pin it in place, we are able to ensure that the majority of the tongue is exposed to the radiation field. In our hands, doses of radiation directed to the tongue ranging from 20 to 40 Gy have resulted in robust ulcerative mucositis that is easily visualized and

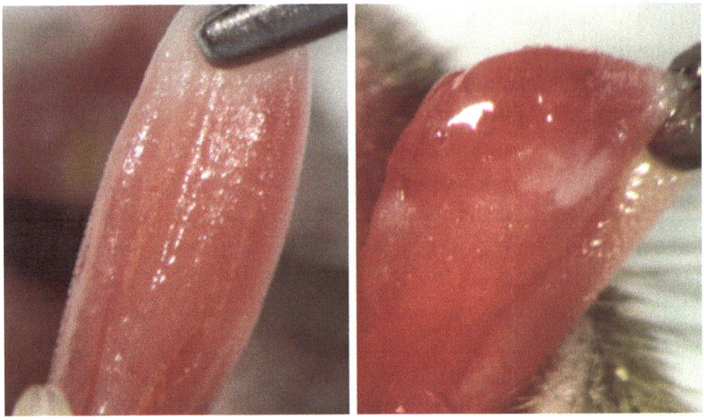

Fig. 1 The clinical appearance of normal mouse tongue (*left*) compared with mouse tongue with ulcerative mucositis (*right*)

disease severity (ulceration/mucosal integrity) can be reliably assessed utilizing stereomicroscopy. Figure 1 shows the mouse tongue in the naïve state as well as how the tongue appears when it presents with ulcerative mucositis. The robust nature of the disease in this model provides an advantage over other mouse models in that persistent ulceration lasts 4–6 days and therefore the efficacy of therapies can be effectively assessed (unpublished observations).

While less common than mice or hamsters (to be discussed later), rats can effectively be used to model radiation- and/or chemotherapy-induced mucositis. Similar to the mouse model, the ventral tongue is the most common area of disease induction/assessment and provides a larger working area (mucosal surface) than in mice. Although this anatomy provides fewer technical challenges with disease induction, the use of rats remains limited by the low availability of transgenics and fewer reagents. The most commonly used rat mucositis models involve fluorouracil (5-FU) and methotrexate therapy often accompanied by a superficial surgically induced lesion that leads to an ulcerative mucositis (Lee et al. 2007; Vivela-Goulat et al. 2008). Lesions in these models tend to be localized to the area of injury and tend to provide relatively short windows of active disease.

While mouse and rat models have been very useful to define responses to various chemotherapy and/or radiation regimens on both a cellular and molecular level, the limited anatomical area available for evaluation, the challenges associated with use of topical formulations, and the subtlety of clinical changes have limited their applicability in interventional studies for drug development.

Given some of the limitations of the murine and rat models of oral mucositis, the hamster was evaluated as a potential species to model this disease. The development of the hamster mucositis model was based on a number of factors, but the most important was the physical structure of the hamster cheek pouch which provides a larger mucosal surface that is easily accessible for disease examination and is ideally suited for the study of topical therapies. In addition, the hamster cheek pouch consists of a renewing squamous epithelium, which is similar to that of the human. From a modeling perspective, the pouch can be easily isolated with a lead shield from the rest of the animal's head thereby permitting targeted radiation with no systemic toxicity. Furthermore, the hamster's oral bacterial flora is similar to that of the human in that gram positive bacteria are the dominant species. Alterations in hamster oral flora following cancer therapy mirror those described in humans—both quantitatively and qualitatively (McMillan and Lowell 1985). Most importantly, mucositis development in the hamster is robust and follows a predictable course as erythema, superficial necrosis, and frank ulceration are followed by spontaneous healing. With all the benefits of the hamster mucositis model, there remains a shortage in the molecular biological/genetic techniques/reagents that can be used with the hamster compared to those available for the rat and mouse models. Even with this deficiency, the predictive value of the hamster model has resulted in it being the most widely used mucositis model for the development of new pharmacological therapies.

When utilizing an animal model of oral mucositis, it is extremely important to shield the entire animal from radiation while providing a targeted dose to the mucosa to prevent systemic complications. To ensure focused exposure, Sonis et al. developed a technique that used lead to shield the entire hamster with the exception of a small slit that allowed the cheek pouch to be isolated and receive radiation to the entire mucosa (Sonis et al. 2000). Following anesthesia with an intraperitoneal injection of ketamine (160 mg/ml) and xylazine (8 mg/ml), the left buccal pouch is everted, fixed, and isolated using a lead shield. Mucositis is induced with a single dose of radiation (40 Gy/dose) targeting the buccal mucosa at a rate of 1–2.5 Gy/min. The radiation dose is incredibly important in this model as there is a necessity to induce disease that is clinically significant, yet not so severe that you are not able to show a therapeutic response. In order to consistently evaluate disease severity in this model, a scoring scale was developed to grade mucositis using outcomes that are analogous to clinical scoring. The mucositis scoring scale is described descriptively in Table 1, while images of a naïve hamster cheek pouch as well as a cheek pouch with ulcerative mucositis are shown in Fig. 2.

In this model, clinically significant mucositis, characterized by erythema and superficial sloughing, is seen by day 10–12 and ulceration (the most significant clinically relevant endpoint) begins by day 14 and lasts for approximately 4–6 days. Typically, clinically significant mucositis in this model begins to resolve by day 24 and completely healed by day 35. When utilizing this model to assess a pharmacological intervention, the most translatable/predictive endpoint is not peak disease severity, but the number of days that the hamsters exhibit ulceration (or scores greater than or equal to 3 as defined by the above scale). With duration of ulceration as a primary endpoint, a 25% or greater reduction in the number of days of ulcerative mucositis days has been a consistent predictor of a robust clinical effect.

Radiation treatment for human head and neck cancer is a consistent cause of ulcerative mucositis. In this patient population, radiation is typically given in small, daily fractionated doses of 2 Gy, 5 days/week, for cumulative doses of 60–70 Gy. This fractionated radiation regimen induces a predictable pattern of mucositis with

Table 1 Scoring scale used to grade mucositis in the hamster model using outcomes that are analogous to clinical scoring

Score	Description
0	Pouch completely healthy. No erythema or vasodilation
1	Light to severe erythema and vasodilation. No erosion of mucosa
2	Severe erythema and vasodilation. Erosion of superficial aspects of mucosa leaving denuded areas. Decreased stippling of mucosa
3	Formation of off-white ulcers in one or more places. Ulcers may have a yellow/gray color due to pseudomembrane. Cumulative size of ulcers should equal less than or equal to ¼ of the pouch. Severe erythema and vasodilation
4	Cumulative size of ulcers should equal about ½ of the pouch. Loss of pliability. Severe erythema and vasodilation
5	Virtually all of pouch is ulcerated. Loss of pliability (pouch can only partially be extracted from mouth)

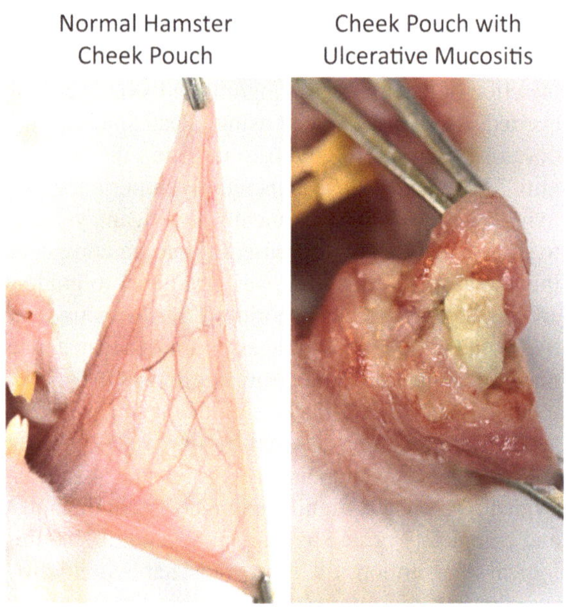

Fig. 2 The clinical appearance of normal hamster cheek pouch (*left*) compared with hamster cheek pouch with ulcerative mucositis (*right*)

erythematous changes occurring at cumulative doses of 10–20 Gy and ulceration developing at doses of 30 Gy and higher that persists for up to a month following completion of radiation therapy. The timing of the radiation therapy in the clinical setting necessitated the development of additional animal models that accurately recapitulated the regimen in patients so that optimal dose scheduling of candidate agents designed to treat the head and neck cancer population could be determined. Consequently, Sonis et al. developed a model in which animals were irradiated for four consecutive days with daily doses ranging from 5 to 15 Gy and cumulative doses of 20–60 Gy per animal. This model was later modified to more closely mimic the patient treatment paradigm by using 7.5 Gy radiation fractions delivered for four consecutive days, followed by a 2-day rest period, after which the 4-day regimen is repeated (Ara et al. 2008; Watkins et al. 2010). The use of two radiation cycles permits the study of interventional agents in schedules that are similar to humans and allows for a better understanding of the importance of dose schedule prior to entering patient trials.

Hamster Models for Chemotherapy-Induced Mucositis with or Without Concomitant Radiation

Radiation is rarely given without concomitant chemotherapy. This combination approach in anticancer therapy results in a synergy in their ability to produce mucosal injury, therefore an animal model combining the two is an important component in the development of any anti-mucositis agent (Alvarez et al. 2008).

Onset and severity of oral mucosal toxicity is dependent upon the chemotherapeutic agent selected, its dose and frequency, interval between dosing, and animal age. One example is the chemotherapeutic agent Fluorouracil (5-FU). Previous studies have determined the optimal dose and schedule for 5-FU administration. In a hamster model of oral mucositis, two intraperitoneal (IP) doses of 60 mg/kg of 5-FU were administered on days 0 and 2. Since the cheek pouch mucosa is anatomically protected from functional trauma as occurs in humans, superficial irritation of the mucosa was performed using an 18-gauge needle on day 4 to model the functional trauma that occurs in human mucosa (Sonis et al. 1990, 1995). In this model, disease peaks approximately 10 days following the first dose of 5-FU and is all but resolved by day 16–18.

Concomitant chemotherapy and radiation is significantly stomatotoxic. As such, two primary protocols have been developed in the hamster to model mucositis progression in patients receiving concomitant therapy. As described by Alvarez et al. (2003), mucositis is induced using 5-FU delivered as single IP doses (60 mg/kg) on days −4 and −2 coupled with a single dose of radiation (30 Gy/dose) administered on day 0. Another derivation of this model has utilized a single dose of 35 Gy of radiation on day 0, coupled with a 5 mg/kg dose of cisplatin on day 1 (Watkins et al. 2010). In both models, ulcerative mucositis is observed in 16–18 days following therapy and has proven effective in the development of potential novel therapeutics.

Gastrointestinal Mucositis Models

It has long been recognized that intestinal complications are a major limiting factor for radiotherapy when it is used to treat pelvic and abdominal cancer (Linard 2003; Richter 1996); likewise chemotherapy, particularly drugs such as 5-fluorouracil (5-FU) and irinotecan, causes significant mucosal damage. Unlike oral mucositis where the signs of mucosal erythema and ulceration are obvious and easy to identify, there are inherent difficulties in visualizing the clinical damage that occurs in the gastrointestinal tract. There are also obvious difficulties in retrieving specimens for analysis from patients for clinical and histological studies. As a result, investigation into the pathobiology of *gastrointestinal* mucositis, perhaps more so than oral mucositis, has heavily relied on the use of animal models. Most commonly, rat and murine models have been used to examine not only various features of the pathobiology of gastrointestinal mucositis, but also potential interventions to ameliorate the effects of mucotoxic agents (Logan 2007; Keefe 2004). The toxicities of radiotherapy and chemotherapy, including targeted treatments, have been investigated with animal models. The development of suitable animal models and subsequent characterization of changes that occur in the tissues has resulted in subsequent studies that have examined potential interventions to prevent or treat mucositis (Bowen 2011) and subsequently led to improved management strategies for mucositis in the clinical setting (for example, keratinocyte growth factor).

Radiation

Various animal models have been used to study damage to the gastrointestinal tract following radiation extending back to the beginning of the twentieth century (Bowen 2011). Damage caused by radiation includes both acute and chronic effects. Acute radiation toxicity is characterized by decreased intestinal crypt depth as well as villus height along with basement membrane degradation (Strup-Perrot 2005) whilst chronic radiation toxicity is associated largely with fibrosis. Murine (Kawashima 2006; Li 2005; Polistena 2008; Ponemone 2010) and rat (including Sprague–Dawley [Richter 1996; Allgood 1996; Landberg 1996; Langberg 1996b; Landberg 1994; Landberg 1994b; Landberg 1996; Landberg 1994; Richter 1997; Scott 1995; Hauer-Jensen 1988], Wistar [Strup-Perrot 2005; Cameron 2011], and Dark Agouti [Ong 2010;Yeoh 2007]) have been used to demonstrate radiation-induced damage to different sites in the gastrointestinal tract. As well as the animal used, the techniques of intestinal irradiation studied have varied including direct radiation in the form of whole body irradiation (Garg 2010) or selective radiation to specific areas of the bowel (Polistena 2008). Studies have also investigated the role of scattered radiation in causing gastrointestinal mucosal toxicity (Cameron 2011) which is clinically relevant if the intestinal mucosa is inadvertently irradiated when other abdominal organs/structures are treated.

Early studies referred to radiation enteropathy or late intestinal radiation injury, which is characterized by mucosal ulceration, inflammation, vascular sclerosis, and reactive fibrosis (Richter 1996). These histological changes that were described essentially appear to reflect what is now recognized as gastrointestinal mucositis; furthermore it was shown early on that that pro-inflammatory cytokines might play a role in the pathobiology of mucosal damage in irradiated intestinal mucosa. Increased immunohistochemical staining for transforming growth factor-β1 (TGF-β1), platelet derived growth factor-AA (PDGF-AA), and interleukin-1α (IL-1α) was demonstrated early on (Landberg 1994). Subsequent studies have demonstrated also that interleukin-1 (IL-1), tumor necrosis factor (TNF), and interleukin-6(IL-6) levels are elevated in the small intestine following radiation exposure as well as activation of nuclear factor-κB (NF-κB) (Linard 2003). This is in keeping with the current understanding of the pathobiology of mucositis.

Murine Models

Murine models were used in the early 1980s to examine the effect of radiation on intestinal crypt cells, specifically the stem cell population in the crypts (Potten 1981). These early studies used the "Mouse crypt survival model," developed by Potten et al. After total body irradiation, the entire small intestine is removed and following fixation in Carnoy's solution prepared as "gut bundles" where ten segments of ileum are bundled together embedded in paraffin and stained with hematoxylin and eosin. The number of regenerating crypts within each of the ten circumferences is counted which is compared to controls and from this a crypt survival curve is generated. Mice have been used in intervention experiments for preventing radiation-induced gastrointestinal toxicity; these agents included streptomycin, keratinocyte growth factor (Farrell 1998), transforming growth factor-β3 (Booth 2000), IL-11 (Potten 1995, 1996), and teduglutide (Booth 2004).

Another model used in recent studies has used an "exteriorized ileum" model to demonstrate effects of radiation on the intestinal mucosa (Polistena 2008). This model involves anesthetizing mice and surgically exposing a section of the ileum that can be then irradiated. Following irradiation the exposed loop of intestine is replaced into the abdomen (Polistena 2008). Polistena et al. demonstrated using this model that a single dose of either 19 or 38 Gy caused damage that was dose dependent. The mucosal changes included increased apoptosis of epithelial cells within the crypts and lymphocytic infiltration of the underlying tissue.

Animal models have been used to examine the effect of radiation on other cell populations in the abdomen which are thought to play a role in mucosal maintenance rather than specifically the intestinal mucosa; for example studies have suggested that temporal alterations in intestinal mucosal immune cells and the cytokine environment in the gastrointestinal tract may influence the development of gastrointestinal toxicity (Garg 2010).

Rat Models

Early animal models using male Sprague–Dawley rats for investigating radiation injury associated with accelerated fractionation were developed and involved the creation of an "artificial hernia" with a loop of small bowel sutured to the inside of the left scrotum (Hauer-Jensen 1988). This model allowed accurate reproducibility of the radiation field whilst not disturbing the intestinal transit or causing morphological changes in the intestinal mucosa. This model has been used to determine optimum interfraction intervals in order to minimize the risk of chronic gastrointestinal toxicity (Landberg 1996) as well as determine the influence of radiation fraction size on the development of chronic toxicity (Landberg 1996b). Further use of this model has found that local administration of interleukin-11 ameliorates intestinal injury due to radiation (Boerma 2007).

Wistar rats have been used to investigate the role of matrix metalloproteinases (MMP) and their inhibitors (TIMPS) in the remodeling that occurs in the extracellular matrix following radiation exposure (Strup-Perrot 2005). In this model, the entire abdomen was irradiated with the head and chest (to just below the ribs) shielded. Examination of the irradiated ileum showed an increase in MMP and TIMP activity associated with a corresponding increase in IL-1β and TNF.

More recently, the Dark Agouti rat mammary adenocarcinoma (DAMA) model, previously used in the context of chemotherapy-induced damage (see next section), has been adapted further to examine the effects of radiation on the gut (Ong 2010; Yeoh 2007). The use of a tumor-bearing model is especially important when evaluating potential interventions for mucositis. It is imperative that the intervention for mucositis does not have any effect that promotes tumor growth or indeed, impedes the response of the tumor to cancer treatment. In these studies, tumor-bearing dark agouti rats received a total radiation dose of up to 45 Gray (Gy), 18 fractions over a 6-week period at a dose of 2.5 Gy per fraction. Even after a week, damage was evident histologically. Structurally, the jejunum and colon demonstrated various atypical features such as disorganization of goblet cells and general degenerative changes. Furthermore, p53, NF-κB, and cyclooxygenase 1 and 2 (COX-1 and 2) levels were increased in the tissues. In a follow-up study, radiation-induced damage was shown to be associated with upregulation of IL-1β, IL-6, and TNF mRNA in the jejunum and colon as had been previously reported in the context of chemotherapy, highlighting the fact that despite different modalities of treatment, the pathobiology of mucositis may be similar (Ong 2010).

Chemotherapy

In recent years, perhaps more so than radiotherapy, animal studies have been increasingly used to study the effects of chemotherapy on the gastrointestinal tract. The effects of chemotherapy on the gastrointestinal tract include inflammatory cell infiltration of the mucosa, loss of crypt architecture, and villus blunting.

Murine Models

Murine models have been developed to investigate single and multiple cycle chemotherapy as well as continuous administration of drugs (Xiang 2011). Xiang et al. demonstrated that interleukin-1 receptor antagonist (IL-1Ra) protected the murine gastrointestinal tract from damage caused by administration of cyclophosphamide in models which replicated these three clinically relevant situations. Likewise, Wu et al. demonstrated in similar animals that a mucosal damage from 5-FU was reduced by IL-1Ra administration prior to chemotherapy (Wu 2011).

Farrell et al. demonstrated in a "murine tumor-bearing model" subjected to either multiple dose chemotherapy, chemoradiation or radiation alone that KGF administration resulted in improved crypt survival, amelioration of weight loss, and improved weight gain during recovery (Farrell 1998). Multiple dose chemotherapy was modeled by injecting mice with 5-FU for four consecutive days, whereas in chemoradiation models, mice were given a single dose of methotrexate followed by a single 6 Gy fraction of radiation 1 h later.

Rat Models

The DAMA model has been used extensively to characterize damage in various sites in the gastrointestinal tract including the small intestine and colon (Keefe 2004). In the DAMA model, tumor cells are injected subcutaneously into each flank; the tumor is then allowed to grow before chemotherapy agents are administered.

Early on, the DAMA model was used to investigate the effect of methotrexate on the small intestine and colon (Gibson 2005). It was shown in these early studies that in the small intestine, methotrexate administration caused a reduction in villous area and the length of crypts as well as reducing the number of cells undergoing mitosis. These morphological changes were accompanied by increased apoptosis and increased expression of p53 in the tissues. The study also demonstrated the increased resistance of the colon to damage by methotrexate.

The use of the DA rat model has been extended and modified in some instances to use non-tumor bearing animals to examine the toxicities of other drugs on the gastrointestinal tract including irinotecan and 5-fluorouracil (5-FU) (Gibson 2007; Logan 2008, 2009). In the clinical setting, both irinotecan and 5-fluorouracilcan cause significant gastrointestinal toxicity. In the DA rat model, it was shown that irinotecan caused widespread gastrointestinal damage, particularly in the jejunum and colon. Crypt degeneration and villous atrophy were described in the small intestine, and in the colon unlike the rats that received methotrexate, significant crypt degeneration was observed (Gibson 2007). Similar damage was seen in the model following administration of 5-fluorouracil (Logan 2009). Further work demonstrated that irinotecan caused alteration in cell kinetics of the intestinal epithelium as well as alterations in components of the extracellular matrix (Al-Daooqi 2011). These studies highlighted the fact that different drugs cause different degrees of damage in the tissues.

In broad terms, the damage caused by the drugs in the DAMA animal model essentially mirrors that described in humans, namely an increase in apoptosis which is followed by crypt hypoplasia and reduction in mitosis followed by a rebound hyperplasia of the crypts (Keefe 2000). There are, of course, differences when extrapolating results from animal studies to the clinical situation. The main difference between the changes observed in the DAMA rat model and the damage that is observed in humans is that it occurs over a much shorter time course in the rats, taking a 5-day period to "normalize," at least morphologically, compared to up to 16 days in humans (Keefe 2004).

As well as looking at morphological changes in the gastrointestinal tract, the use of models using the DA rats have been developed further to examine more global changes that occur following administration of cytotoxic agents as well as changes in the luminal environment rather than the structural changes in the mucosa. A study looking at serum levels of pro-inflammatory cytokines demonstrated that changes in interleukin-1β, interleukin-6, and tumor necrosis factor could be detected in the serum of the rats just prior to the development of changes in the alimentary tract (Logan 2008). Furthermore, the DAMA rat model has been used to demonstrate changes that occur in gene expression following the administration of irinotecan, in particular MAP kinase signaling (Bowen 2007).

Potential interventions explored using the DAMA rat model included interleukin-11 (Gibson 2002) and velafermin (fibroblast growth factor-20) (Gibson 2005) both of which were found to have some beneficial effect in reducing the severity of intestinal damage seen after cytotoxic drug administration. Of importance has been the investigation of keratinocyte growth factor (KGF) in this model, also known as palifermin (Farrell 1998; Gibson 2005; Gibson 2002) as results from animal studies proved to be important in the translation of palifermin into the clinical setting.

Examination of microflora in the gastrointestinal tract in the DA rat has demonstrated that administration of irinotecan also affect the composition of bacteria in the stomach, jejunum, and colon and it has been suggested that this may also contribute to the signs of gastrointestinal mucositis such as diarrhea (Stringer 2009; Stringer 2007). Furthermore, functional changes such as changes in mucin secretion have been shown in the rats following both irinotecan and 5-FU administration (Stringer 2009).

Other studies using Wistar rats have also shown that changes occur in small intestine function including altered electrolyte secretion and disruption of intestinal permeability following administration of methotrexate that correlate with histological changes (Carneiro-Filho 2004). Recent studies have also used a Wistar rat model to investigate the gastrointestinal toxicities associated with targeted therapies using small-molecule tyrosine kinase inhibitors (TKI) (Bowen 2010; Bateman 2011). This model of lapatinib-induced has been used to examine changes along the length of the gastrointestinal tract and preliminary results have demonstrated that diarrhea is dose dependent and it has been suggested that diarrhea is secretory in nature associated with imbalances in electrolytes which occur secondary to epithelial changes such as increased proliferative index and crypt depth in the jejunum [50]. This model highlights the importance of animal selection in the development of appropriate models to study toxicity with respect to strain and sex differences in metabolic enzyme profiles. TKI are predominantly metabolized through the CYP3A4 pathway and in this example, Wistar rats have an appropriate drug metabolic profile for investigating TKI (Bowen 2011).

Radiation-Induced Proctitis

Overview of the Condition

Radiation-induced proctitis is a common complication associated with radiation directed at the lower abdomen or pelvis typically associated with rectal, prostate, or cervical malignancies. There are both acute and delayed forms of radiation proctitis that may present following radiation therapy. The symptoms of acute proctitis have been reported as a complication in as many as 75% of patients undergoing pelvic radiation (Counter et al. 1999) with symptoms beginning almost immediately following therapy and persisting for up to 3 months. Delayed or late radiation proctitis is a chronic disease that can occur well after the completion of the radiation therapy and result in symptoms that may persist for decades. Disease symptomology usually begins in the acute phase during the first or second week following radiation, with presentation of rectal bleeding that can be associated with diarrhea and/or discharge of mucus. Late proctitis symptoms appear anywhere from a few months to several years following the radiation therapy and as with the acute disease, chronic rectal bleeding is the most common late symptom. Other late symptoms may include a discharge of mucus from the rectum and more instances of tenesmus, or a feeling or inability to empty the bowel upon defecation, which is likely a result of extensive rectal tissue fibrosis or possibly the formation of rectal strictures.

Diagnosis and treatment of radiation proctitis usually follows a colonoscopy to exclude the possibility of any other underlying pathology. Current treatment options for radiation proctitis vary greatly in method as well as rate of success (Leiper and Morris 2007). While there are few, if any, compelling clinical studies in the treatment of radiation proctitis, the majority of therapies are based on the results of small unblinded studies with mixed results. The most common first-line therapies have been adapted from the treatments used in inflammatory bowel disease (IBD), including 5-ASA, steroids, sucralfate, and metronidazole. Increasingly, endoscopic therapies are being employed to control bleeding associated with radiation proctitis which includes heat probes, lasers, and most commonly argon plasma coagulation (APC). APC involves the flow ionized argon gas and can target small or large areas of bleeding without making physical contact with the tissue. While APC has been shown to be quite effective in a number of small studies (Leiper and Morris 2007), there are no large clinical studies to help support these results.

The Biology of Radiation-Induced Proctitis

The overall pathogenesis of mucositis described for the mouth and GI tract is likely similar to that of proctitis, although the mechanisms of radiation-induced proctitis have not been as aggressively studied. In the days to months following radiation the onset of acute proctitis begins with early apoptosis, disruptions of mitosis, and fibroblastic proliferation that leads to swelling and sloughing of the rectal mucosa (Haboubi et al. 1988; Brunn and Flethcher 2006), while the later changes are primarily vascular and fibrotic in nature (Fajardo 2005). The neovascularization and telangiectasias that occur in the late phases of disease lead to the persistent bleeding while increased fibrosis may lead to ischemia and eventual necrosis of the bowel tissue (Fajardo 2005; Haselton et al. 1985). The exact mechanisms for the late changes in vascularity and fibrosis have yet to be elucidated; however there is evidence that several growth factors including platelet-derived growth factor, vascular endothelial growth factor, and fibroblast growth factor play key roles in the pathology (Brunn and Flethcher 2006) as well as more recent evidence of mast cell involvement (Blirando et al. 2011). Furthermore, radiation-induced increased expression of $\alpha_v\beta_3$ integrin, which is known to have potent angiogenic activity, may contribute to the vascular changes observed in radiation proctitis (Abdollani et al. 2005). In addition, lesions resulting from radiation-induced proctitis are associated with increased angiogenesis and vascular expression of CD39 which promotes inflammation. Taken together, these mechanisms may individually or collectively contribute to long-term defects in vascular integrity associated with this disease (Sheth et al. 2009). Additional studies are needed to further understand the mechanisms underlying the induction of radiation-induced proctitis to provide better avenues for effective interventions.

Animal Model of Radiation-Induced Proctitis

Recent modifications of rat models of radiation-induced proctitis have proven to be effective in creating a clinically relevant model of the condition (Kang et al. 2000). A single 17.5 Gy dose of radiation is directed at the rectum of a rat to induce proctitis. More recently, we have developed fractionated radiation-induced models of proctitis where radiation doses of 4–6 Gy are administered for four consecutive days, followed by a 2-day rest period, after which the 4-day regimen is repeated (Lyng et al. 2012). In all of these models, radiation is performed following the administration of ketamine and xylazine anesthesia. Animals are placed on a silicon rubber sheet and lead shielding is used to isolate the lower abdomen. The animals are placed at an angle to isolate the rectum and minimize the possibility of irradiating the small intestine. The rectum is irradiated with a single, acute dose or multiple daily fractions of radiation using a 160 kVp (15-ma) source at a focal distance of 30 cm, hardened with a 0.35 mm Al filtration system at a rate of 2.5 Gy/min. Disease induction occurs within 5 days of radiation and peaks at approximately 7–10 days post radiation exposure in the acute model and slightly later in the fractionated

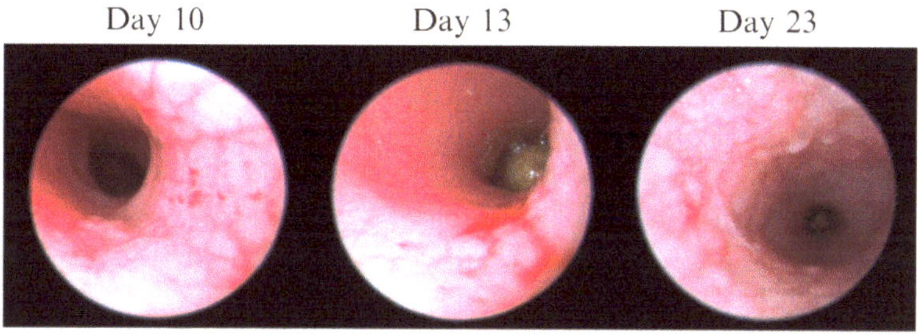

Fig. 3 Endoscopic appearance of the progression of proctitis following fractionated radiation

radiation model (day 12–14). In an effort to overcome the necessity of relying on histological endpoints alone to assess mucosal damage and permit in situ assessment of proctitis progression, we have applied video endoscopy to models of radiation-induced proctitis (Lyng et al. 2012). Video endoscopy allows for daily visual assessment of the extent and severity of disease using clinically relevant endpoints as well as the tracking of mucosal healing following any potential therapeutic interventions. To conduct endoscopy the rats are anesthetized with isoflurane and the endoscope is slowly inserted into the rectum while the colon is insufflated with air. This method allows for clear imaging of the colon mucosal surface and both video and still images are recorded to help assess disease severity. Severity of disease is scored on 0–4 scale with 0 = normal; 1 = loss of vascularity; 2 = loss of vascularity and friability; 3 = friability and erosions; 4 = ulcerations and bleeding. Changes in the colon observed endoscopically include modification of the vascular pattern, often accompanied by friability, erosion, and active bleeding. The images in (Fig. 3) show the progression of proctitis, as visualized endoscopically following a regimen of fractionated radiation.

While endoscopic data adds an important visual and clinical component to the animal models of proctitis, the use of histology can still add valuable confirmatory information several weeks following the study completion. Endoscopic correlation with pathologic findings has been confirmed using models for inflammatory bowel disease (Lyng et al. 2008; Hamilton et al. 2011).

Radiation-Induced Dermatitis

Overview of the Condition

Radiation-induced dermatitis is injury to the skin as a result of exposure to a radiation source, generally associated with receiving radiation therapy for cancer treatment of cutaneous neoplasms, head and neck, and breast cancers. Dermatitis may result from either a single high dose of radiation or a series of fractionated doses.

Radiation-induced dermatitis occurs in approximately 95% of patients receiving radiation therapy (Rosen et al. 1999; Fujishiro et al. 2000). The condition is unpleasant, painful, and may contribute to poor quality of life and in some cases, radiation dermatitis may become so severe that it becomes necessary to interrupt or discontinue treatment. The disease onset is usually characterized by the reddening of the skin, but may also include epilation. As the disease progresses dry desquamation, wet desquamation, decreased sweating, edema, ulceration, and bleeding can occur. Areas of telangiectasias and necrosis are observed in the most severe cases. Symptom progression depends on both treatment-related factors, such as the total radiation dose, fractionation, total duration of treatment, volume of tissue irradiated, and type of radiation delivered; as well as patient-related factors such as age, menopausal state, and pre-disposing genetic factors (Isomura et al. 2008).

Current treatment options for radiation-induced dermatitis are similar to those associated with thermal burns, are primarily palliative, and include a wide variety of products. In general, emollients or hydrating lotions are used during the early stages of the condition. If symptoms progress, topical corticosteroids, dressings, and/or radiation dose reduction may be necessary (Hymes et al. 2006; McQuestion 2006).

The Biology of Radiation-Induced Dermatitis

Radiation dermatitis differs from thermal dermatitis in its pathophysiology and therefore there is an unmet clinical need to develop specific, mechanistically based interventions. Although the mechanism has not been well studied, it is likely that the pathoetiology of radiation-induced dermatitis is similar to other epithelial toxicities. DNA damage caused by ionizing radiation generates free radicals, which in turn activate IκB kinase (IκK) inducing NF-κB translocation to the nucleus, and subsequently, transcription of pro-inflammatory mediators. This leads to a complex pattern of tissue injury and recruitment of inflammatory cells (Hymes et al. 2006). The p53 pathway may be activated and cytokines such as interleukin 1 (IL-1), interleukin 6 (IL-6), and transforming growth factor β (TGF-β) are secreted by keratinocytes. Histological analysis of the injured tissue reveals severe inflammation with an associated loss of epithelial architecture, a loss of sebaceous and sweat glands, destruction of hair follicles, and necrosis.

Animal Model of Radiation-Induced Dermatitis

As presented by Mancini et al. (2012) the model originally described by Abe and Urano (1990) has been modified to accommodate both acute and fractionated radiation disease induction protocols. In the acute model, a single 30 Gy dose of radiation is directed to the dorsal skin of the mouse. Two days prior to irradiation the skin on the backs of animals is prepared by removal of the hair covering the entire back of each animal using an electric shaver. The animals are then anesthetized using ketamine and xylazine, placed on a silicon rubber sheet and the loose dorsal skin is gently stretched and secured with two 25-gauge needles. A lead shield is then placed over the animal so that an area of skin about 2 cm × 4 cm in size was exposed.

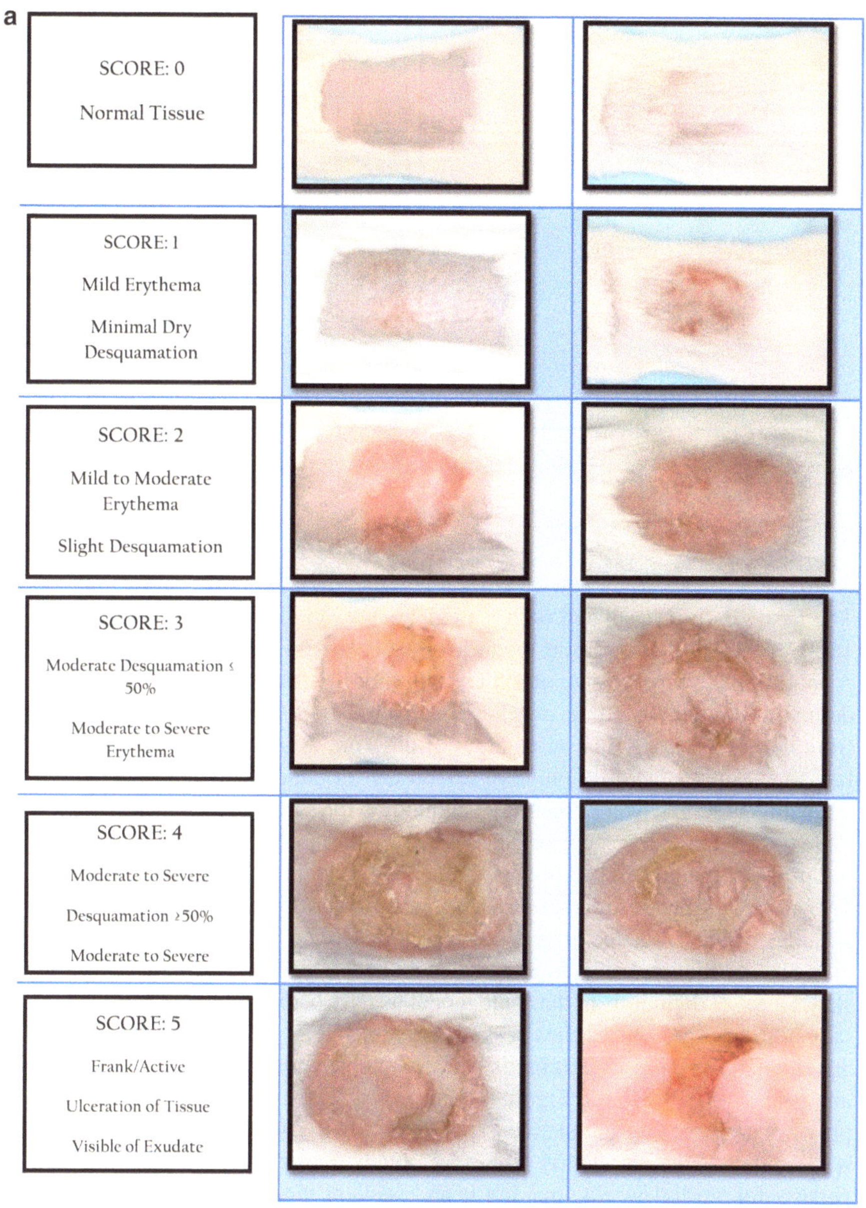

Fig. 4 Scoring scale for an animal model of radiation-induced dermatitis (**a**). Clinical and histological features of moderate to severe dermatitis following 60 Gy of radiation compared with no radiation exposure (NR)

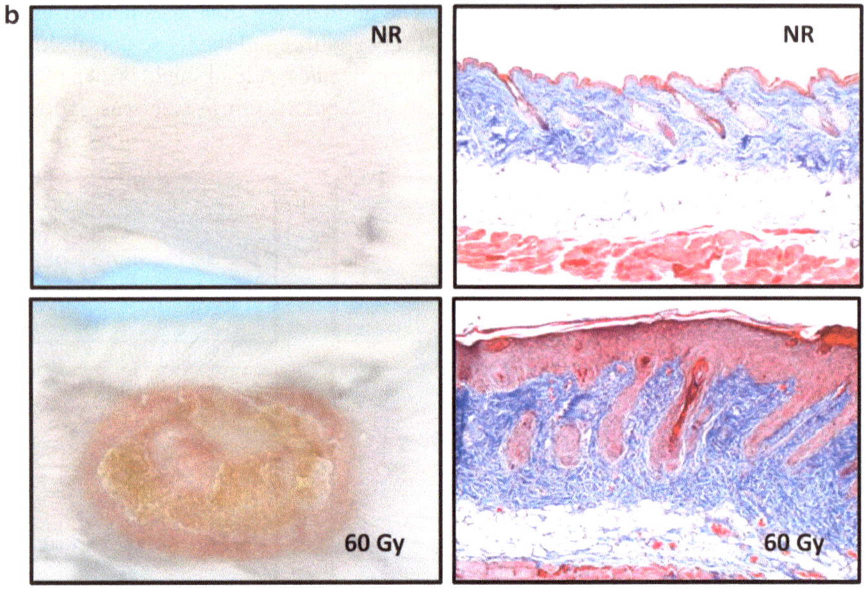

Fig. 4 (continued)

The tissue is then irradiated with a single, acute, 30 Gy dose of radiation using a 160 kVp (15-ma) source at a focal distance of 30 cm, hardened with a 0.35 mm Al filtration system at a rate of 1.5 Gy/min. For the fractionated radiation-induced dermatitis model, radiation doses are delivered in six smaller fractions ranging from 6 to 10 Gy each on a schedule of 3-days on, 2-days off, 3-days on resulting in a total cumulative dose range of 36–60 Gy. In either radiation-induced model, dermatitis is scored every other day beginning 6 days after radiation using a 6 point scale (0–5) in which a score of 0 is defined by normal skin; a score of 1 is defined by mild erythema; a score of 2 is defined by moderate to severe erythema with or without slight desquamation; a score of 3 is defined by desquamation of 25–50% of the irradiated area; a score of 4 is defined by desquamation of greater than 50% of the irradiated area; and a score of 5 is defined frank ulceration of the skin. Onset of dermatitis in the acute model usually begins around 1 week following radiation exposure and reaches peak severity another 5–7 days later. In the fractionated model, the onset of dermatitis occurs at approximately 10 days post-radiation and peaks 7–10 days later. Figure 4 outlines the scoring scale in greater detail and provides an example of moderate to severe dermatitis. One of the major benefits of both models is that dermatitis scores are obtained every other day and the disease persists for several weeks, making it an excellent model system to test potential therapeutics.

References

1. Linard C, et al. Abdominal irradiation increases inflammatory cytokine expression and activates NF-kappaB in rat ileal muscularis layer. Am J Physiol Gastrointest Liver Physiol. 2003;285(3):G556–65.
2. Richter KK, et al. Association of transforming growth factor beta (TGF-beta) immunoreactivity with specific histopathologic lesions in subacute and chronic experimental radiation enteropathy. Radiother Oncol. 1996;39(3):243–51.
3. Logan RM, et al. The role of pro-inflammatory cytokines in cancer treatment-induced alimentary tract mucositis: pathobiology, animal models and cytotoxic drugs. Cancer Treat Rev. 2007;33(5):448–60.
4. Keefe DM. Gastrointestinal mucositis: a new biological model. Support Care Cancer. 2004;12(1):6–9.
5. Bowen JM, Gibson RJ, Keefe DM. Animal models of mucositis: implications for therapy. J Support Oncol. 2011;9(5):161–8.
6. Strup-Perrot C, et al. Expression of matrix metalloproteinases and tissue inhibitor metalloproteinases increases in X-irradiated rat ileum despite the disappearance of CD8a T cells. World J Gastroenterol. 2005;11(40):6312–21.
7. Kawashima R, et al. IL-13 receptor alpha2 promotes epithelial cell regeneration from radiation-induced small intestinal injury in mice. Gastroenterology. 2006;131(1):130–41.
8. Li J, et al. Loss of caveolin-1 causes the hyper-proliferation of intestinal crypt stem cells, with increased sensitivity to whole body gamma-radiation. Cell Cycle. 2005;4(12):1817–25.
9. Polistena A, et al. Local radiotherapy of exposed murine small bowel: apoptosis and inflammation. BMC Surg. 2008;8:1.
10. Ponemone V, et al. Effect of adiponectin deficiency on intestinal damage and hematopoietic responses of mice exposed to gamma radiation. Mutat Res. 2010;690(1–2):102–7.
11. Allgood JW, et al. Timing of concomitant boost irradiation affects incidence and severity of intestinal complications. Int J Radiat Oncol Biol Phys. 1996;34(2):381–7.
12. Langberg CW, Hauer-Jensen M. Optimal interfraction interval to minimize small bowel radiation injury in treatment regimens with two fractions per day: an experimental study in a rat model. Radiother Oncol. 1996;41(3):249–55.
13. Langberg CW, Hauer-Jensen M. Influence of fraction size on the development of late radiation enteropathy. An experimental study in the rat. Acta Oncol. 1996;35(1):89–94.
14. Langberg CW, et al. Expression of fibrogenic cytokines in rat small intestine after fractionated irradiation. Radiother Oncol. 1994;32(1):29–36.
15. Langberg CW, et al. Influence of fractionation schedule on development of intestinal complications following localized irradiation. An experimental study in the rat. Acta Oncol. 1994;33(4):403–8.
16. Langberg CW, et al. Relationship between intestinal fibrosis and histopathologic and morphometric changes in consequential and late radiation enteropathy. Acta Oncol. 1996;35(1): 81–7.
17. Langberg CW, et al. Significance of overall treatment time for the development of radiation-induced intestinal complications. An experimental study in the rat. Cancer. 1994;73(10): 2663–8.
18. Richter KK, et al. Increased transforming growth factor beta (TGF-beta) immunoreactivity is independently associated with chronic injury in both consequential and primary radiation enteropathy. Int J Radiat Oncol Biol Phys. 1997;39(1):187–95.
19. Scott BR, Langberg CW, Hauer-Jensen M. Models for estimating the risk of ulcers in the small intestine after localized single or fractionated irradiation. Br J Radiol. 1995;68(805):49–57.
20. Hauer-Jensen M, Poulakos L, Osborne JW. Effects of accelerated fractionation on radiation injury of the small intestine: a new rat model. Int J Radiat Oncol Biol Phys. 1988;14(6): 1205–12.

21. Cameron S, et al. Radiation-induced damage in different segments of the rat intestine after external beam irradiation of the liver. Exp Mol Pathol. 2012;92(2):243–58.
22. Ong ZY, et al. Pro-inflammatory cytokines play a key role in the development of radiotherapy-induced gastrointestinal mucositis. Radiat Oncol. 2010;5:22.
23. Yeoh AS, et al. A novel animal model to investigate fractionated radiotherapy-induced alimentary mucositis: the role of apoptosis, p53, nuclear factor-kappaB, COX-1, and COX-2. Mol Cancer Ther. 2007;6(8):2319–27.
24. Garg S, et al. Influence of sublethal total-body irradiation on immune cell populations in the intestinal mucosa. Radiat Res. 2010;173(4):469–78.
25. Potten CS, et al. The correction of intestinal microcolony counts for variation in size. Int J Radiat Biol Relat Stud Phys Chem Med. 1981;40(3):321–6.
26. Farrell CL, et al. Keratinocyte growth factor protects mice from chemotherapy and radiation-induced gastrointestinal injury and mortality. Cancer Res. 1998;58(5):933–9.
27. Booth D, et al. Transforming growth factor-B3 protects murine small intestinal crypt stem cells and animal survival after irradiation, possibly by reducing stem-cell cycling. Int J Cancer. 2000;86(1):53–9.
28. Potten CS. Interleukin-11 protects the clonogenic stem cells in murine small-intestinal crypts from impairment of their reproductive capacity by radiation. Int J Cancer. 1995;62(3): 356–61.
29. Potten CS. Protection of the small intestinal clonogenic stem cells from radiation-induced damage by pretreatment with interleukin 11 also increases murine survival time. Stem Cells. 1996;14(4):452–9.
30. Booth C, et al. Teduglutide ([Gly2]GLP-2) protects small intestinal stem cells from radiation damage. Cell Prolif. 2004;37(6):385–400.
31. Boerma M, et al. Local administration of interleukin-11 ameliorates intestinal radiation injury in rats. Cancer Res. 2007;67:9501–6.
32. Xiang D, et al. Interleukin-1 receptor antagonist attenuates cyclophosphamide-induced mucositis in a murine model. Cancer Chemother Pharmacol. 2011;67(6):1445–53.
33. Wu ZQ, et al. Interleukin-1 receptor antagonist reduced apoptosis and attenuated intestinal mucositis in a 5-fluorouracil chemotherapy model in mice. Cancer Chemother Pharmacol. 2011;68(1):87–96.
34. Gibson RJ, et al. Relationship between dose of methotrexate, apoptosis, p53/p21 expression and intestinal crypt proliferation in the rat. Clin Exp Med. 2005;4(4):188–95.
35. Gibson RJ, et al. Establishment of a single-dose irinotecan model of gastrointestinal mucositis. Chemotherapy. 2007;53(5):360–9.
36. Logan RM, et al. Serum levels of NFkappaB and pro-inflammatory cytokines following administration of mucotoxic drugs. Cancer Biol Ther. 2008;7(7):1139–45.
37. Logan RM, et al. Is the pathobiology of chemotherapy-induced alimentary tract mucositis influenced by the type of mucotoxic drug administered? Cancer Chemother Pharmacol. 2009;63(2):239–51.
38. Al-Dasooqi N, et al. Irinotecan-induced alterations in intestinal cell kinetics and extracellular matrix component expression in the dark agouti rat. Int J Exp Pathol. 2011;92(5):357–65.
39. Keefe DM, et al. Chemotherapy for cancer causes apoptosis that precedes hypoplasia in crypts of the small intestine in humans. Gut. 2000;47(5):632–7.
40. Gibson RJ, et al. Effect of interleukin-11 on ameliorating intestinal damage after methotrexate treatment of breast cancer in rats. Dig Dis Sci. 2002;47(12):2751–7.
41. Gibson RJ, et al. Velafermin improves gastrointestinal mucositis following irinotecan treatment in tumor-bearing DA rats. Cancer Biol Ther. 2007;6(4):541–7.
42. Gibson RJ, Bowen JM, Keefe DM. Palifermin reduces diarrhea and increases survival following irinotecan treatment in tumor-bearing DA rats. Int J Cancer. 2005;116(3):464–70.
43. Gibson RJ, et al. The effect of keratinocyte growth factor on tumour growth and small intestinal mucositis after chemotherapy in the rat with breast cancer. Cancer Chemother Pharmacol. 2002;50(1):53–8.

44. Stringer AM, et al. Irinotecan-induced mucositis manifesting as diarrhoea corresponds with an amended intestinal flora and mucin profile. Int J Exp Pathol. 2009;90(5):489–99.
45. Stringer AM, et al. Chemotherapy-induced diarrhea is associated with changes in the luminal environment in the DA rat. Exp Biol Med (Maywood). 2007;232(1):96–106.
46. Stringer AM, et al. Gastrointestinal microflora and mucins may play a critical role in the development of 5-Fluorouracil-induced gastrointestinal mucositis. Exp Biol Med (Maywood). 2009;234(4):430–41.
47. Stringer AM, et al. Irinotecan-induced mucositis is associated with changes in intestinal mucins. Cancer Chemother Pharmacol. 2009;64(1):123–32.
48. Carneiro-Filho BA, et al. Intestinal barrier function and secretion in methotrexate-induced rat intestinal mucositis. Dig Dis Sci. 2004;49(1):65–72.
49. Bowen J, et al. Characterisation of EGF receptor tyrosine kinase inhibitor-induced diarrhoea. Support Care Cancer. 2010;18 Suppl 3:S107.
50. Bateman E, et al. Further characterisation of epidermal growth factor receptor (EGFR) tyrosine kinase inhibitor-induced diarrhoea. Support Care Cancer. 2011;19 Suppl 2:S250.

Part II
The Pathobiology of Specific Toxicities

Nausea and Vomiting

Pernille H. Hansen, Jesper Palshof, and Jørn Herrstedt

"One thing only I know, and that is that I know nothing."

Socrates (philosopher, born 469 BC)

Abstract Despite the fact that research in the mechanism of vomiting goes back more than 300 years, there are still major gaps in our knowledge. The first report of the clinical use of an antineoplastic drug was published in 1942, but it was not until cisplatin, the most emetogenic of all drugs, was approved in 1978, the necessity of focusing on antiemetic research became evident.

This chapter will focus on the biological basis of nausea and vomiting from the time of the very simple animal experiments, suggesting the existence of a vomiting center, to current research methodology enabling precise subdivision of receptors involved in the emetic reflex arch.

P.H. Hansen
Laboratory of Neuropsychiatry/Molecular Neuropharmacology Laboratory, Department of Neuroscience and Pharmacology, University of Copenhagen, Copenhagen, Denmark

J. Palshof
Department of Oncology, Herlev Hospital, University of Copenhagen, Copenhagen, Denmark

J. Herrstedt (✉)
Department of Oncology, Odense University Hospital, and University of Southern Denmark, Sdr. Boulevard 29, 5000 Odense C, Denmark
e-mail: herrstedt@ouh.regionsyddanmark.dk

S.T. Sonis and D.M. Keefe (eds.), *Pathobiology of Cancer Regimen-Related Toxicities*,
DOI 10.1007/978-1-4614-5438-0_6, © Springer Science+Business Media New York 2013

Introduction

Nausea and vomiting are components of a natural mechanism that protects the human body from ingested toxins. Antineoplastic agents are probably the most toxic medication offered to patients. It is therefore not a surprise that the majority of these agents induce significant nausea and vomiting if patients are not offered proper antiemetic prophylaxis. This chapter will focus on the mechanism of nausea and vomiting induced by chemotherapy.

Pathophysiology: The Early Days

The mechanism of nausea and vomiting has been of interest to scientists long before the development of antineoplastic agents. The first experimental work was probably done by Johann Jakob Wepfer who in 1679 described the involvement of the cat and dog pylorus and of the pyloric part of the stomach during vomiting [1]. For many years the mechanism was therefore thought to be due to contractions of the stomach only, but in 1881 [2] experiments demonstrated that also respiratory muscles were essential in generating the abdominal pressure needed for expulsion of stomach content. The famous American physiologist, Walter Bradford Cannon, in 1898 described the activity of various muscle groups during vomiting by the use of X-rays [3]. Giannuzzi in 1865 [4] was probably the first to propose the existence of a vomiting center, but it was not until Professor Lazarus Thumas published his experiments in the dog brain in 1891, that the existence of a central emetic reflex arch (the vomiting center) was further explored [5]. He proposed the vomiting center to be located in the medulla oblongata. Hatcher and Weiss in 1923 did a number of ablation experiments in dogs and cats and concluded that the vomiting center was more precisely situated in the sensory nuclei of the vagi nerves [6]. In 1949 Borison and Wang published their findings from experiments done by direct electrical stimulation of the medulla of cats. They used decerebrate cats in order to avoid anesthesia that would blur the response to electrical stimulation. They confirmed the existence of a vomiting center localized to the solitary fasciculus and nucleus and to the dorsal border of the lateral reticular formation [7]. In other studies, they used the gastric irritant, copper sulfate, to study the upper pathways from the gastrointestinal tract in dogs [8], and the loci of the central emetic action of cardiac glycosides [9] and apomorphine [10]. In a comprehensive review from 1953, Borison and Wang summarized the results of the pathophysiology of vomiting and concluded that two central areas are involved: the well-known vomiting center, functioning as a coordinating zone for peripheral and central receptor sites and the chemoreceptor trigger zone (CTZ) located in the area postrema, outside the blood–brain barrier, and as such able to sense both blood-borne and cerebrospinal fluid-borne emetic stimuli [11].

The existence and location of a vomiting center and of the CTZ have been intensively debated for many years. In support of the existence of a CTZ in humans, Lindstrom and Brizzee managed to relieve intractable vomiting in five patients by superficial thermal lesions in the area postrema [12]. On the other hand, in an electrical stimulation study from 1983 [13], Miller and Wilson found no support for a restricted localization of a vomiting center in cats, a result the group confirmed a decade later using a refined technique [14]. Today it is recognized that a vomiting center in the sense of a well-defined anatomical region does not exist. More likely coordination of the emetic reflex is a result of intertwined neural networks penetrating to the nucleus of the solitary tract and receiving input from the area postrema and the abdominal vagus nerve [14, 15].

Mechanism of Chemotherapy-Induced Nausea and Vomiting

Carl et al. proposed three hypothetical models for antineoplastic drug-induced emesis [16]. All models involved the CTZ in the area postrema, but speculated in different ways of what this area of the brain sensed.

The first model proposed that receptors in the area postrema would directly detect antineoplastic drugs. Peroutka in 1987 investigated a possible interaction of cisplatin, 5-fluorouracil, doxorubicin, mitomycin, carmustine, cyclophosphamide, methotrexate, and thio-Thepa with dopamine D_2, histamine H_1, and muscarinic cholinergic receptors and found no binding of any of these antineoplastic agents at the receptors [17]. Herrstedt et al. used radioligand binding techniques to investigate the affinity of cisplatin (high emetic risk), carboplatin and doxorubicin (moderate emetic risk), etoposide (low emetic risk), and vincristine (minimal emetic risk) at dopamine D_2, histamine H_1, muscarinic cholinergic, α_1-adrenergic, and 5-hydroxytryptamine $(5\text{-HT})_3$ receptors. None of the agents had any detectable affinity for the receptors, except for vincristine which showed nanomolar affinity for muscarinic cholinergic receptors. An antagonist effect at muscarinic cholinergic receptors could explain constipation, a frequent side effect of vincristine [18].

The second model proposed that antineoplastic drugs directly or by release of humoral factors could trigger neural input to the CTZ and/or other brain vomiting centers. This model assumes that the CTZ and/or other brain centers detect nerve impulses not humoral factors.

The third model hypothesized that antineoplastic agents induce nausea and vomiting by release of endogenous factors which reach the CTZ and/or other emetic sensing centers via the circulation. Carl et al. did a number of experiments using cisplatin as the emetic stimulus, but found no support for this model [16].

The most plausible model is therefore number 2, but subsequent studies of neurotransmitters and receptors have led to significant modifications of the original

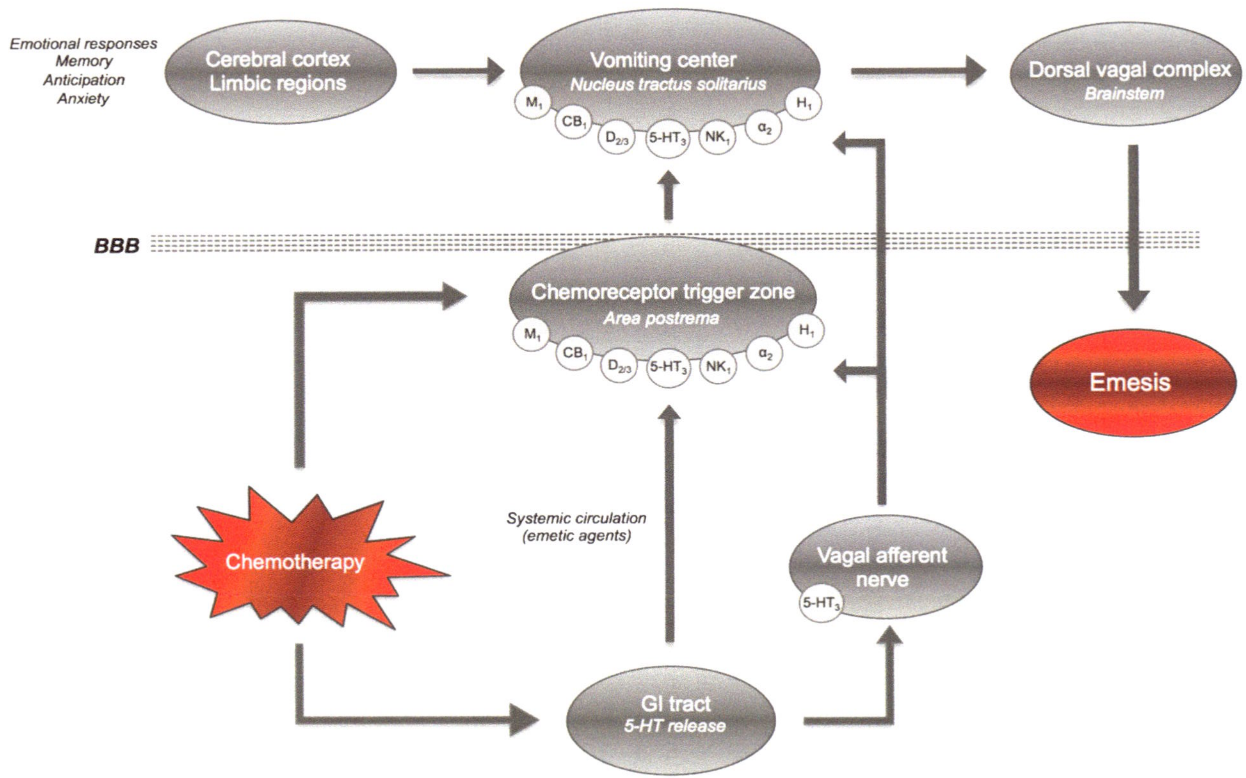

Fig. 1 Neuronal circuits involved in chemotherapy-induced nausea and vomiting

hypothesis. Although many aspects of the pathophysiology of chemotherapy-induced nausea and vomiting are still unclear, we know that a number of transmitters such as dopamine, serotonin, and substance P are involved and binding of these transmitters at their receptors in the gut and/or in areas in the medulla oblongata initiates an agonist effect resulting in nausea and vomiting (Fig. 1). The pathophysiology of chemotherapy-induced nausea and vomiting has previously been reviewed focusing serotonin and serotonin receptors plus substance P and neurokinin receptors [19]. In the following serotonin, substance P, and the corresponding receptors will be summarized, whereas the involvement of dopamine and dopamine receptors will be described in more detail.

Transmitters and Receptors Involved in the Emetic Reflex Arch

Dopamine and Dopamine Receptors

More than 50 years ago scientists believed that 3-hydroxytyramine (dopamine), a metabolite of the amino acid tyrosine, merely existed as a metabolic precursor for noradrenaline in the brain. In 1957 Carlsson and coworkers elegantly showed how reserpine, a vesicular monoamine transporter blocker, could cause depletion of dopamine and concomitant loss of sympathetic nerve stimulation (Table 1). These effects could be rescued by administering 3,4-dihydroxyphenylalanine (L-DOPA) that restored dopamine levels but not noradrenaline levels, demonstrating an independent role of dopamine [20]. Subsequently dopamine has received much attention for its important modulator role in the central nervous system controlling critical aspects of motor activity, cognition, and reward mechanisms, i.e., dysfunction of the dopaminergic system has been associated with several neuropsychiatric diseases [21].

Both in the CNS and periphery, dopamine synthesis occurs in a two-step process, where the initial and rate-limiting step is hydroxylation of tyrosine to the dopamine precursor L-DOPA catalyzed by tyrosine hydroxylase. The second step is decarboxylation of L-DOPA to form dopamine. Synthesized, dopamine is transported into secretory vesicles for storage and is then ready for rapid release into the cleft of dopaminergic synapses where it can recognize dopamine receptors located both at the postsynaptic as well as the presynaptic neuron. A rapid reuptake of dopamine from the extracellular space facilitates a homeostatic balance in the synaptic cleft, thereby regulating dopaminergic neuronal activity. Once dopamine is back in the axon terminal, it can either be reloaded into secretory vesicles or degraded to the major metabolites 3,4-dihydroxyphenyl-acetic acid (DOPAC) and homovanillic acid (HVA), respectively. Dopaminergic innervations are most prominent in the CNS, but also exist in the periphery [22, 23]. In the human brain, the vast majority of dopaminergic innervations originates in the midbrain (substantia nigra and ventral tegmental area) and reach out to different parts of the forebrain; the nigrostriatal pathway that drives the motor system; the mesolimbic pathway which drives the pleasure center; and the mesocortical pathway, which drives our attention, concentration,

Table 1 History of DA and DA receptors

Year of publication	Identification and development of DA and its receptors	References
1957	Discovery of DA as a brain neurotransmitter	Carlsson et al. [20]
1963	Metopimazine exhibit antiemetic effect	Moertel et al. [33]
1964	Metoclopramide exhibit antiemetic effect	Justin-Besancon and Laville [119]
1972	First demonstration of DA-sensitive adenylate cyclase	Kebabian et al. [25]
1977	Two different classes of DA receptors described	Katzman et al. [27]
1979		Kebabian and Calne [26]
1983	DA receptor sensitivity in the CTZ	Shen et al. [34]
1988	Cloning of the DA D_2R and identification of two isoforms	Bunzow et al. [31]
1989		Giros et al. [32]
1990	DA D_3R cloning	Sokoloff et al. [36]
1995	The emetic pathway has a DA D_3R component	Yoshida et al. [37]
1996		Yoshikawa et al. [38]
2000	First demonstration of a selective DA D_3R antagonist	Stemp et al. [39]
2012	DA D_3R antagonist activity against cisplatin-induced emesis	Herrstedt et al. [124]

CTZ chemoreceptor trigger zone; *DA* dopamine; D_2R dopamine D_2 receptor; D_3R dopamine D_3 receptor

and relatedness to others. The nigrostriatal innervations account for a main part of the total dopamine in the brain, and play a crucial role in the regulation of motor control, e.g., the progressive loss of these projections is one of the most noticeable pathological features of Parkinson's disease. Conversely, this pathway is responsible for mediating many of the neurological adverse effects associated with antipsychotic drugs [24].

Once dopamine is released from presynaptic terminals in response to neuronal firing, it exerts its action by activating metabotropic guanosine nucleotide-binding protein (G-protein) coupled receptors (GPCR) designated D_1 to D_5. The first biochemical evidence for a dopamine receptor was the identification of dopamine-stimulated adenylate cyclase activity [25]. Dopamine receptors were originally divided into two different classes according to their different G-protein coupling and pharmacological specificity, the D_1-like (D_1 and D_5 subtypes) and the D_2-like (D_2, D_3, and D_4 subtypes) receptors [26, 27]. While D_1-like receptors are found exclusively on postsynaptically dopamine-receiving neurons, D_2-like receptors are expressed both postsynaptically on dopamine targets and presynaptically on dopaminergic neurons [28].

All subtypes have been expressed in varying densities in a range of organs including the gastrointestinal tract [21, 29, 30]. In the brain, the D_1 receptor is the most abundant dopamine receptor having a broad expression pattern throughout the forebrain with the highest density found in the striatum and substantia nigra. D_5 receptors are typically expressed at lower levels than D_1 receptors, and are mainly detected in the hippocampus and hypothalamus. After the initial cloning of the D_2 receptor, it became evident that the receptor existed both as a short and a long isoform [31, 32]. D_2 receptors are widely distributed in all areas receiving dopaminergic innervation and are particularly found at high densities in the striatum. D_3 receptors have a more limited expression pattern mostly of limbic origin and generally present at lower levels than D_2 receptors. In turn, D_4 receptors are the type expressed at the lowest level in the brain most apparent in limbic and cortical areas.

The discovery of dopamine's independent role as a neurotransmitter in the brain [20] opened up the first antiemetic research involving D_2 antagonists back in the 1960s [33]. The idea, that dopamine receptor antagonists could serve as potential antiemetics, was based on pharmacological evidence demonstrating that agents with the ability to activate dopamine D_2 receptors could elicit emesis, strongly suggesting a role for dopaminergic neurons in the initiation of nausea and vomiting. This is further supported by the fact that dopamine receptor agonists and other pharmacological agents that increase dopamine concentrations (e.g., L-DOPA) are associated with adverse effects such as nausea and vomiting. Elevated levels of dopamine close to the CTZ have been associated with the genesis of emesis, and the presence of D_2 receptors in this area has led to the hypothesis that dopamine's contribution in the initiation of nausea and vomiting is related to this specific part of the medulla [34]. Therefore it was believed that antiemetics act as antagonists at D_2 receptors probably located in the CTZ, whereas the dopaminergic neurotransmission within the basal ganglia, the brain region comprising the highest density of dopamine receptors, is likely responsible for many of the extrapyramidal side effects associated with these drugs. This is in concordance with the fact that metopimazine is the dopamine antagonist with the lowest risk of causing extrapyramidal side effects and in the same time metopimazine and its acid metabolite have the lowest penetration of the blood–brain barrier as compared to other dopamine antagonists such as metoclopramide, chlorpromazine, and domperidone [35]. Only a few years after the cloning of the D_3 receptor in 1990 [36], it became evident that the receptor participated in elicit of emesis. Experiments involving the D_3 receptor agonist, 7-hydroxy-2 (*N*,*N*-di-*n*-propyl-amino) tetraline (7-OH-DPAT), demonstrated that the compound was able to induce emesis in a dose-dependent manner in dogs [37] and ferrets [38]. Because several of the available dopamine antagonists often cause extrapyramidal side effects, there has been an interest in targeting the much more weakly expressed D_3 receptors, as a means of avoiding these adverse effects. With the development of a selective D_3 receptor antagonist [39], a new tool for investigating D_3 receptors role in emesis was provided. The history of dopamine and dopamine receptors is summarized in Table 1.

Serotonin and Serotonin Receptors

The isolation of a vasoactive substance from beef serum was reported by Rapoport and coworkers in 1948 [40] and further characterized as 5-hydroxy-3-β-aminoethylindole

(5-hydroxytryptamine) called serotonin in 1949 [41]. In the following years these findings were confirmed and explored in more detail by the group [42]. 5-hydroxytryptamine (serotonin) is synthesized from tryptophan that is converted to 5-hydroxytryptophan which in turn is metabolized to serotonin. The total conversion of serotonin in a time period can be evaluated by measurement of the main metabolite of serotonin, 5-hydroxyindole-acetic acid in the urine.

Gaddum and Hameed did some experiments in 1954 with 5-hydroxytryptamine antagonists and concluded that serotonin produces some of its effects by acting at specific receptors [43]. A few years later (1957) Gaddum and Picarelli identified two different receptors for serotonin in the guinea-pig ileum. They named them M receptors and D receptors because they could be blocked with morphine and dibenzyline respectively [44]. More than 20 years later (1979) Peroutka and Snyder described two binding sites for serotonin in brain tissue, namely 5-HT_1 and 5-HT_2 receptors [45]. New drug tools made it possible to characterize receptors in more detail and in 1986 enabled Bradley et al. to publish their proposal for the classification of functional receptors of serotonin [46]. According to this classification serotonin receptors could be divided into 5-HT_1, 5-HT_2, and 5-HT_3 receptors. The 5-HT_3 receptor broadly encompasses the M receptor and the 5-HT_2 receptor the D receptor previously identified by Gaddum and Picarelli [44].

Recent developments and in particular the capability of doing receptor cloning studies identified a number of additional serotonin receptors and in 1994 a group of scientists with a combined total experience of more than 100 years in serotonin and serotonin receptor research published a new classification dividing serotonin receptors into seven different groups 5-HT_1-5-HT_7 [47]. The number of receptor subgroups has continued to increase and constituted a total of at least 14 in 2002 [48]. Of these, the 5-HT_{1A}, 5-HT_{2A}, 5-HT_{2C}, 5-HT_3, and 5-HT_4 receptors are all known to be involved in the emetic reflex arch. With the exception of the 5-HT_3 receptor, 5-HT receptors are members of the G-protein-coupled superfamily. The 5-HT_3 receptor, which from an antiemetic point of view, is by far the most interesting of the 5-HT receptors, and is a pentameric, ligand-gated ion channel of the cysteine-loop receptor family (Fig. 2). It occurs with the highest frequency in the gut and in the nucleus of the tractus solitarius [49, 50]. In 1995–2003 the human genes for five subunits (A-E) of the 5-HT_3 receptor were cloned [51–54] and it is now evident that 5-HT_3 receptors can be subdivided into 5-HT_{3A}, 5-HT_{3B}, 5-HT_{3C}, 5-HT_{3D}, and 5-HT_{3E} [55]. The knowledge about receptor pharmacology continues to grow and recently it was published that the brain specific 5-HT_{3B} receptor is heterogeneous and includes several subunits, the role of which is unclear [56]. It has been speculated that these isoforms are individual and can be used for personalized therapy [57].

One of the true breakthroughs in antiemetic research was the finding (1981) that high doses of metoclopramide was effective in preventing emesis induced by cisplatin [58]. In 1978 it had been demonstrated that metoclopramide is a weak 5-HT_3 receptor antagonist [59] and it could be hypothesized that the antiemetic effect of high doses of metoclopramide was due to an antagonist effect at 5-HT_3 receptors, whereas the majority of side effects was connected to dopamine D_2 receptor antagonism.

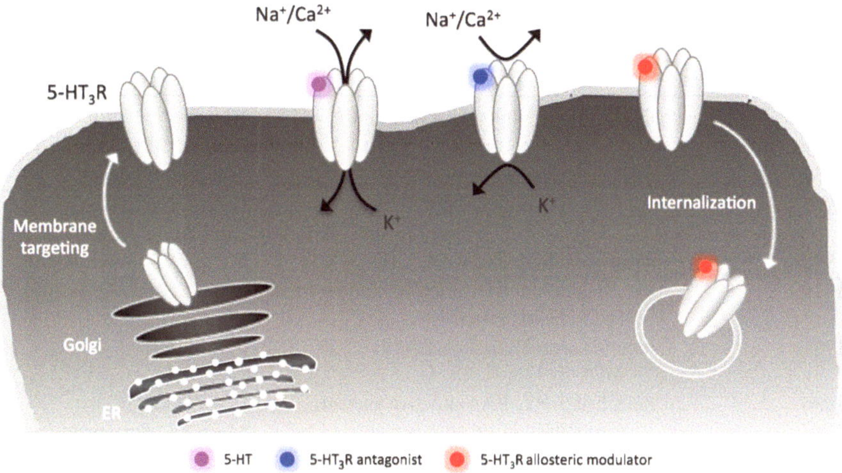

Fig. 2 Differential effects of antiemetics on 5-HT$_3$ receptors

The finding of the first selective 5-HT$_3$ receptor antagonist, MDL72222, in 1984 [60], was followed by several animal studies demonstrating antiemetic effect of MDL2222 and other 5-HT$_3$ receptor antagonists [61, 62] and the first two clinical studies were published in The Lancet in May [63] and June [64] 1987, respectively. The history of serotonin and serotonin receptors is summarized in Table 2.

Substance P and Neurokinin Receptors

In 1931 von Euler and Gaddum followed up on the finding of Dale and Dudley 2 years earlier, that acetylcholine is present in the fresh spleen of the ox and horse [65]. During this research in atropinized rabbit intestine, they found a new depressor substance that could not be identified as acetylcholine, histamine, adenosine, or kallikrein [66]. This new substance has later been confirmed to be substance P. Lembeck in 1953 made his experiments on the only method available at this time, the isolated guinea pig ileum and compared ventral and dorsal root extracts. He demonstrated that dorsal root extract contained ten times more of a contracting compound (substance P) than ventral roots [67]. It took almost 40 years from the findings of von Euler and Gaddum before substance P was isolated and completely purified [68]. Chang and Leeman in 1970 showed that substance P is an undecapeptide composed of Lys$_1$, Arg$_1$, Glx$_2$, Pro$_2$, Gly$_1$, Leu$_1$, Met$_1$, and Phe$_2$. They found the NH$_2$-terminal residue to be arginine. In 1983 two additional mammalian tachykinins were identified [69, 70]. These were named neurokinin A (also known as substance K) and neurokinin B (also known as neuromedin K). The three tachykinins share the

Table 2 History of 5-HT and 5-HT receptors

Year of publication	Identification and development of 5-HT and its receptors	References
1948	Discovery of 5-hydroxytryptamine (5-HT)	Rapport et al. [40]
1949		Rapport [41]
1954	5-HT acts at specific receptors	Gaddum and Hameed [43]
1957	Identification of two different receptors for 5-HT	Gaddum and Picarelli [44]
1978	Metoclopramide is a weak 5-HT_3 RA	Fozard and Mobarok [59]
1979	Identification of two different binding sites for 5-HT in brain tissue	Peroutka and Snyder [45]
1984	Identification of the first selective 5-HT_3 RA, MDL 72222	Fozard [60]
1986	Selective 5-HT_3 RAs effective against cisplatin-induced emesis	Costall et al. [61] Miner and Sanger [62]
1986	Classification of serotonin receptors into three groups	Bradley et al. [46]
1987	First clinical trials demonstrating antiemetic effect of a 5-HT_3 RA	Leibundgut and Lancranjan [63] Cunningham et al. [64]
1990	Consensus meeting agrees that the highest number of 5-HT receptors is in the NTS	Pratt [50]
1994	Classification of serotonin receptors into seven groups and several subgroups	Hoyer [47]
1995	Cloning of the human gene for 5-HT_3 receptor subunit A	Belelli et al. [51] Miyake et al. [52]
1999	Cloning of the human gene for 5-HT_3 receptor subunit B	Davies et al. [53]
2007	Cloning of the human genes for 5-HT_3 receptor subunits C, D, and E	Niesler et al. [54]

RA receptor antagonist; *NTS* nucleus tractus solitarius

common C-terminal sequences Phe, X, Gly, Leu, Met, NH_2, which is characteristic for the tachykinin family of peptides [71]. In the following years accumulating evidence appeared that these peptides acted as transmitters at specific receptors and in the period 1984–1986 three receptors were identified and named NKA (substance K) receptor, NK_1 (substance P) receptor, and NKB (neuromedin K) receptor respectively [72–74]. A few years later (1987–1990) cloning of the receptor genes was completed, see Table 3 [75–77]. Today the presence of substance P, NKA, and NKB has been verified in the CNS and in the peripheral nervous system as well [78]. The first NK_1 receptor antagonist was developed in 1981 [79], but as a peptide it was unable to penetrate the blood–brain barrier due to the large size of these compounds. It took another 10 years before Snider et al. published the finding of the first non-peptide NK_1 receptor antagonist, CP 96,345 [80] and this was much more important for the potential clinical development of agents acting on the NK_1 receptor, because penetration of the blood–brain barrier seems to be a necessity for clinical effect. It soon became evident that the emetic pathway include a tachykinin component

Table 3 History of substance P and NK receptors

Year of publication	Identification and development of substance P and NK receptors	References
1931	Discovery of substance P	Von Euler and Gaddum [66]
1953	Substance P acts as a neurotransmitter	Lembeck [67]
1970	Substance P purified and sequenced	Chang and Leeman [68]
1981	First peptide NK_1 receptor antagonist	Leander et al. [79]
1983	Discovery of NKA and NKB	Kangawa et al. [69]
1983		Kimura et al. [70]
1984		Buck et al. [72]
1986	Three different neurokinin receptors, NK_1, NK_2, and NK_3	Laufer et al. [73]
1986		Lee et al. [74]
1987	Cloning of the substance K (NKA) receptor	Masu et al. [75]
1989	Cloning of the substance P (NK_1) receptor	Yokota et al. [76]
1990	Cloning of the neuromedin K (NKB) receptor	Shigemoto et al. [77]
1991	First non-peptide NK_1 receptor antagonist	Snider et al. [80]
1993	Substance P, NKA, and NKB present in CNS and PNS	Maggi et al. [78]
1993	The emetic pathway has a tachykinin component	Tattersall et al. [81]
1997	First clinical trial demonstrating antiemetic effect of a NK_1 RA	Kris et al. [82]

RA receptor antagonist; *NK* neurokinin

[81] and a natural development was the design of several specific NK_1 receptor antagonists leading to the publication of the first clinical trial demonstrating antiemetic effect of one of these agents in 1987 [82].

Other Transmitters and Receptors

Although we know a good deal about the involvement of dopamine, serotonin, and substance P in the emetic pathway, these transmitters cannot entirely explain the mechanism resulting in nausea and vomiting induced by chemotherapy. Other transmitters such as enkephalins [83], arginine vasopressin [84], peptide YY [85], ghrelin [86], and endocannabinoids [87] have been proposed to be involved in the emetic pathway, but little is known about the significance of these transmitters. Contrary to dopamine, serotonin, and substance P, endocannabinoids have a protective effect against emesis, when blocking cannabinoid $(CB)_1$ receptors. This has been taken advantage of and synthetic cannabinoid drugs have been developed and used as antiemetics [88]. The side effects of these agents limit their use in clinical practice.

Histamine and acetylcholine are important transmitters in the pathway of motion sickness-induced nausea and vomiting, but do not seem to be of important in chemotherapy-induced nausea and vomiting. Noradrenaline is the principal neurotransmitter of the sympathetic nervous system and one of the key monoamines in the central nervous system. Serving as the main output for the sympathetic nervous system, noradrenaline plays an essential role in preparing the body for "fight-or-flight" responses to external stressors. Noradrenaline is synthesized from dopamine in one enzymatic step by dopamine β-hydroxylase and stored in secretory vesicles until they are released in response to neuronal firing. Noradrenaline is widely distributed in the brain, but most noradrenergic neurons are found in the brain stem. Noradrenaline is, similar to dopamine and 5-HT, a neuromodulatory neurotransmitter that can alter glutamatergic and GABAergic-mediated synaptic transmission. It exerts its actions by binding to two different classes of metabotropic G-protein coupled receptors: the α- and β-adrenergic receptors [89]. Pharmacological experiments have demonstrated that noradrenaline and agonists with the ability to activate central α-adrenergic receptors can produce emesis [90, 91]. High levels of noradrenaline in the area postrema and the presence of adrenergic receptors in this region but also in the nucleus of the tractus solitarius further suggest a role for noradrenaline in the multifactorial mechanisms related to chemotherapy-induced nausea and vomiting. Experiments involving central infusion of noradrenaline or administration of α_2-receptor agonists demonstrated that the agents were able to induce an emetic response that could be blocked by α_2-receptor antagonists in dogs [92] and cats [93], suggesting a role for α_2-receptors in emesis. Furthermore, urinary noradrenaline levels have been suggested to predict acute (0–24 h after chemotherapy) and delayed (24–48 h after chemotherapy) chemotherapy-induced nausea and vomiting in cancer patients, since relatively higher noradrenaline levels were positively correlated with an increased intensity of delayed nausea following treatment [94]. More recent studies have demonstrated that the selective noradrenaline reuptake inhibitor, (*R*)-sila-venlafaxine, could attenuate morphine- and cisplatin-induced emesis in the ferret [95, 96]. The antiemetic actions of (*R*)-sila-venlafaxine were hypothesized to involve an extracellular elevation of noradrenaline that in turn could activate presynaptic α_2-receptors on neuronal circuits involved in emesis, in order to reduce additional release of noradrenaline and/or prevent the release of potential other emetic agents.

Mechanism of Antiemetic Effect

The mechanism of effect of antiemetics is in a way very simple, but at the same time extremely complicated. The simple explanation is that serotonin receptor antagonists act by blockade of serotonin (primarily 5-HT_3) receptors, dopamine receptor antagonists by blockade of dopamine (D_2 and D_3) receptors, and neurokinin receptor antagonists by blockade of neurokinin (NK_1) receptors. Focus has primarily been on the mechanism of action of serotonin receptor antagonists. Treatment with antineoplastic agents, e.g., cisplatin, will induce damage to the enterochromaffin cells in the gut, causing release of serotonin, chromogranin A, substance P, and other transmitters. Serotonin will block 5-HT_3 receptors in the gut (located on the vagus nerve) and result in initiation of the emetic reflex arch.

The Serotonin and 5-HT_3 Receptor Hypothesis

Unfortunately the mechanism is not as simple as described above. We know that excess amounts of serotonin and the urine metabolite 5-hydroxyindole-acetic acid (5-HIAA) is released in the first 4–12 h after the initiation of high-dose cisplatin [97, 98] and dacarbazine [98] therapy, but this is not the case after treatment with low doses of cisplatin [98] and with other antineoplastic agents such as, e.g., cyclophosphamide [98]. Despite the lack of an excess level of serotonin after treatment with cyclophosphamide, serotonin receptor antagonists are very effective in preventing cyclophosphamide-induced emesis. Contrary, in cisplatin- and dacarbazine-induced emesis, there is a parallel increase in 5-HIAA, and the numbers of emetic episodes [97, 98]. The limited effect of serotonin receptor antagonists in the delayed phase of chemotherapy-induced nausea and vomiting (24–120 h after chemotherapy) is in good correlation with the lack of excess serotonin release in the period 24–96 h after high-dose cisplatin [99]. In subsequent cycles of cisplatin the peak levels of 5-HIAA are higher compared to the first cycle and elevations are more sustained [99]. Again, this is in good correlation with the fact that the effect of serotonin receptor antagonists (and of other antiemetics) obtained in the first cycle of chemotherapy is not maintained in subsequent cycles [100].

The most likely receptor binding mechanism is a competitive inhibition, although the existence of noncompetitive antagonists has been asserted [101]. Recently it has been proposed that the 5-HT_3 receptor antagonist, palonosetron, has a somewhat different binding mechanism. In a radioligand binding assay, ondansetron and granisetron were competitive inhibitors, whereas palonosetron exhibited allosteric binding and positive cooperativity [102]. Furthermore Rojas and coworkers made a number of independent experiments and concluded that palonosetron triggers 5-HT_3 receptor internalization and causes prolonged inhibition of the receptor function [103]. This could explain the approximately 10% increase in effect on delayed nausea and vomiting induced by cisplatin or anthracycline/cyclophosphamide chemotherapy as compared to another 5-HT_3 receptor antagonist, granisetron [104].

Differential Time Course of Chemotherapy-Induced Emesis and of the Action of Antiemetics

Old data from patients receiving chemotherapy without antiemetic protection and from placebo-controlled trials has provided us with some knowledge about the time to onset of emesis induced by various antineoplastic agents [105–109]. We do not know the mechanism behind this delay in emesis. The time to onset of emesis is: cisplatin 2–4 h [106], high-dose cyclophosphamide 6–13 h [107, 108], carboplatin 4–9 hours [109], and doxorubicin 4–6 h [105]. In a comparative trial including a placebo arm, the 5-HT_3 receptor antagonist, ondansetron, significantly prolonged the time to onset of emesis induced by cisplatin from a median of 2.8 to 11.6 h [106].

Also a differential time course of action of antiemetics has been suggested [110, 111]. In a randomized, double-blind trial comparing the use of adding the dopamine receptor antagonist, metopimazine to ondansetron plus placebo, the combination was significantly superior, but there seemed to be little effect of the combination as compared to ondansetron alone in the first 8–12 h after cisplatin [110]. Hesketh et al. summarized four randomized, double-blind studies, three in cisplatin and one in anthracycline/cyclophosphamide (AC) chemotherapy. In all studies patients received antiemetic therapy with the 5-HT_3 receptor antagonist, ondansetron plus dexamethasone, and were randomized to additional therapy with the NK_1 receptor antagonist, aprepitant or placebo [111]. In the cisplatin studies, the risk of first emesis was seen after 15–18 h, whereas this was the case after 3 h in the AC study, suggesting a differential time course of action of aprepitant in these two different settings. It should, however, be recalled that the median time to first emesis in cisplatin-treated patients is almost 12 h [106], when ondansetron alone is used as antiemetic protection, meaning that in this period, it would be difficult to improve the effect on emesis by the addition of a second or even a third antiemetic.

Interactions and Cross-Talking Between Transmitter Systems

A major part of our knowledge about interactions and cross-talking between transmitter systems comes from the field of neuropsychiatric research [112–114].

Jovanovic-Micic and coworkers produced dose-dependent emesis in cats by intracerebroventricular injection of dopamine [115]. They tested eight different dopamine receptor antagonists to prevent the dopamine-induced emesis, but noticed that only haloperidol was effective. Because haloperidol is also an α-adrenergic receptor antagonist, they hypothesized that dopamine-induced emesis is due to the release of noradrenaline and to an action on α_2-adreno-receptors within the area postrema.

Recently a number of electrophysiological and biochemical trials have suggested that receptor crosstalk occurs between 5-HT_3 receptors and NK_1 receptors. In theory, co-activation of either of these receptors will enhance the activity of the other to produce a response. For example, release of serotonin and binding to 5-HT_3 receptors will enhance the response of NK_1 receptors to substance P and increase the risk of emesis. The 5-HT_3 receptor antagonist, palonosetron has in electrophysiological trials been able to inhibit substance P-mediated responses, probably by interactions with the 5-HT_3 receptor, supporting the hypothesis about crosstalk between 5-HT_3 and NK_1 receptors [103, 116]. In a well-described animal model of vomiting, the last shrew, Darmani et al. was able to demonstrate that the emetic response to a selective 5-HT_3 receptor agonist and to a selective NK_1 receptor agonist was synergistically increased by the combination of the two. Furthermore pretreatment with one of the selective antagonists to 5-HT_3 receptors (tropisetron) or NK_1 receptors (CP 99,994) each reduced the emetic response not only to the corresponding agonist but to both agonists [117].

Conclusion

The first report of the clinical use of an antineoplastic drug was published in 1942 [118]. During the next 30 years a number of antineoplastic drugs were FDA approved, and implemented in clinical practice. Antiemetic therapy was non-existing or limited to single agent therapy with a dopamine D_2 antagonist, typically a phenothiazine or metoclopramide [119]. When cisplatin, the most emetogenic of all antineoplastic agents, was approved in 1978, the necessity of focusing on antiemetic research became evident. Today we can rely on evidence-based recommendations such as those published by the Multinational Association of Supportive Care in Cancer and the European Society for Medical Oncology [120]. Despite more than 30 years of antiemetic research, there continue to be major gaps in our knowledge, restricting the possibility to prevent nausea and vomiting in all patients.

One gap is the lack of understanding of the mechanism leading to antiemetic effect of corticosteroids. Corticosteroids seem to improve the activity of almost any other antiemetic and are well established as useful antiemetics. Recently the presence of functional glucocorticoid receptors located in the nucleus of the tractus solitarius has been verified in cats [121] and also corticosteroids have been shown to antagonize 5-HT_{3A} receptors in an experimental model [122].

A more significant gap is the lack of understanding about the mechanism of nausea. If we look for support in clinical trials, one gets the impression that it is necessary to target numerous receptors to prevent nausea. Serotonin- and neurokinin receptor antagonists, the only two classes of drugs specifically developed as antiemetics, are highly selective for 5-HT_3 and NK_1 receptors respectively, but primarily effective in the prophylaxis of vomiting. Actually the most effective antinausea agents are the "dirty drugs" like metopimazine [110] and olanzapine [123], targeting multiple receptors. A recent pilot study showed promising results with APD 421, an agent targeting dopamine D_2 but also central D_3 receptors. APD 421 in combination with the 5-HT_3 receptor antagonist ondansetron completely prevented vomiting induced by cisplatin in 15/17 patients (88%) in the first 24 h and nausea was less than 10 mm on a 100 mm VAS in all patients [124].

Antiemetic research has been an extraordinary example of the use of translational research, but has also emphasized the distance between the laboratory and the clinic and between different animal models and patients. Today the human genome has been cloned and modern technology will soon enable such analyses to be implemented in the daily clinic. We already know of several polymorphisms that can explain differences in antiemetic effect [125], and maybe gene expression profiles could become a future instrument in individual patient-oriented prophylaxis of nausea and vomiting. In the meantime we have to take comfort in the fact that despite the major gap in our knowledge, 30 years of research has made it much easier for patients to complete chemotherapy without suffering from intractable nausea and vomiting.

References

1. Wepfer JJ. Historia cicutae aquaticae. Basel. 1679;152.
2. Mellinger C. Beitrage zur Kenntniff des Erbrechens. Pfluegers Arch. 1881;24:232–40.
3. Cannon WB. The movement of the stomach studied by means of the röntgen rays. Am J Physiol. 1898;1:359–82.
4. Giannuzzi G. Untersuchungen über die Organe, welche an dem Brechaet theilnehmen, und über die Physiologische Wirkung de Tartarus stibiatic. Zent Med Wizz. 1865;3:129–31.
5. Thumas LJ. Über das Brechcentrum und über die Wirkung einiger pharmakologischer Mittel auf dasselbe. Arch für Patol Anat. 1891;123:44–69.
6. Hatcher RA, Weiss S. Studies on vomiting. J Pharmacol Exp Ther. 1923;22:139–93.
7. Borison HL, Wang SC. Functional location of central coordinating mechanism for emesis in cat. J Physiol. 1949;12:305–13.
8. Wang SC, Borison HL. Copper sulphate emesis: a study of afferent pathways from the gastrointestinal tract. J Physiol. 1951;14:520–6.
9. Borison HL, Wang SC. Locus of the central emetic action of cardiac glycosides. J Physiol. 1951;76:335–8.
10. Wang SC, Borison HL. A new concept of organization of the central emetic mechanism: recent studies on sites of action of apomorphine, copper sulphate and cardiac glycosides. Gastroenterology. 1952;22:1–12.
11. Borison HL, Wang SC. Physiology and pharmacology of vomiting. Pharmacol Rev. 1953;5:193–230.
12. Lindstrom PA, Brizze KR. Relief of intractable vomiting from surgical lesions in the area postrema. J Neurosurg. 1962;19:228–36.
13. Miller AD, Wilson VJ. "Vomiting center" reanalyzed: an electrical stimulation study. Brain Res. 1983;270:154–8.
14. Miller AD, Nonaka S, Jakûs J. Brain areas essential or non-essential for emesis. Brain Res. 1994;647:255–64.
15. Carpenter DO. Neural mechanisms of emesis. Can J Physiol Pharmacol. 1990;68:230–6.
16. Carl PL, Cubeddu LX, Lindley C, Myers RD, Rezvani AH. Do humoral factors mediate cancer chemotherapy-induced emesis? Drug Metab Rev. 1989;21:319–33.
17. Peroutka SJ. Chemotherapeutic agents do not interact with neurotransmitter receptors. Cancer Chemother Pharmacol. 1987;19:131–2.
18. Herrstedt J, Hyttel J, Pedersen J. Interaction of the antiemetic metopimazine and anticancer agents with brain dopamine D_2, 5-hydroxytryptamine$_3$, histamine H_1, muscarine cholinergic and α1-adrenergic receptors. Cancer Chemother Pharmacol. 1993;33:53–6.
19. Herrstedt J. Antiemetics: an update and the MASCC guidelines applied in clinical practice. Nat Clin Pract Oncol. 2008;5:32–43.
20. Carlsson A, Lindqvist M, Magnusson T. 3,4-Dihydroxyphenylalanine and 5-hydroxytryptophan as reserpine antagonists. Nature. 1957;180:1200.
21. Iversen SD, Iversen LL. Dopamine: 50 years in perspective. Trends Neurosci. 2007;30:188–93.
22. Anden NE, Carlsson A, Dahlstroem A, Fuxe K, Hillarp NA, Larsson K. Demonstration and mapping out of nigro-neostriatal dopamine neurons. Life Sci. 1964;3:523–30.

23. Dahlstroem A, Fuxe K. Evidence for the existence of monoamine- containing neurons in the central nervous system. I. Demonstration of mono- amines in the cell bodies of brain stem neurons. Acta Physiol Scand. 1964;232(Suppl):231–55.
24. Bunney B. Antipsychotic drug effects on the electrical activity of dopaminergic neurons. Trends Neurosci. 1984;7:212–5.
25. Kebabian JW, Petzold GL, Greengard P. Dopamine-sensitive adenylate cyclase in caudate nucleus of rat brain, and its similarity to the "dopamine receptor". Proc Natl Acad Sci. 1972;69:2145–9.
26. Kebabian JW, Calne DB. Multiple receptors for dopamine. Nature. 1979;277:93–6.
27. Katzman R, Markman MH, Ahn HS, Mishra RK, Gardner E. Effects of drugs and lesions on dopamine-stimulated adenylate cyclase: evidence for different classes of dopamine receptors. Trans Am Neurol Assoc. 1977;102:76–9.
28. Beaulieu JM, Gainetdinov RR. The physiology, signaling and pharmacology of dopamine receptors. Pharmacol Rev. 2011;63:182–217.
29. Missale C, Nash SR, Robinson SW, Jaber M, Caron MG. Dopamine receptors: from structure to function. Physiol Rev. 1998;78:189–225.
30. Carlsson A. A paradigm shift in brain research. Science. 2001;294:1021–4.
31. Bunzow JR, Van Tol HH, Grandy DK, et al. Cloning and expression of a rat D2 dopamine receptor cDNA. Nature. 1988;336:783–7.
32. Giros B, Sokoloff P, Martres MP, Riou JF, Emorine LJ, Schwartz JC. Alternative splicing directs the expression of two D2 dopamine receptor isoforms. Nature. 1989;342:923–6.
33. Moertel CG, Reitemeier RJ, Gage RP. A controlled clinical evaluation of antiemetic drugs. JAMA. 1963;186:116–8.
34. Shen WW, Baig MS, Sata LS, Hofstatter L. Dopamine receptor supersensitivity and the chemoreceptor trigger zone. Biol Psychiatry. 1983;18:917–21.
35. Jolliet P, Nion S, Allain-Veyrac G, et al. Evidence of the lowest brain penetration of an antiemetic drug, metopimazine, compared to domperidone, metoclopramide and chlorpromazine, using an in vitro model of the blood–brain barrier. Pharmacol Res. 2007;56:11–7.
36. Sokoloff P, Giros B, Martres M-P, Bouthenet M-L, Schwartz J-C. Molecular cloning and characterization of a novel dopamine receptor (D3) as a target for neuroleptics. Nature. 1990;347(6289):146–51.
37. Yoshida N, Yoshikawa T, Hosoki K. A dopamine D3 receptor agonist, 7-OH-DPAT, causes vomiting in the dog. Life Sci. 1995;57:PL347–50.
38. Yoshikawa T, Yoshida N, Hosoki K. Involvement of dopamine D3 receptors in the area postrema in R(+)-7-OH-DPAT induced emesis in the ferret. Eur J Pharmacol. 1996;301:143–9.
39. Stemp G, Ashmeade T, Branch CL, et al. Design and synthesis of trans-N-[4-[2-(6-cyano-1,2,3, 4-tetrahydroisoquinolin-2-yl)ethyl]cyclohexyl]-4-quinolinecarboxamide (SB-277011): a potent and selective dopamine D3 receptor antagonist with high oral bioavailability and CNS penetration in the rat. J Med Chem. 2000;43:1878–85.
40. Rapport MM, Green AA, Page IH. Partial purification of the vasoconstrictor in beef serum. J Biol Chem. 1948;176:735–41.
41. Rapport MM. Crystalline Serotonin. Science. 1948;108:329–30.
42. Freyburger WA, Graham BE, Rapport MM, et al. The pharmacology of 5-hydroxytryptamine (serotonin). J Pharmacol Exp Ther. 1952;105:80–6.
43. Gaddum JH, Hameed KA. Drugs which antagonize 5-hydroxytryptamine. Br J Pharmacol. 1954;9:240–8.
44. Gaddum JH, Picarelli ZP. Two kinds of tryptamine receptors. Br J Pharmacol. 1957;12:323–8.
45. Peroutka SJ, Snyder SH. Multiple serotonin receptors: differential binding of ^{3}H-5-hydrxytryptamine, ^{3}H-lysergic acid diethylamide and ^{3}H-spiropidol. Mol Pharmacol. 1979;16:687–9.
46. Bradley PB, Engel G, Feniuk W, et al. Proposals for the classification and nomenclature of functional receptors for 5-hydroxytryptamine. Neuropharmacology. 1986;25:563–76.
47. Hoyer D, Clarke DE, Fozard JR, et al. International union of pharmacology classification of receptors for 5-hydroxytryptamine (serotonin). Pharmacol Rev. 1994;46:157–203.

48. Hoyer D, Hannon JP, Martin GR. Molecular, pharmacological and functional diversity of 5-HT receptors. Pharmacol Biochem Behav. 2002;71:533–54.
49. Kilpatrick GJ, Jones BJ, Tyers MB. Identification and distribution of 5-HT_3 receptors in the brain using radioligand binding. Nature. 1987;330:746–8.
50. Pratt GD, Bowery NG, Kilpatrick GJ, et al. Consensus meeting agrees about distribution of 5-HT_3 receptors in mammalian hindbrain. TIPS. 1990;11:135–7.
51. Belelli D, Balcarek JM, Hope AG, et al. Cloning and functional expression of a human 5-hydroxytryptamine type 3AS receptor subunit. Mol Pharmacol. 1995;48:1054–62.
52. Miyake A, Mochizuki S, Takemoto Y, Akuzawa S. Molecular cloning of human 5-hydroxytryptamine3 receptor: heterogeneity in distribution and function among species. Mol Pharmacol. 1995;48:407–16.
53. Davies PA, Pistis M, Hanna MC, et al. The 5-HT3B subunit is a major determinant of serotonin-receptor function. Nature. 1999;397:359–63.
54. Niesler B, Frank B, Kapeller J, Rappold GA. Cloning, physical mapping and expression analysis of the human 5-HT3 serotonin receptor-like genes HTR3C, HTR3D and HTR3E. Gene. 2003;310:101–11.
55. Niesler B, Kapeller J, Hammer C, Rappold G. Serotonin type 3 receptor genes: HTR3A, B, C, D, E. Pharmacogenomics. 2008;9:501–4.
56. Jensen AA, Davies PA, Braüner-Osborne H, Krzywkowski K. 3B but which 3B? And that's just one of the questions: the heterogeneity of human 5-HT_3 receptors. TIPS. 2008;29:437–44.
57. Niesler B. 5-HT3 receptors: potential of individual isoforms for personalized therapy. Curr Opin Pharmacol. 2011;11:81–6.
58. Gralla RJ, Itri LM, Pisko SE, et al. Antiemetic efficacy of high-dose metoclopramide: randomized trials with placebo and prochlorperazine in patients with chemotherapy-induced vomiting. N Engl J Med. 1981;305:905–9.
59. Fozard JR, Mobarok ALIA. Blockade of neuronal tryptamine receptors by metoclopramide. Eur J Pharmacol. 1978;49:109–12.
60. Fozard JR. MDL 72222, a potent and highly selective antagonist at neuronal 5-hydroxytryptamine receptors. Naunyn Schmiedebergs Arch Pharmacol. 1984;326:36–44.
61. Costall B, Domeney AM, Naylor RJ, Tattersall FD. 5-hydroxytryptamine M-receptor antagonism to prevent cisplatin-induced emesis. Neuropharmacology. 1986;25:959–61.
62. Miner WD, Sanger GJ. Inhibition of cisplatin-induced vomiting by selective 5-hydroxytryptamine M-receptor antagonism. Br J Pharmacol. 1986;88:497–9.
63. Leibundgut U, Lancranjan I. First results with ICS 205–930 (5-HT3 receptor antagonist) in prevention of chemotherapy-induced emesis. Lancet. 1987;329:1198.
64. Cunningham D, Hawthorne J, Pople A, et al. Prevention of emesis in patients receiving cytotoxic drugs by GR38032F, a selective 5HT 3 receptor antagonist. Lancet. 1987;329:1461–3.
65. Dale HH, Dudley HW. The presence of histamine and acethylcholine in the spleen of the ox and the horse. Physiology. 1929;68:97–123.
66. Von Euler VS, Gaddum JH. An unidentified depressor substance in certain tissue extracts. J Physiol (Lond). 1931;1931:577–83.
67. Lembeck F. Zur Frage der zentralen Ubertragung afferenter Impulse III. Mitteilung. Das Vorkommen und die Bedeutung der Substanz P in den dorsalen Wurzeln des Ruckenmarks. Arch Exp Pathol Pharmakol. 1953;219:197–213.
68. Chang M, Leeman SE. Isolation of sialogogic peptide from bovine hypothalamic tissue and its characteristic as substance P. J Biol Chem. 1970;245:4784–90.
69. Kangawa H, Minamino N, Fukuda A, Matsuo H. Neuromedin K: a novel mammalian tachykinin identified in porcine spinal cord. Biochem Biophys Res Commun. 1983;114:533–40.
70. Kimura S, Okada M, Sugita Y, Kanzawa I, Munekata E. Novel neuropeptides, neurokinin A and B isolated from porcine spinal cord. Proc Jpn Acad. 1983;59B:101–4.
71. Erspamer V. The tachykinin peptides family. Trends Neurosci. 1981;4:267–9.
72. Buck SH, Burcher E, Shults CW, Lovenberg W, O'Donohue TL. Novel pharmacology of substance K–binding sites: a third type of tachykinin receptor. Science. 1984;226:987–9.
73. Laufer R, Gilon C, Chorev M, Selinger Z. Characterization of a neurokinin B receptor site in rat brain using a highly selective radioligand. J Biol Chem. 1986;261:10257–63.

74. Lee C-M, Campbell NJ, Williams BJ, Iversen LL. Multiple tachykinin binding sites in peripheral tissues and in the brain. Eur J Pharmacol. 1986;130:209–17.
75. Masu Y, Nakayama K, Tamaki H, Harada Y, Kuno M, Nakanishi S. cDNA cloning of bovine substance-K receptor through oocyte expression system. Nature. 1987;329:836–8.
76. Yokota Y, Sasai Y, Tanaka K, et al. Molecular characterization of a functional cDNA for rat substance P receptor. J Biol Chem. 1989;264:17649–52.
77. Shigemoto R, Yokota Y, Tsuschida K, Nakanishi S. Cloning and expression of a rat neuromedin K receptor cDNA. J Biol Chem. 1990;265:623–8.
78. Maggi CA, Patacchini R, Roveto P, Giachetti A. Tachykinin receptors and receptor antagonists. J Auton Pharmac. 1993;13:23–93.
79. Leander S, Håkanson R, Rosell S, Folkers K, Sundler F, Tornquist K. A specific substance P antagonist blocks smooth muscle contractions induced by non-cholinergic, non-adrenergic nerve stimulation. Nature. 1981;294:467–9.
80. Snider RM, et al. A potent nonpeptide antagonist of the substance P (NK1) receptor. Science. 1991;251:435–7.
81. Tattersall FD, et al. The tachykinin NK1 receptor antagonist CP-99,994 attenuates cisplatin induced emesis in the ferret. Eur J Pharmacol. 1993;250:R5–6.
82. Kris MG, et al. Use of a NK1 receptor antagonist to prevent delayed emesis after cisplatin. J Natl Cancer Inst. 1997;89:817–8.
83. Harris HL. Cytotoxic therapy-induced vomiting is mediated via enkephalin pathways. Lancet. 1982;1:714–6.
84. Fisher RD, Rentschler RE, Nelson JC, Godfrey TE, Wilbur DW. Elevation of plasma antidiuretic hormone (ADH) associated with chemotherapy-induced emesis in man. Cancer Treat Rep. 1982;66:25–9.
85. Perry MR, Rhee J, Smith WL. Plasma levels of peptide YY correlate with cisplatin-induced emesis in dogs. J Pharm Pharmacol. 1994;46:553–7.
86. Rudd JA, Ngan MP, Wai MK, et al. Anti-emetic activity of ghrelin in ferrets exposed to the cytotoxic anticancer agent cisplatin. Neurosci Lett. 2006;392:79–83.
87. Walsh D, Nelson KA, Mahmoud FA. Established and potential therapeutic applications of cannabinoids in oncology. Support Care Cancer. 2003;11:137–43.
88. Rocha M, Stéfano SC, De Cássia HR, Oliveira R, Da Silveira DX. Therapeutic use of cannabis sativa on chemotherapy-induced nausea and vomiting among cancer patients: systematic review and meta-analysis. Eur J Cancer Care. 2008;17:431–43.
89. Ahlquist RP. A study of adrenotropic receptors. Am J Physiol. 1948;153:586.
90. Jenkins LC, Lahay D. Central mechanisms of vomiting related to catecholamines response: anaesthetic implication. Can Anaesth Soc J. 1971;18:434–41.
91. Borison HL. Area postrema: chemoreceptor circumventricular organ of the medulla oblongata. Prog Neurobiol. 1989;32:351–90.
92. Lang IM, Sarna SK. The role of adrenergic receptors in the initiation of vomiting and its gastrointestinal motor correlates. J Pharmacol Exp Ther. 1991;263:395–403.
93. Beleslin DB, Strbac M. Noradrenaline-induced emesis: alpha-2 adrenoreceptor mediation in the area postrema. Neuropharmacology. 1987;26:1157–65.
94. Fredrikson M, Hursi T, Steineck G, Fürst CJ, Börjesson S, Peterson C. Delayed chemotherapy-induced nausea is augmented by high levels of endogenous noradrenaline. Br J Cancer. 1994;70:642–5.
95. Showell GA, Barnes MJ, Daiss JO, et al. (R)-sila-venlafaxine: a selective noradrenaline reuptake inhibitor for the treatment of emesis. Bioorg Med Chem Lett. 2006;16:2555–8.
96. Warneck JB, Cheng FH, Barnes MJ, et al. Action of (R)-sila-venlafaxine and reboxetine to antagonize cisplatin-induced acute and delayed emesis in the ferret. Toxicol Appl Pharmacol. 2008;232:369–75.
97. Du Bois A, Kriesinger-Schroeder H, Meerpohl H-G. The role of serotonin as a mediator of emesis induced by different stimuli. Support Care Cancer. 1995;3:285–90.
98. Cubeddu LX, Hoffmann IS, Fuenmayor NT, Malave JJ. Changes in serotonin metabolism in cancer patients: its relationship to nausea and vomiting induced by chemotherapeutic drugs. Br J Cancer. 1992;66:198–203.

99. Cubeddu LX, Hoffmann IS. Participation of serotonin on early and delayed emesis induced by initial and subsequent cyles of cisplatinum-based chemotherapy: effects of antiemetics. J Clin Pharmacol. 1993;33:691–7.
100. De Wit R, Schmitz PIM, Verweij J, et al. Analysis of cumulative probabilities show, that the efficacy of 5-HT_3 antagonist prophylaxis is not maintained. J Clin Oncol. 1996;14:644–51.
101. Csillik-Perczel V, Bakonyi A, Yemane T, et al. GYKI-46903, a non-competitive antagonist for 5-HT3 receptors. Pharmacol Toxicol. 1996;79:32–9.
102. Rojas C, Stathis M, Thomas AG, et al. Palonosetron exhibits unique molecular interactions with the 5-HT_3 receptor. Anaesth Analg. 2008;107:469–78.
103. Rojas C, Slusher BS. Pharmacological mechanisms of 5-HT_3 and tachykinin NK_1 receptor antagonism to prevent chemotherapy-induced nausea and vomiting. Eur J Pharmacol. 2012;684:1–7.
104. Saito M, Aogi K, Sekine I, et al. Palonosetron plus dexamethasone versus granisetron plus dexamethasone for prevention of nausea and vomiting during chemotherapy: a double-blind, double-dummy, randomized, comparative Phase III trial. Lancet Oncol. 2009;10:115–24.
105. Borison HL, McCarthy LE. Neuropharmacology of chemotherapy-induced emesis. Drugs. 1983;25 Suppl 1:8–17.
106. Cubeddu LX, Hoffmann IS, Fuenmayor NT, Finn AL. Efficacy of ondansetron (GR 38032) and the role of serotonin in cisplatin-induced nausea and vomiting. N Engl J Med. 1990;322:810–6.
107. Fetting J, Grochow LB, Folstein MF, Ettinger DS, Colvin M. The course of nausea and vomiting after high-dose cyclophosphamide. Cancer Treat Rep. 1982;66:1487–93.
108. Beck TM. The pattern of emesis following high-dose cyclophosphamide and the antiemetic efficacy of ondansetron. Anti Cancer Drugs. 1995;6:237–42.
109. Martin M, Diaz-Rubio E, Sánchez A, Almenarez J, López-Vega JM. The natural course of emesis after carboplatin treatment. Acta Oncol. 1990;29:593–5.
110. Herrstedt J, Sigsgaard T, Handberg J, Schousboe BMB, Hansen M, Dombernowsky P. Randomized, double-blind comparison of ondansetron versus ondansetron plus metopimazine as antiemetic prophylaxis during platinum-based chemotherapy in patients with cancer. J Clin Oncol. 1997;15:1690–6.
111. Hesketh PJ, Warr DG, Street JC, Carides AD. Differential time course of action of the 5-HT_3 and NK_1 receptor antagonists when used with highly and moderately emetogenic chemotherapy (HEC and MEC). Support Care Cancer. 2011;19:1297–302.
112. Blier P. Crosstalk between the norepinephrine and serotonin systems and its role in antidepressant response. J Psychiatry Neurosci. 2001;26(Suppl):S3–10.
113. Di Giovanni G, Esposito E, Di Matteo V. Role of serotonin in central dopamine dysfunction. CNS Neurosci Ther. 2010;16:179–94.
114. Werkman TR, McCreary AC, Kruse CG, Wadman WJ. NK_3 receptors mediate an increase in firing rate of midbrain dopamine neurons of the rat and guinea pig. Synapse. 2011;65:814–26.
115. Jovanovic-Micic D, Samardzic R, Beleslin DB. The role of α-adrenergic mechanisms within the area postrema in dopamine-induced emesis. Eur J Pharmacol. 1995;272:21–30.
116. Rojas C, Li Y, Zhang J, et al. The antiemetic 5-HT_3 receptor antagonist palonosetron inhibits substance P-mediated responses in vitro and in vivo. J Pharmacol Exp Ther. 2010;335:362–8.
117. Darmani NA, Chebolu S, Amos B, Alkam T. Synergistic antiemetic interactions between serotonergic 5-HT_3 and tachykininergic NK_1-receptor antagonists in the last shrew (Cryptotis parva). Pharmacol Biochem Behav. 2011;99:573–9.
118. Goodman LS, Wintrobe MM, Dameshek W, et al. Nitrogen mustard therapy; use of methyl-bis (beta-chloroethyl) amine hydrochloride and tris (beta-chloroethyl) amine hydrochloride for Hodgkin's disease, lymphosarcoma, leukemia and certain allied and miscellaneous disorders. J Am Med Assoc. 1946;132:126–32.
119. Justin-Besancon L, Laville C. Antiemetic action of metoclopramide with respect to apomorphine and hydergine. C R Seances Soc Biol Fil. 1964;158:723–7.
120. Roila F, Herrstedt J, Aapro M, et al. Guideline update for MASCC and ESMO in the prevention of chemotherapy- and radiotherapy-induced nausea and vomiting: results of the Perugia consensus conference. Ann Oncol. 2010;21 Suppl 5:v232–43.

121. Ho CM, et al. Dexamethasone has a central antiemetic mechanism in decerebrated cats. Anesth Analg. 2004;99:734–9.
122. Suzuki T, et al. Inhibitory effect of glucocorticoids on human-cloned 5-hydroxy- tryptamine$_{3A}$ receptor expressed in xenopus oocytes. Anesthesiology. 2004;101:660–5.
123. Navari RM, Gray SE, Kerr AC. Olanzapine versus aprepitant for the prevention of chemotherapy-induced nausea and vomiting: a randomized phase III trial. J Support Oncol. 2011;9:188–95.
124. Herrstedt J, Summers YJ, Daugaard G et al. The dopamine D_2/D_3 receptor antagonist APD421 in combination with ondansetron effectively prevents acute cisplatin-induced nausea and vomiting (CINV). Ann Oncol. 2012;23(suppl 9):ix507.
125. Kaiser R, Sezer O, Papies A, et al. Patient-tailored antiemetic treatment with 5- hydroxytryptamine type 3 receptor antagonists according to cytochrome P-450 2D6 genotypes. J Clin Oncol. 2003;20:2805–11.

Mucositis

Noor Al-Dasooqi, Dorothy M. Keefe, and Stephen T. Sonis

Abstract Alimentary tract (AT) mucositis is a serious and debilitating side effect of cancer therapy, primarily characterised by damage to the mucous membranes throughout the AT, and associated with mouth ulcers and pain, vomiting, diarrhoea, rectal bleeding and abdominal cramps to name a few. Considerable progress has been made in our understanding of the biological basis for mucositis, and the understanding of pathobiology continues to evolve. In recent years, the biggest advance has come from the 5-phase model of mucositis development. The main pathobiological features of mucositis include changes to tissue structure, and surface area, microbiome changes and inflammation. With a better understanding of mucositis pathogenesis, the focus of more recent research is shifting to targeted therapy-induced toxicity, toxicity clustering and the investigation of genetic polymorphisms in toxicity prediction. The ultimate goal of mucositis researchers is to identify the most appropriate targets for therapeutic interventions and to be able to predict toxicity risk and personalise interventions to genetically suitable patients. Continuing research efforts are needed to further our understanding of mucositis pathobiology and the pharmacogenomics of toxicity. This book chapter will update our understanding of the recent concepts of mucositis pathogenesis.

N. Al-Dasooqi (✉)
Department of Medicine, University of Adelaide, Adelaide, SA, Australia
e-mail: noor.al-dasooqi@adelaide.edu.au

D.M. Keefe
University of Adelaide and Adjunct Professor, Sansom Institute for Health Research, University of South Australia, Adelaide, SA, Australia
e-mail: Dorothy.keefe@health.sa.gov.au

S.T. Sonis
Brigham and Women's Hospital, 75 Francis Street, Boston, MA 02115, USA
e-mail: ssonis@partners.org

S.T. Sonis and D.M. Keefe (eds.), *Pathobiology of Cancer Regimen-Related Toxicities*,
DOI 10.1007/978-1-4614-5438-0_7, © Springer Science+Business Media New York 2013

Introduction

Oral mucositis (OM) is a frequent complication of anti-neoplastic therapy. Clinically, this condition is associated with significant pain, and hence, can significantly impair quality of life. The manifestations of OM form a continuum, where in its mild form it presents erythematous atrophic lesions and in its severe form presents ulcerative lesions which penetrate the submucosa. Loss of mucosal layer integrity in OM, especially in patients undergoing myeloablative therapy, represents a clinically significant risk factor for bacteraemia, fungaemia and sepsis [31, 111, 113]. Furthermore, OM can become a dose-limiting toxicity in some patients resulting in dose reductions or treatment breaks.

Anti-cancer therapies have also been shown to affect mucosal surfaces in the rest of the alimentary tract. Gastrointestinal (GI) mucositis has become an increasingly recognised and better characterised toxicity of chemotherapy and radiotherapy. It is most likely to affect the small intestine, but can also occur in the oesophagus, stomach and large intestine. Symptoms include pain, ulceration, nausea and vomiting, diarrhoea and rectal bleeding [24, 58] and depend on the area affected. Proctitis following radiation is also considered a manifestation of GI mucositis. Systemic effects include infection, fatigue, malnutrition and even occasionally death [5, 58, 77].

The frequency of oral and GI mucositis varies significantly amongst different treatment regimens. Anthracycline-, taxane- and platinum-based regimens are associated with relatively low level of toxicity where approximately 1–10% of patients will experience OM. In contrast, regimens containing 5-fluorouracil (5-FU) or irinotecan (CPT-11) induce grade 3–4 oral and GI mucositis, respectively, in more than 15% of patients. The addition of radiation to these treatment regimens will further increase the risk of grade 3–4 oral and GI mucositis to approximately 30% of patients. Patients undergoing radiotherapy for head and neck cancer are amongst those at highest risk of developing OM with 85% of patients presenting with symptoms [7, 33, 111, 113, 131]. Limited epidemiological data is available on targeted therapies for cancer. However, recently, oral ulceration was reported in 66% of patients receiving mTOR inhibitors for various solid tumours [118]. These ulcers resembled those of aphthous stomatitis and were distinct from those seen in conventional mucositis. These observations suggest differences in pathology of mucositis as a result of conventional vs. targeted anti-cancer treatment.

Although the pathobiology of mucositis is complex, much progress continues to be made in understanding the pathological and signalling alterations leading to tissue injury. Our current understanding of this condition is based on early radiation animal OM models which highlight the multifactorial nature of the condition and implicates a cascade of events in multiple tissue regions [106, 107, 115–117]. This observation gave rise to the 5-phase model of OM, which describes the sequence of genetic and histopathological events following cytotoxic treatment. This early OM model questioned the simplistic, historical view of mucositis being an epithelium-based condition and prompted the need for investigation of the intricate functional interaction of genes, cells and tissues. Based on this, subsequent studies implicated microvascular injury [85], pro-inflammatory cytokines [66, 70, 114, 137], microbiome–host interactions [120–122, 124] and extracellular alterations [1, 3] in the development of mucositis.

This chapter will outline the oral and GI toxicities associated with commonly used cytotoxic and targeted therapies, and give an insight into the pathobiology of regimen-related mucositis.

The Normal Alimentary Tract

Understanding pathobiology required understanding of the normal structures and function. The alimentary tract is a hollow tube that extends from the mouth to anus. It has similar structural organisation throughout its length [75], and like all body cavities which open to the exterior it is lined by mucosa. All mucosae are covered by a layer of epithelial cells lining the lumen, supported by a lamina propria. The primary functions of the mucosa include protection, secretion (of digestive enzymes, hydrochloric acid, mucins and antibodies), absorption of metabolic substrates, serving as a barrier from pathogens and forming an immunologic barrier [80].

Oral Mucosa

The oral cavity is the start of the GI tract, and is lined by oral mucosa, a stratified squamous epithelium, which may or may not be keratinised (this is highly dependent on regions of high friction) [22, 119]. Within the epithelium itself there are four layers of cells: the basal cell layer (which contains the stem cells) and the prickle cell layer which is found in both keratinised and non-keratinised tissue. Keratinised epithelium has a granular layer followed by a keratinised layer, whereas non-keratinised epithelium has an intermediate layer and a superficial layer [119].

Oesophagus and Stomach

Essentially the oesophagus is a long muscular tube which functions to transport material from the oral cavity to the stomach. It is lined with a non-keratinised stratified squamous epithelium. The stomach functions as a "blender" and reservoir for food preparation and storage prior to presentation to the small intestine. The stomach is lined with a simple columnar epithelium, which invaginates into the lamina propria forming gastric pits and gastric glands. A number of specialised cells line the gastric glands and these are responsible for specific gastric secretions.

Intestinal Anatomy and Histology

The small intestinal epithelia can simplistically be divided into two units: proliferative units known as crypts and functional units known as villi [134]. The epithelium lining the small intestine is simple columnar epithelium and is constantly replaced as cells are sloughed into the lumen [6]. The large intestine is also composed of simple columnar epithelium; however there are no villi postnatally [127]. All intestinal cells arise from stem cells found in the base of the crypts with each cell dividing four to six times before reaching its ultimate destination [6]. Within the intestinal epithelium, there are four main cell lineages namely columnar, mucin-secreting (Goblet cells), endocrine and Paneth cells.

Cell Renewal

A hallmark of AT mucosal layer maintenance is the high cell turnover rate which allows for the necessary continuous renewal of the epithelium. Rapidly dividing epithelial progenitor cells, located in the proliferative zones of crypts in the small intestine and colon, undergo continuous, regulated apoptosis, proliferation and differentiation into a range of cell types with specialised functions as they migrate towards the lumen [89, 90]. This process is tightly regulated by mucosal as well as submucosal signalling in order to maintain homeostasis and to compensate for disturbances which may occur in the AT [17].

Mucositis

The Historic Paradigm for Mucositis Pathogenesis

The historic paradigm for the development of mucositis proposed that radiotherapy and chemotherapy caused clonogenic cell death in rapidly proliferating cells of the basal epithelium leading to epithelial atrophy, breakage and ulceration [15, 45, 62, 108, 112], and resulting in oral symptoms. A number of observations supported this view: firstly, children and young animals (who have a faster rate of proliferation than adults) experience increased rates of mucositis [104]; secondly, epidermal growth factor (EGF) exposure prior to chemotherapy increased the occurrence of mucositis [109, 110]; and thirdly, inhibiting proliferation with TGF-β3 prior to chemotherapy proved to have protective effects [108, 112]. However, observations in the clinic suggested a more complex mechanism of injury. This view was challenged in the last decade as more data has become available on cellular and tissue responses to anti-cancer agents.

The first animal models for investigation of chemotherapy- [116, 117] and radiation-induced [114] OM were developed by Sonis. These animal models

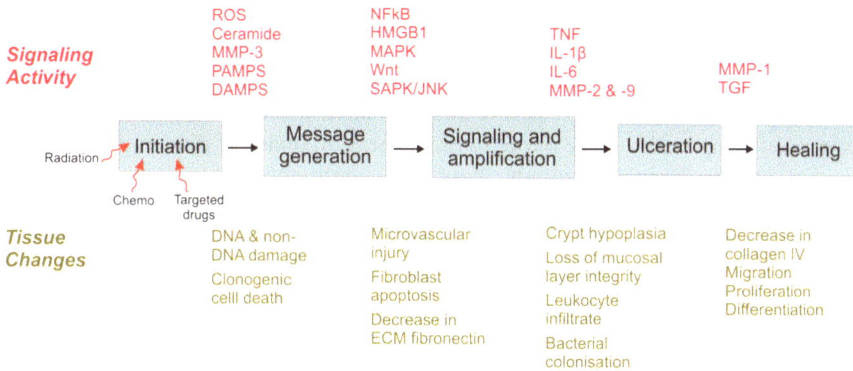

Fig. 1 The 5-phase model for the progression of AT mucositis

were the first to successfully capture the clinical and histological features of mucositis and were of much significance for investigation of mucositis pathogenesis and assessment of anti-mucotoxic interventions. These models implicated molecular and cellular markers with the onset of mucositis and highlighted the need for more studies in the area of mucositis pathogenesis. In 2001, Paris and colleagues demonstrated radiotherapy to cause extensive microvascular injury in mice. In particular, they noted that endothelial and fibroblast apoptosis in the submucosa precedes epithelial tissue damage and clinical signs [85]. Further work confirmed the occurrence of early and delayed changes in gene expression in the irradiated oral mucosa [115] and indicated that damage occurs to all mucosal tissue compartments, not just the epithelium, and is a result of up-regulation of a number of stress response genes associated with the endothelium, muscle, inflammatory infiltrate and epithelial cells in these tissues [115]. This formed the basis for the 5-phase model of mucositis.

5-Phase Model for OM

The 5-phase model for mucositis progression, developed by Sonis [106, 107], divides the biological sequences of mucositis into 5 phases (Fig. 1): *initiation*, encompassing the primary damage response and is a result of DNA and non-DNA damage with generation of reactive oxygen species (ROS); *message generation*, involving the up-regulation of transcription factors including NFκB and subsequent activation of cytokine and stress response mediators including MAPK, cyclooxygenase-2 (COX-2), Wnt, SAPK/JNK and ceramide signalling; *signalling and amplification*, producing proteins, such as TNF, IL-1β and IL-6, which cause direct tissue damage and provide positive feedback to amplify the process; *ulceration*, resulting in painful ulcers, bacterial penetration into injured mucosa and an influx of macrophages and other inflammatory cells and finally *healing*, which spontaneously occurs upon cessation of chemotherapy [106, 107]. Ongoing refinement of the model continues as further pieces of the jigsaw fit into place. Recent studies have found matrix metalloproteinases (MMPs), originally proposed to have a role in the ulceration phase of mucositis, to be *key mediators of damage* through their early influence of cell kinetics and inflammatory mediators and later the extracellular matrix (ECM) proteins [1, 3]. Changes in MMP and ECM proteins are now thought to have a direct impact on the loss of mucosal layer integrity seen in mucositis.

Evolution of GI Mucositis

The first attempt to characterise the effect of cytotoxics on the histopathological features of the gut was made by Keefe [52], who investigated the effect of chemotherapy on the mucosa of the small intestine in both humans and rats. For the first time, it was shown that small intestinal mucositis occurs symptomatically in a significant number of patients, and peaks 3–7 days after treatment. Early studies in this field suggested that the principal mechanism for small intestinal mucositis was apoptosis of the intestinal crypts that result in intestinal hypoplasia [52, 53, 89, 90]. From then on, studies demonstrated the effects of cytotoxic agents on the large intestine and implicated intestinal bacterial flora and mucus secretions in GI mucositis [49, 125].

Over the past 15 years, mucositis research has evolved such that oral and GI mucositis are now seen as a continuum of pathobiologic changes over time. Keefe [54] proposed that oral and GI mucositis should both be considered to be elements of alimentary mucositis (AM), which covers mucosal damage through the entire alimentary tract, with regional differences being due to the specialised needs of each area [54]. Up until now, the future directions for GI mucositis research are informed by the progress already made in the study of OM. Many elements of the 5-phase model of OM have now been proven true for GI mucositis [1, 3, 18, 40, 137]. These will be discussed further below.

Underlying Causes for Mucositis Symptoms

The clinical symptoms of mucositis are a result of direct epithelial injury; however, the development of this condition is a consequence of a plethora of biological events in multiple cells and tissue compartments. The symptoms manifesting depend on the area of the alimentary tract most affected. For the purposes of this chapter four key events will be discussed: cell kinetic changes, inflammation, gut bacterial changes and surface area changes.

Cell Kinetic Changes

Cytotoxic agents induce cell death in both tumour and healthy tissues via apoptosis [43]. Rapidly proliferating cells are more susceptible so the oral and GI mucosa is particularly vulnerable [59]. There are differences in susceptibility along the alimentary tract with early damage occurring in the small intestine (day 2–3), less damage in the colon and later damage in the oral cavity (day 7–10). There is an increase in the number of cells undergoing apoptosis throughout the alimentary tract corresponding with a decrease in the number of cells that are mitotically active [19, 40, 53, 109, 110]. Early studies investigating the effect of EGF on mucositis have shown that EGF stimulated epithelial cell proliferation and significantly increased the severity of mucositis [109, 110] indicating the importance of renewal rate of the epithelium in determining susceptibility of the mucosa to injury. More recently, it was shown that a novel peptide derived from the 18-kDA antrum mucosal protein (AMP-18) attenuated radiation-induced OM and this was attributed in part to the antiapoptotic properties of the peptide [135]. In the small intestine, studies have demonstrated that chemotherapy causes crypt hypoplasia, followed by rebound crypt hyperplasia and finally restoration of normal tissue [25, 39, 40]. Gibson et al. [41] demonstrated that IL-11 was effective at attenuating methotrexate-induced mucositis by inducing a compensatory crypt cell proliferation [41]. These studies highlight the importance of intricate regulating of cell kinetics in maintaining mucosal layer integrity and demonstrate their contribution to mucosal injury following cytotoxics.

Inflammation

In recent years it has been shown that pro-inflammatory cytokines play a key role in both oral and GI toxicities. Early studies demonstrated that cytokines targeting epithelial cell proliferation including EGF [109, 110] or transforming growth factor-beta 3 (TGF-β3) [109, 110] modified the course of mucositis in animal models. More recently, gene expression and tissue levels of TNF and IL-1β were shown to closely correlate with oral mucosa injury following radiation [114, 115]. The pro-inflammatory cytokines TNF, IL-1β and IL-6 were also associated with the development of chemotherapy-induced GI mucositis [63–67]. Furthermore, elevated levels of pro-inflammatory cytokines, in particular TNF, IL-1β and IL-6, have been identified as being excellent markers of the inflammatory response induced by chemotherapy [63, 65].

According to the 5-phase model for mucositis development, it has been suggested that the transcription factor NF-κB acts as a 'gatekeeper' for various pathways (one of which are the pro-inflammatory cytokines). Logan et al. [63, 65] demonstrated that peaks in NF-κB expression preceded peaks in pro-inflammatory cytokines in oral and intestinal tissue following chemotherapy [63, 65]. Once pro-inflammatory cytokines are up-regulated, they have been suggested to provide feedback by activating NF-κB and amplifying injurious pathways [105–107]. NF-κB has also been shown to up-regulate cyclooxygenase-2 (COX-2), an inducible enzyme involved in inflammation through its role in prostaglandin production. COX-2 expression was shown to dramatically increase, particularly in submucosal fibroblasts and endothelial cells of the oral mucosa, 10 and 16 days after targeted radiation [111, 113]. Furthermore, changes in COX-2 expression paralleled with the development of ulcerative mucositis, suggesting an amplification role for this enzyme [111, 113]. These results have also been demonstrated clinically where a statistically significant increase in NF-κB and COX-2 was noted in the oral mucosal of patients undergoing chemotherapy [64, 67].

Few studies have looked specifically at anti-inflammatory cytokines and toxicities, but these have demonstrated a shift in the balance between pro- and anti-inflammatory cytokines with a net shift towards pro-inflammatory cytokines [35]. IL-10 is one of the more researched anti-inflammatory cytokines in mucositis. DeKoning and colleagues demonstrated that IL-10-deficient mice experience an increased weight loss and more severe intestinal damage following methotrexate treatment in comparison to wild-type controls [29]. Furthermore, cytokines directed at inhibiting pro-inflammatory cytokines have proven useful as interventions for mucositis [21]. IL-11, a cytokine secreted by bone marrow stromal cells, has demonstrated ability to reduce the severity of GI mucositis [41]. The mechanism of action of IL-11 is partly through its ability to inhibit cytokine release and down-regulate inflammatory mediators such as TNF (Fig. 2). Another anti-inflammatory agent known as anti-inflammatory amino acid decapeptide (RDP58) also inhibits the production of certain cytokines and has demonstrated effectiveness in ameliorating clinical symptoms (primarily diarrhoea) and intestinal inflammation in mice following chemotherapy [138].

Microbiome Changes

Recent studies demonstrated a shift in the ecological balance in oral and gut flora following anti-cancer treatment [79, 101, 120–124, 129]. In the mouth, this shift in bacterial flora is suggested to occur as a result of neutropenia and mucosal surface changes after treatment [73]. Furthermore, OM has been identified as an independent risk factor for the development of infections [16, 97]. A recent study by Napenas et al. [79] determined the profile of the oral bacterial flora following anthracycline therapy in an outpatient cancer population. They found that over 60% of bacterial flora identified in the buccal mucosa of patients was exclusively present post-chemotherapy suggesting an alteration in the nature of oral flora [79]. Bacterial substitution (of mainly coagulase-negative staphylococci for streptococci) has also been reported on the oral mucosa after hematopoietic cell transplantation and this was clearly associated with OM [103]. The observations of shifts in oral bacterial flora and infection development following cancer therapy have been attributed not only to the development of mucositis but also to secondary factors including the use of antibiotics [103] and the development of neutropenia [128] and xerostomia [103].

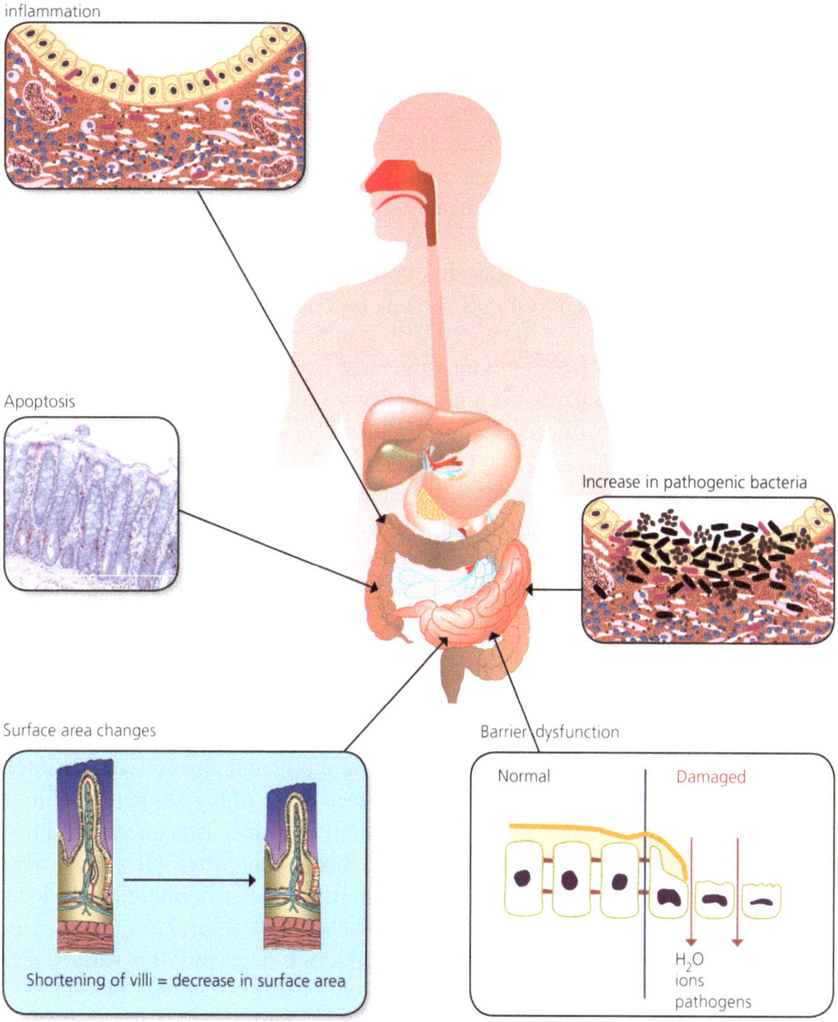

Fig. 2 The histopathological features underpinning GI toxicity

The gut microbiota and its influence on GI mucositis is becoming an increasingly recognised field of research. According to the 5-phase model for mucositis development, the gut microbiota plays little role in the pathobiology of mucositis where bacterial translocation is an outcome of mucosal layer degradation following inflammation and apoptosis. However, recent studies have shown a shift from commensal bacteria, in particular *Bifidobacterium spp.*, towards *Salmonella spp.* and *E. coli* following chemotherapy treatment [120, 121, 124]. Stringer et al. [120, 121, 124] demonstrated that decreases in commensal bacteria, represented by *Bifidobacterium spp.*, inversely followed the pattern of diarrhoea induced by chemotherapy. These results strongly suggest a role for gut microbiome changes in diarrhoea induction

(Fig. 7.2). It is not yet clear how commensal bacteria influence the different stages of mucositis development; however, it has been suggested that these are capable of influencing inflammatory processes, intestinal permeability, mucus layer composition, epithelial repair mechanisms and regulation of immune effector molecules including TLRs [129]. Intervention studies using probiotic factors to treat 5-FU-induced mucositis have proven effective in the pre-clinical setting and provide further evidence for the involvement of intestinal bacterial flora in the development of mucositis [91].

Recently, the occurrence of mucositis has been recognised as a risk factor for infection [14]. In the clinic, it is a well known observation that chemotherapy cycles complicated by mucositis are associated with significantly higher risk of infections and use of antibiotics [31]. Furthermore, the most frequently occurring infection in cycles complicated by mucositis is a gram-positive cocci, an infection which interestingly accounts for 3 of every 4 episodes of bacteraemia in patients treated for haematological malignancies [133]. The ulcerative phase of mucositis has been suggested to be a major contributor, allowing the translocation of residential microorganisms and their products from the alimentary tract into the bloodstream. This is reported to most probably occur at gingival borders, soft tissues of the oral cavity, oesophagus, ileum, cecum and rectum [14].

Surface Area Changes

This is particularly important in the GI tract where efficient absorptive function is strongly reliant on availability of a large surface area. Many studies have used surface area changes, particularly within the small intestine, as markers of GI toxicity [40, 41, 53]. All of these studies have shown that when GI surface area decreases, GI toxicities occur including malnutrition. Furthermore, well-documented studies are indicative of a relationship between structural changes and the incidence of malabsorption [94, 102, 136]. Reduction in crypt/villus surface area is a hallmark of small intestinal damage following cancer treatment [40, 41, 53]. After a short delay this is followed by a rebound hyperplasia, with the epithelium proliferating first to replace damaged stem cells and then to repopulate the crypt and villi to restore normal structure. The hyperplastic villi are functionally immature, however, without full activity of brush border enzymes, which contributes to the incidence of malabsorption [94, 102, 136]. Transient lactose intolerance is common (Fig. 2).

Gut Toxicity of Commonly Used Anti-cancer Agents

While all chemotherapeutic agents can cause mucositis, there is a hierarchy of drugs that do most commonly (Table 1). The incidence, severity and location of mucositis depend on the drug/radiotherapy used, the site and type of tumour, as well as host factor.

Table 1 Grade 3–4 toxicities induced by commonly used chemotherapeutic agents

Chemotherapeutic drugs	Uses	Grade 3–4 GI toxicity	Incidence of grade 3–4 GI toxicity	References
Covalent DNA-binding drugs				
Cyclophosphamide	Leukaemia, malignant and lymphocytic lymphomas, breast carcinoma	None	None	
Cisplatin	Metastatic ovarian and testicular carcinomas, bladder cancer	Nausea and vomiting	7%	Ridwelski et al. [93]
Melphalan	Multiple myeloma, ovarian carcinoma and breast cancer	Mucositis, nausea and vomiting	28%	Palumbo et al. [81]
Antimetabolites				
Methotrexate	Breast, head and neck, lung carcinomas, cutaneous T cell lymphoma	Mucositis, diarrhoea	10%	Guardiola et al. [42]
5-Fluorouracil	Colon, rectum, breast, stomach and pancreas cancers	Stomatitis, diarrhoea	20%	Paolo et al. [82]
Mercaptopurine	Acute lymphatic leukaemia	Mucositis	46%	Figliolia et al. [34]
Antitumor antibiotics and related compounds				
Doxorubicin	Leukaemia, neuroblastoma, soft tissue and bone sarcomas, breast, ovarian, bladder, thyroid, gastric carcinoma	Nausea and vomiting, mucositis	32%	Hortobagyi et al. [48]
Microtubule-targeting drugs				
Docetaxel	Head and neck, gastric, breast and prostate cancer	Mucositis	9%	Esteban et al. [32]; Guardiola et al. [42]
Paclitaxel	Ovarian, breast and non-small cell lung cancers	Nausea and vomiting, diarrhoea	11%	
DNA topoisomerase inhibitors				
Irinotecan	Colorectal cancer	Diarrhoea	45%	Rothenberg et al. [96]

Radiotherapy

Radiotherapy is commonly used to treat cancer, either alone or in combination with chemotherapy and/or targeted agents with approximately 50% of all cancer patients receiving radiotherapy at some stage during their cancer journey [12]. Toxicity from radiotherapy is dependent on the normal tissue tolerance of the site irradiated. Mucositis is extremely common with head and neck cancer (almost 100%), and whenever the treatment fields include areas of the GI tract. These toxicities are classified as *acute*, occurring a few hours after treatment; *early-delayed*, beginning 1–2 weeks after treatment and *late-delayed*, occurring months after treatment.

Acute toxicities following oral cavity radiotherapy include xerostomia, bacterial, fungal or viral infections, dental caries, loss of taste and osteoradionecrosis while those induced following radiation to the gastrointestinal tract symptoms include nausea and vomiting, pain, diarrhoea, constipation and rectal bleeding. The pathobiology of acute radiation toxicity has been extensively researched. It is well established that radiation triggers a sequence of signals that ultimately result in cell damage. Early tissue culture models reveal the induction of various early response genes upon irradiation including c-jun, c-fos, Erg-1 and NFkB [132] as well as vascular adhesion genes [92]. Sonis et al. [115] confirmed the occurrence of early (MAPK, PCAF, serum amyloid A1, Hsp70 and keratin-14) and delayed (proteasome 26S, keratin 6b, adenosine deaminase and SCC AG-1) changes in gene expression in the irradiated oral mucosa. This indicates that damage occurs to all mucosal tissue compartments, not just the epithelium, and is a result of up-regulation of a number of stress response genes associated with the endothelium, muscle, inflammatory infiltrate and epithelial cells in these tissues [115]. While the majority of radiotherapy damage occurs in or close to the field of treatment, with sufficient dosing systemic effects can occur. An example of these systemic effects is seen in patients receiving radiation treatment for lung cancer. These patients often suffer from mouth and throat soreness and experience diarrhoea [55]. Such symptoms are indicative of damage to segments of the alimentary tract which are not in the treatment field and indicate a systemic effect [55].

An increasingly recognised issue of radiotherapy is the chronic side effects that it is capable of inducing months after the treatment has stopped. Late onset bowel dysfunction (previously labelled 'chronic radiation proctitis') results from dysfunction of the GI tract at numerous points. Approximately 90% of patients develop permanent changes in their bowel habits after pelvic radiation, 50% of which experience a reduction in their quality of life [46]. One of the earlier studies implicated bacterial invasion- and inflammation-induced IL-1 production in chronic radiation toxicity [74]. Since then, we now understand that radiation-induced acute injury (extending from the epithelium to the submucosa) is capable of triggering a 'regenerative response' which either leads to repair or development of ulceration, fibrosis and stricturing [46]. Risk factors which can hamper the proper regeneration of gut tissue include radiotherapy-specific factors (volume of tissue irradiated, radiation dose and the type of radiotherapy), previous surgery, concomitant chemotherapy cycles and genetic susceptibility [4]. Oral complications due to radiotherapy, including xerostomia, have also been described as chronic toxicities often being sustained up to 6 months after treatment [51, 100].

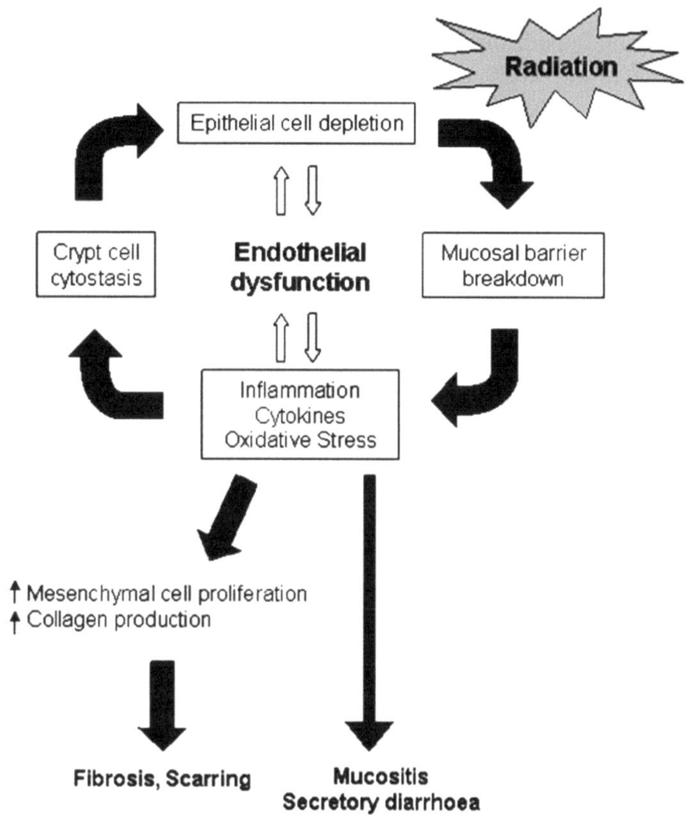

Fig. 3 The interactions between tissue factors affected by radiation in the intestine which leads to acute intestinal radiation toxicity and subsequently sustains the chronicity of radiation-induced fibrosis

A hallmark of the late toxicity is fibrosis [44, 60]. In the gut and other mucosal tissues, fibrosis is characterised by (1) thickening of the smooth muscle wall due to excessive transmural deposition of ECM components, (2) damage to the myenteric nerve plexus and (3) eventual long-term injury to epithelial, endothelial and mesenchymal stem cell compartments (Fig. 3). One of the leading causes of this fibrosis is the prolonged up-regulation of fibrogenic cytokines, such as transforming growth factor-β1 and its downstream effector connective tissue growth factor and their dynamic interactions with the ECM [37, 60].

Irinotecan

The topoisomerase-I inhibitor irinotecan is commonly used for the treatment of patients with advanced colorectal carcinoma [39, 69]. Although effective, irinotecan can be highly toxic, causing grade 4 diarrhoea in up to 45% of patients [96]. An animal model of irinotecan damage demonstrates extensive structural and morphological alteration in both small and (unusually) large intestine [39]. Necropsies on rats administered with a 250 mg/kg dose of irinotecan revealed that many had developed duodenal perforations and fatty lysis consistent with peritonitis. Histopathological examination confirmed the presence of lesions characterised by crypt degeneration (consistent with radiomimetic damage to the proliferative compartment), resulting in villus atrophy. Crypt degeneration was generally more severe and extensive in the large intestine [39].

Delayed diarrhoea is the most prominent toxicity associated with irinotecan use in the clinic. Past studies have suggested this toxicity to be a result of gastrointestinal enterocolitis, with accompanying water, electrolyte and mucus secretion in the large intestine. However, current evidence attributes irinotecan-induced diarrhoea to changes in intestinal bacterial flora, particularly implicating elevated levels of β-glucuronidase-producing bacteria. A recent study by Stringer et al. [123] demonstrated an increase in β-glucuronidase levels in the small and large intestines following irinotecan [123]. β-glucuronidase is a member of the glycosidase family of enzymes that catalyse the breakdown of complex carbohydrates. In the gut, this enzyme is capable of hydrolysing the less toxic form of irinotecan (which is secreted into the gut via bile), rendering it toxic (SN-38) and causing damage. Irinotecan has also been shown to increase submucosal collagen deposits [1] and induce smooth muscle over-contractility associated with elevated levels of tissue nitric oxide [61] during the period of maximal tissue injury. NFkB [63, 65], the pro-inflammatory cytokines [63, 65] and MMPs [1] are also affected by irinotecan and contribute directly to the associated inflammation occurring in the gut following its administration. Irinotecan has also been shown to cause damage to components of the tight junction including claudin-1 and occludin in the large intestine resulting in bacterial translocation [78].

Methotrexate

This is an anti-folate commonly used in cancers including breast and lung cancer, non-Hodgkin lymphoma, head and neck cancers and leukaemia. It is structurally altered following administration to become a tight inhibitor of dihydrofolate reductase, a key enzyme in intracellular folate homeostasis [26]. This causes a partial depletion of reduced folate substrate and directly inhibits folate-dependent enzymes and synthesis of nucleotide precursors of DNA [26]. Methotrexate induces enterocolitis in animal models [40, 83, 84]. Furthermore, when used as part of an MACC regimen for non-small cell lung cancer, methotrexate induced gastrointestinal toxicities of various grades in 64% of patients [20]. Furthermore, approximately 10% of those are grade 3–4 GI toxicities [42]. Methotrexate works by decreasing the proliferation of tumour and normal cells [130]. The most pronounced damage resulting from methotrexate occurs in the small intestine [83, 84]. Although the exact mechanism of methotrexate-induced gastrointestinal injury is not known, it is thought that a combination of progenitor cell loss, reduced proliferative capacity and inflammation in the small intestine leads to disruption of barrier integrity and toxicity [66]. In animal models of mucositis, methotrexate induces dose-dependent villus atrophy and elongation of crypts in the small intestine along with increased secretion of electrolytes and a reduction of mucosal surface area as indicated by decreased mannitol excretion ratio [25]. Trefoil factors, which are well known for their ability to regulate cell migration, apoptosis and proliferation, have been implicated in methotrexate-induced mucositis. Particularly, intestinal trefoil factor-deficient mice were shown to be more susceptible to chemotherapy- and radiation-induced mucositis (as evident by increased weight loss and more severe diarrhoea) indicating the protective properties of trefoil factors [11]. Lymphocytes and macrophages from methotrexate-treated mice were shown to secrete enhanced amounts of cytokines upon stimulation, in particular TNF and IL-1β, suggesting a contributive role for both the adaptive and innate immune system in methotrexate-induced mucositis, respectively [29].

5-Fluorouracil

This antimetabolite is commonly used to treat colon, rectum, breast, stomach and pancreatic cancers. 5-FU metabolism in the anabolic pathway prevents the methylation reaction of deoxyuridylic acid to thymidylic acid and interferes with the synthesis of DNA and the formation of RNA hence inducing anti-tumour activity. The gastrointestinal toxicities associated with this drug include stomatitis, esophagopharyngitis, diarrhoea, nausea and vomiting. Furthermore, grade 3–4 GI toxicities occur in 20% of patients receiving 5-FU and include stomatitis and diarrhoea [82]. Following intravenous injection, 5-FU distributes into tumours, oral and intestinal mucosa, bone marrow, liver and other tissues throughout the body. In animal models of 5-FU-induced toxicity, it has been shown to cause significant thinning of the epithelium throughout the alimentary tract [66]. Intestinal histopathologic alterations include blunting and fusion of the villi, enterocyte hyperplasia as well as significant changes in intestinal flora and mucin secretion [66, 120, 121, 124]. These changes can potentially contribute to the systemic effects of 5-FU particularly to the development of chemotherapy-induced mucositis.

The use of 5-FU in continuous infusion regimens in the clinic is associated with toxicity as well as other complicating factors including the potential of infection, bleeding and thromboembolism [50]. Capecitabine, a prodrug of 5-FU, was developed in an attempt to overcome these issues as it is enzymatically converted within tumour, where it inhibits DNA synthesis and slows growth of tumour tissue. Capecitabine is used for the treatment of colorectal and breast cancers. In recent clinical trials, capecitabine was found to be noninferior and has an improved side effect profile in comparison to 5-FU as part of a chemoradiotherapy regimen in patients with rectal cancer [47]. However, capecitabine can induce severe diarrhoea which is dose limiting in many cases. The mechanisms underlying this toxicity are not well understood.

Targeted Therapies

The early promise that targeted therapies would increase cure rates while minimising toxicity has not been completely fulfilled [56]. Toxicities include those due to the presence of target receptors on healthy tissues, the promiscuous cross-reactivity of tyrosine kinase inhibitors and the adverse effects of immunostimulation. AT side effects are extremely common, with oral problems and diarrhoea being particularly common. The mechanisms are not necessarily the same as for other cancer treatments [57, 58, 99, 126]. There are an ever increasing number of targeted agents in use, so only three will be discussed.

Trastuzumab is a monoclonal antibody directed at the HER2 receptor [10, 27, 88]. The AT is a major site of toxicity [58, 99]. Trastuzumab is used as a single agent or in combination with cytotoxic chemotherapy or radiotherapy in patients with HER2-overexpressing breast cancer [9, 13, 30]. Single agent trastuzumab induces gastrointestinal toxicities following 12% of administrations. These include diarrhoea, abdominal pain and vomiting [2]. Moreover, phase II studies investigating the efficacy and tolerability of trastuzumab in combination with paclitaxel have shown diarrhoea to be the most frequent toxicity manifesting in 30% of patients [36]. Toxicity profiling for newer combination regimens where trastuzumab is given with carboplatin or capecitabine has also shown significant induction of mucositis, diarrhoea and vomiting at grade 3–4 intensity [8, 98].

HER2 is expressed by many normal tissues including neural, gut and cardiac tissue. Monoclonal antibodies, such as trastuzumab, are extremely specific to the ligand–receptor interaction for which they are designed, for example, species specificity of trastuzumab for human or primate HER2 but not rodent *neu* [86]. As such, most animal models of toxicity are inappropriate to test humanised or fully human monoclonal antibodies due to conformational disparity. Trastuzumab has also been shown to have a much shorter half life in mice (~7 days) compared to humans (~21 days). Hence, the lack of well-designed and reported animal models sufficient to capture toxicity of anti-cancer agents makes it difficult to understand the underlying pathobiology of regimen-related toxicity thereby jeopardising patient safety.

Lapatinib. This small molecule TKI has been recently approved for treatment of advanced and metastatic breast cancer. It is orally administered, dual tyrosine kinase inhibitor which directly targets the tyrosine kinase domain of the EGFR and HER2 receptors by reversibly binding to the cytoplasmic ATP-binding site, blocking receptor phosphorylation and activation, and preventing subsequent downstream signalling [76]. Diarrhoea is the most frequently reported adverse effect of lapatinib monotherapy (42%) [23]. Moreover, a phase III trial showed a significant increase in diarrhoea with lapatinib plus capecitabine verses capecitabine alone, with 60% of patients experiencing diarrhoea in the combination group [38].

The mechanisms of lapatinib-induced diarrhoea have recently been investigated. Unpublished observations in a hamster model of TKI-induced toxicity indicate lapatinib treated jejunal tissue exhibits a higher proliferative index and crypt depth compared to controls. High-dose lapatinib also caused an increase in the serum anion gap. Lapatinib-induced diarrhoea is suggested to be secretory in nature, due to electrolyte imbalance secondary to epithelial changes.

Links to Other Toxicities

To date, studies investigating the biological basis of therapy-related toxicity have focused on studying the various toxicities in isolation from each other. It is well known that the response to anti-cancer agents differs across species and tissues. It is now thought that regimen-related toxicities in different tissues might share a common pathobiology. Recently, it has been recognised that toxicities arise in patients in clusters rather than in isolation [5]. Toxicities associated with cytotoxic treatment include localised (ulceration, abdominal pain, diarrhoea, malabsorption) as well as systemic (fatigue, nausea, infection) toxicities. The recognition of concurrent occurrence of tissue-based and systemic toxicities has caused a shift in the potential applications of toxicity models to investigate regimen-related toxicity rather than tissue-specific toxicity.

Biomarkers of Mucositis

Although AT mucositis is highly prevalent, there are currently no diagnostic biomarkers for this condition and the only method by which it can be diagnosed is through assessment of symptoms and clinical signs according to clinical mucositis grading scales. However, there is no established scale that is universally accepted or meets all the criteria a clinical mucositis scoring system should achieve [111, 113]. Two of the most commonly used scales are the NCI-CTCAE and the WHO. Furthermore, accurate evaluation of mucositis is impossible due to the difficulty of obtaining sequential biopsies before, during and after treatment and the inaccessibility of important segments of the AT [111, 113]. Also, while the physical damage to the gut is short-lived, functional damage can persist for days or weeks. This has prompted the need for a biomarker of gut toxicity in response to cytotoxic anti-cancer treatment. Lutgens et al. [68] identified two criteria for a biomarker of gastrointestinal toxicity (1) that it is easily accessible to clinicians and (2) that it must work independently of each patient's medication and metabolic activities such as individual diets and nutrition [68]. Furthermore, an ideal biomarker could predict before treatment whether a patient will get a particular toxicity, whether a given toxicity will respond to a particular intervention, or measure the severity of that toxicity when it cannot be accurately assessed by history and examination. In order to be valuable, a biomarker has to pass the common sense test: Is it better than just asking the patient how they feel, or examining them?

A wide range of biomarkers have been investigated including citrulline, calprotectin and the pro-inflammatory cytokines [63, 65, 68, 95]. These biomarkers have been shown to parallel with mucositis development, but the degree to which they predict the condition is yet unknown. Findings have highlighted the difficulties in identifying a biomarker for mucositis, i.e. (1) our lack of understanding its pathobiology (2) the fact that mucositis is multifactorial, often arising from a combination of pathologies and (3) links between toxicities: since mucositis does not occur in isolation, systemic biomarkers may be marking a different toxicity (or more than one).

An emerging class of biomarkers for mucositis is the pro-inflammatory cytokines and MMPs. These have augmented expression profiles in the AT in animal models of mucositis [3, 64, 67]. Preliminary findings from a pilot human study suggest that TNF, IL1β, MMP-3 and MMP-9 are potential biomarkers of gastrointestinal toxicity induced by 5-fluorouracil, capecitabine or irinotecan. Prospective studies are now required to assess the feasibility of biomarkers for mucositis in particular and regimen-related toxicity in general.

Models for Studying Mucositis

Toxicology is an area of rapid growth, particularly in the field of oncology drug development. Modelling and experimentation with cell-free, in vitro and in vivo combinations are used to best predict toxicity potential due to drug–drug interactions, off target effects and mechanisms of action of novel agents [28]. Pre-clinical formulation and toxicity testing is the final step of pre-clinical development of new compounds before phase I clinical trials begin. As such, the animal models used for toxicity testing are vitally important.

In recent years, our knowledge of the pathobiology of mucositis has increased. However, the complexity of the condition in terms of timing and sequence of injury events is still not fully understood. Animal models of AT mucositis have been heavily utilised to explore that pathobiological basis of the condition and assess the appropriateness of anti-mucotoxic agents. The key advantage in animal models is the ability to thoroughly and accurately investigate any region of the AT in a highly homogeneous population. Furthermore, animal models of AT mucositis have been utilised to examine anything from basic survival endpoint to highly specialised biomarkers of damage.

Numerous animal models have been established for the investigation of chemotherapy- and radiotherapy-induced AT injury. One of the first models to investigate chemotherapy-induced gastrointestinal mucositis was developed by Keefe and colleagues in the mid-1990s (Keefe thesis). The dark agouti rat mammary adenocarcinoma (DAMA) model of gastrointestinal mucositis remodels the structural, morphological and functional changes which are seen in humans following cytotoxic treatment [53]. The presence of the isogenic tumour in this model allows for the simultaneous assessment of tumour protection when assessing the efficacy of anti-mucotoxic treatments.

Specific Challenges in Animals Models Investigating Regimen-Related Toxicity

A major challenge in the interpretation of toxicity results in animal models is the substantial differences between species (including man) in enzyme and transporter activity leading to variation in drug pharmacokinetics, metabolic and clearance pathways [71]. There are important anatomical and physiological differences between humans and smaller animals, including relatively higher amounts of hepatic enzymes in small animals, so that in general small animals tend to eliminate drugs more rapidly than humans when compared on a weight-normalised basis [72].

The necessity for animal models to investigate multiple toxicities has presented new challenges in establishing and applying these models. Existing AT mucositis models reflect very specific toxicities, in response to specific agents, in a specific animal strain and are not optimised to allow for the investigation of toxicities in multiple tissues. Furthermore, the extent of other non-AT toxicities in these models has not been investigated. Animal models are established following extensive validation, where the dose and scheduling of cytotoxic agents are set to induce the specific toxicity of interest. For example, in the first chemotherapy-induced mucositis model, developed by Sonis and colleagues, oral mucositis is induced by administering a specific dose of 5-Floururacil (5-FU) following buccal cheek pouch scratching [116, 117]. In this model oral mucositis is induced as indicated by the presence of oral lesions similar to those noted in patients. However, this model is not optimised to reflect the other AT or non-AT toxicities which 5-FU is known to induce including diarrhoea and dermatitis.

The specificity by which certain animal models respond to experimental treatments or conditions and the failure to capture the full toxicity profile seen in humans can be due to a number of things: (1) the animal dosing is not comparable to that used in the clinic, (2) there are substantial differences between species in enzymes and transporters leading to variations in drug pharmacokinetics, metabolism and clearance [71]. Hence, we need to carefully consider and select the most appropriate animal model for the compound under investigation.

Future Directions

Mucositis is a major oncological problem which has been clinically observed for decades. But it was only recently that we had comprehensive advances in our knowledge regarding the pathobiology and possible molecularly targeted treatments. The bulk of investigations into mucositis mechanisms have been in relation to the oral cavity and digestive tract; with little or no research in other mucosal surfaces (such as respiratory and genitourinary tract) which are generally less susceptible (the exception being bladder toxicity and haemorrhagic cystitis from cyclophosphamide). The co-occurrence of multiple toxicities following cytotoxic treatment has prompted the need for investigating regimen-related toxicities as a whole rather than in isolation. As a consequence, Peterson and colleagues identified a number of new frontiers in mucosal injury research including the *development of molecularly targeted mucositis therapeutics*, *genomics research*, *development of new types of cancer therapy* and *studies involving non-alimentary tract mucosa* [87]. Better understanding of toxicity clusters in response to certain chemotherapy regimens and cancer types is now required in order to target mechanistic research and allow for advancement of our knowledge of regimen-related mucositis.

References

1. Al-Dasooqi N, Bowen J, Gibson R, Logan R, Stringer A, Keefe D. Irinotecan-induced alterations in intestinal cell kinetics and extracellular matrix component expression in the dark agouti rat. Int J Exp Pathol. 2011;92:357–65.
2. Al-Dasooqi N, Bowen J, Gibson R, Sullivan T, Lees J, Keefe D. Trastuzumab induces gastrointestinal side effects in HER2-overexpressing breast cancer patients. Invest New Drugs. 2008;27:173–8.
3. Al-Dasooqi N, Gibson R, Bowen J, Logan R, Stringer A, Keefe D. Matrix metalloproteinases are possible mediators for the development of alimentary tract mucositis in the DA rat. Exp Biol Med. 2010;235:1244–56.
4. Andreyev H. Gastrointestinal symptoms after pelvic radiotherapy: a new understanding to improve management of symptomatic patients. Lancet Oncol. 2005;8:1007–17.
5. Aprile G, Ramoni M, Keefe DM, Sonis S. Application of distance matrices to define associations between acute toxicities in colorectal cancer patients receiving chemotherapy. Cancer. 2008;112:284–92.
6. Bach SP, Renehan AG, Potten CS. Stem cells: the intestinal stem cell as a paradigm. Carcinogenesis. 2000;21(3):469.
7. Bahri S, Agarwala S, Cano E, Johnson J, Myers E, Jeong J. Phase II study of weekly carboplatin and paclitaxel concurrent with radiation therapy (RT) in patients with locally advanced, inoperable squamous cell carcinoma of the head and neck (HNSCC) [abstract 1667]. Prog/Proc Am Soc Clin Oncol. 2000;19:422a.
8. Bartsch R, Wenzel C, Altorjai G, Pluschnig U, Rudas M, Mader R, et al. Capecitabine and trastuzumab in heavily pretreated metastatic breast cancer. J Clin Oncol. 2007;25:3853–8.
9. Baselga J. Clinical trials of single-agent trastuzumab (Herceptin). Semin Oncol. 2000;27 (5 Suppl 9):20–6.
10. Baselga J. Clinical trials of Herceptin (Trastuzumab). Eur J Cancer. 2001;37:18–24.
11. Beck P, Wong J, Li Y, Swaminthan S, Xavier R, Devaney K, et al. Chemotherapy- and radiotherapy-induced intestinal damage is regulated by intestinal trefoil factor. Gastroenterology. 2004;126:796–808.
12. Begg A, Stewart F, Vens C. Strategies to improve radiotherapy with targeted drugs. Nat Rev Cancer. 2011;11:239–53.
13. Belkacemi Y, Gligorov J, Ozsahin M, Marsiglia H, De Lafontan B, Laharie-Mineur H, et al. Concurrent trastuzumab with adjuvant radiotherapy in HER2-positive breast cancer patients: acute toxicity analyses from the French multicentric study. Ann Oncol. 2008;19(6):1110–6.
14. Blijlevens N, Donnelly J. Mucosal barrier injury and infections. Principles and Practice of Cancer Infectious Diseases, Current Clinical Oncology. S. A, Springer Science; 2011.
15. Bloomer W, Hellman S. Normal tissue responses to radiation therapy. New Engl J Med. 1975;293:80–3.

16. Bochud P, Eggiman CT, Van Melle G, Saghafi L, Francioli P. Bacteremia due to viridans streptococcus in neutropenic patients with cancer: clinical spectrum and risk factors. Clin Infect Dis. 1994;18:25–31.
17. Booth D, Potten C. Protection against mucosal injury by growth factors and cytokines. J Natl Cancer Inst. 2001;29:16–20.
18. Bowen J, Gibson R, Cummins A, Keefe D. Intestinal mucositis: the role of the Bcl-2 family, p53 and caspases in chemotherapy-induced damage. Support Care Cancer. 2006;14:713–31.
19. Bowen J, Gibson R, Keefe D, Cummins A. Cytotoxic chemotherapy up-regulates pro-apoptotic Bax and Bak in the small intestine of rats and humans. Pathology. 2005;37:56–62.
20. Buccheri G, Ferrigno D, Rosso A. A phase II study of methotrexate, doxorubicin, cyclophosphamide, and lomustine chemotherapy in advanced non-small cell lung cancer. Cancer. 1993;72:1564–72.
21. Bultzingslowen I, Brennan M, Spijkervet F, Logan R, Stringer A, Raber-Durlacker J, et al. Growth factors and cytokines in the prevention and treatment of oral and gastrointestinal mucositis. Support Care Cancer. 2006;14:519–27.
22. Burkitt HG, Young B, Heath JW. Oral Tissues. In: Burkitt HG, Young B, Heath JW, editors. Wheater's functional histology: a text and colour atlas. New York: Churchill Livingstone; 1993. p. 235–46.
23. Burris H. Dual kinase inhibition in the treatment of breast cancer: initial experience with the EGFR/ErbB-2 inhibitor lapatinib. Oncologist. 2004;9:10–5.
24. Capp A, Inostroza-Ponta M, Bill D, et al. Is there more than one proctitis syndrome? A revisitation using data from the TROG 96.01 trial. Radiother Oncol. 2009;90:400–7.
25. Carneiro-Filho B, Lima I, Araujo D, Cavalcante M, Carvalho G, Brito G, et al. Intestinal barrier function and secretion in methotrexate-induced rat intestinal mucositis. Dig Dis Sci. 2004;49:65–72.
26. Chabner B, Longo D. Cancer chemotherapy and biotherapy. Philadelphia: Lippincott-Raven; 1996.
27. Cobleigh M, Vogel C, Tripathy D, Robert N, et al. Multinational study of the efficacy and safety of the humanized anti-HER2 monoclonal antibody in women who have HER2-overexpressing metastatic breast cancer that has progressed after chemotherapy for metastatic disease. J Clin Oncol. 1999;17:2639–48.
28. Damia G, D'Incalci M. Contemporary pre-clinical development of anticancer agents—what are the optimal preclinical models? Eur J Cancer. 2009;45(16):2768–81.
29. de Koning B, Dieren J, Lindenbergh-Kortleve D, van der Sluis M, Matsumoto T, Yamaguchi K, et al. Contributions of mucosal immune cells to methotrexate-induced mucositis. Int Immunol. 2006;18:941–9.
30. Di Lauro V, Murrone A, Bidoli E, Magri MD, Crivellari D, Veronesi A. Trastuzumab and vinorelbine as highly effective and safe therapy for HER-2-overexpressing metastatic breast cancer. A single institution experience. Tumori. 2008;94(4):464–8.
31. Elting L, Cooksley C, Chambers M, Cantor S, Manzullo E, Rubenstein E. The burdens of cancer therapy: clinical and economic outcomes of chemotherapy-induced mucositis. Cancer. 2003;98:1531–9.
32. Esteban E, Gonzalez de Sande L, Fernandez Y, Corral N, Fra J, Muniz I, et al. Prospective randomised phase II study of docetaxel versus paclitaxel administered weekly in patients with non-small-cell lung cancer previously treated with platinum-based chemotherapy. Ann Oncol. 2003;14:1640–7.
33. Feher O, Martins S, Lima C, Salvajoli J, Kowalski L. Pilot trial of concurrent chemoradiotherapy with paclitaxel and conventional fractionated irradiation in advanced unresectable squamous cell carcinoma of the head and neck [abstract 1694J]. Prog/Proc Am Soc Clin Oncol. 2000;19:431a.
34. Figliolia S, Oliveira D, Pereira M, Lauris J, Mauricio A, Oliveira D, et al. Oral mucositis in acute lymphoblastic leukaemia: analysis of 169 paediatric patients. Oral Dis. 2008;14:761–6.

35. Fiochhi C. Inflammatory bowel disease: etiology and pathogenesis. Gastroenterology. 1998;115:182–205.
36. Fountzilas G, Tsavdaridis D, Kalogera-Fountzila A, Christodoulou C, Timotheadou E, Kalofonos C, et al. Weekly paclitaxel as first-line chemotherapy and trastuzumab in patients with advanced breast cancer. A Hellenic Cooperative Oncology Group phase II study. Ann Oncol. 2001;12:1545–51.
37. Gervaz P, Morel P, Vozenin-Brotons M. Molecular aspects of intestinal radiation-induced fibrosis. Curr Mol Med. 2009;9:273–80.
38. Geyer CE, Forster J, Lindquist D, Chan S, Romieu CG, Pienkowski T, et al. Lapatinib plus capecitabine for HER2-positive advanced breast cancer. N Engl J Med. 2006;355(26):2733–43.
39. Gibson R, Bowen J, Alvarez E, Finnie J, Keefe D. Establishment of a single-dose irinotecan model of gastrointestinal mucositis. Chemotherapy. 2007;53:360–9.
40. Gibson R, Bowen J, Cummins A, Keefe D. Relationship between dose of methotrexate, apoptosis, p53/p21 expression and intestinal crypt proliferation in the rat. Clin Exp Med. 2005;4:188–95.
41. Gibson R, Keefe D, Thompson F, Clarke J, Goland G, Cummins A. Effect of interleukin-11 on ameliorating intestinal damage after methotrexate treatment of breast cancer in rats. Dig Dis Sci. 2002;47:2751–7.
42. Guardiola E, Peyrade F, Chaigneau L, Cupissol D, Tchiknavorian X, Bompas E, et al. Results of a randomised phase II study comparing docetaxel with methotrexate in patients with recurrent head and neck cancer. Eur J Cancer. 2004;40:2071–6.
43. Hannum Y. Apoptosis and the dilemma of cancer chemotherapy. Blood. 1997;89:1845–53.
44. Hauer-Jensen M. Late radiation injury of the small intestine. Clinical, pathophysiologic and radiobiologic aspects. A review. Acta Oncol. 1990;29:401–15.
45. Hendry J, Potten C, Roberts N. The gastrointestinal syndrome and mucosal clonogenic cells: relationships between target cell sensitivities, LD50 and cell survival and their modification by antibiotics. Radiat Res. 1983;96:100–12.
46. Henson C. Chronic radiation proctitis: issues surrounding delayed bowel dysfunction post-pelvic radiotherapy and an update on medical treatment. Ther Adv Gastroenterol. 2010;3:359–65.
47. Hofheinz R, Wenz F, Post S, et al. Capecitabine (Cape) versus 5-fluorouracil (5-FU)-based (neo)adjuvant chemoradiotherapy (CRT) for locally advanced rectal cancer (LARC): long-term results of a randomized, phase III trial. ASCO Annual Meeting Abstract 3504; 2011.
48. Hortobagyi G, Frye D, Buzdar A, Ewer M, Fraschini G, Hug V, et al. Decreased cardiac toxicity of doxorubicin administered by continuous intravenous infusion in combination chemotherapy for metastatic breast carcinoma. Cancer. 1989;63:37–45.
49. Ikuno N, Soda H, Watanabe M, Oka M. Irinotecan (CPT-11) and characteristic changes in the mouse ileum and cecum. J Natl Cancer Inst. 1995;87:1876–983.
50. Kadoyama K, Miki I, Tamura T, Brown J, Sakaeda T, Okuno Y. Adverse event profiles of 5-Fluorouracil and capecitabine: data mining of the public version of the FDA adverse event reporting system, AERS, and reproducibility of clinical observations. Int J Med Sci. 2012;9:33–9.
51. Kakoei S, Haghdoost A, Rad M, Mohammadalizadeh S, Pourdamghan N, Nakhaei M, et al. Xerostomia after radiotherapy and its effects on quality of life in head and neck cancer patients. Arch Iran Med. 2012;15:214–8.
52. Keefe D. The effect of cytotoxic chemotherapy on the mucosa of the small intestine. Department of Medicine, University of Adelaide. PhD; 1998.
53. Keefe D. Chemotherapy for cancer causes apoptosis that precedes hypoplasia in crypts of the small intestine in humans. Gut. 2000;47:632–7.
54. Keefe D. Gastrointestinal mucositis: a new biological model. Support Care Cancer. 2004;12:6–9.
55. Keefe D. Mucositis and lung cancer. World Congress on Lung Cancer; 2009.
56. Keefe D, Bateman E. Tumor control versus adverse events with targeted anticancer therapies. Nat Rev Clin Oncol. 2011;9:98–109.
57. Keefe D, Gibson R. Mucosal injury from targeted anticancer therapy. Support Care Cancer. 2007;15:483–90.

58. Keefe D, Schubert M, Elting L, Sonis S, Epstein J, Raber-Durlacher J, et al. Updated clinical practice guidelines for the prevention and treatment of mucositis. Cancer. 2007;109:820–31.
59. Kerr J, Winterford C, Harmon B. Apoptosis: its significant in cancer and cancer therapy. Cancer. 1994;73:2013–26.
60. Langberg C, Hauer-Jensen M, Sung C, Kane C. Expression of fibrogenic cytokines in rat small intestine after fractionated irradiation. Radiother Oncol. 1994;32:29–36.
61. Lima-Junior R, Figueiredo A, Freitas H, Melo M, Wong D, Leite C, et al. Involvement of nitric oxide on the pathogenesis of irinotecan-induced intestinal mucositis: role of cytokines on inducible nitric oxide synthase activation. Cancer Chemother Pharmacol. 2012;69:931–42.
62. Lockhart P, Sonis S. Relationship of oral complications to peripheral blood leukocytes and platelet counts in patients receiving cancer chemotherapy. Oral Surg Oral Med Oral Pathol. 1971;48:21–8.
63. Logan R, Gibson R, Bowen J, Stringer A, Sonis S, Keefe D. Characterisation of mucosal changes in the alimentary tract following administration of irinotecan: implications for the pathobiology of mucositis. Cancer Chemother Pharmacol. 2008;62:33–41.
64. Logan R, Gibson R, Sonis S, Keefe D. Nuclear factor-kB (NFkB) and cyclooxygenase-2 expression in the oral mucosa following cancer chemotherapy. Oral Oncol. 2007;43:395–401.
65. Logan R, Stringer A, Bowen J, Gibson R, Sonis S, Keefe D. Serum levels of NFkappaB and pro-inflammatory cytokines following administration of mucotoxic drugs. Cancer Biol Ther. 2008;7:1139–45.
66. Logan R, Stringer A, Bowen J, Gibson R, Sonis S, Keefe D. Is the pathobiology of chemotherapy-induced alimentary tract mucositis influenced by the type of mucotoxic drug administered? Cancer Chemother Pharmacol. 2009;63:239–51.
67. Logan R, Stringer A, Bowen J, Yeoh A, Gibson R, Sonis S, et al. The role of pro-inflammatory cytokines in cancer treatment-induced alimentary tract mucositis: pathobiology, animal models and cytotoxic drugs. Cancer Treat Rev. 2007;33:448–60.
68. Lutgens L, Blijlevens N, Deutz N, Donnelly J, Lambin P, De Pauw B. Monitoring myeloablative therapy-induced small bowel toxicity by serum citruilline concentration: a comparison with sugar permeability tests. Cancer. 2005;103:191–9.
69. Ma M, McLeod H. Lessons learned from the irinotecan metabolic pathway. Curr Med Chem. 2003;10:41–9.
70. Manzano M, Bueno P, Rueda R, Ramirez-Tortosa C, Prieto P, Lopez-Pedrosa J. Intestinal toxicity induced by 5-fluorouracil in pigs: a new preclinical model. Chemotherapy. 2007;53:344–55.
71. Marathe PH, Rodrigues AD. Examination of CYP3A and P-glycoprotein-mediated drug-drug interactions using animal models. Methods Mol Biol. 2010;596:385–403.
72. Martignoni M, Groothuis GM, de Kanter R. Species differences between mouse, rat, dog, monkey and human CYP-mediated drug metabolism, inhibition and induction. Expert Opin Drug Metab Toxicol. 2006;2(6):875–94.
73. Martin M, van Saene H. The role of oral microorganisms in cancer therapy. Curr Opin Dent. 1992;2:81–4.
74. McBride W, Mason K, Withers H, Davis C. Effect of interleukin 1, inflammation, and surgery on the incidence of adhesion formation and death after abdominal irradiation in mice. Cancer Res. 1989;49:169–73.
75. Michael H, Gordon I, Wojciech P. Histology: a text and atlas. Philadelphia: Lippincott Williams & Wilkins; 2003.
76. Moy B, Goss P. Lapatinib-associated toxicity and practical management recommendations. Oncologist. 2007;12:756–65.
77. Murphy B. Clinical and economic consequences of mucositis induced by chemotherapy and/or radiation therapy. J Support Oncol. 2007;5:13–21.
78. Nakao T, Kurita N, Komatsu M, Yoshikawa K, Iwata T, Utusnomiya T, et al. Irinotecan injures tight junction and causes bacterial translocation in rat. J Surg Res. 2010;173:341–7.

79. Napenas J, Brennan M, Coleman S, Kent M, Noll J, Frenette G, et al. Molecular methodology to assess the impact of cancer chemotherapy on the oral bacterial flora: a pilot study. Oral Surg Oral Med Oral Pathol Oral Radiol Endod. 2010;109:554–60.
80. Orga P, Mestecky J, Lamm M, Strober W, Bienenstock J, McGhee J. Mucosal immunology. San Diego: Academic; 1994.
81. Palumbo A, Bringhen S, Bruno B, Falcone A, Liberati A, Grasso M, et al. Melphalan 200 mg/m2 versus melphalan 100 mg/m2 in newly diagnosed myeloma patients: a prospective, multicenter phase 3 study. Blood. 2010;115:1873–9.
82. Paolo A, Danesi R, Falcone A, Cionini L, Vannozzi F, Masi G, et al. Relationship between 5-fluorouracil disposition, toxicity and dihydropyrimidine dehydrogenase activity in cancer patients. Ann Oncol. 2001;12:1301–6.
83. Papaconstantinou H, Xie C, Zhang W, Andsari N, Hellmich M, Townsent C, et al. The role of caspases in methotrexate-induced gastrointestinal toxicity. Surgery. 2001;130:859–65.
84. Papaconstantinou H, Xie C, Zhang W, Ansari N, Hellmich M, Townsend C, et al. The role of caspaces in methotrexate-induced gastrointestinal toxicity. Surgery. 2001;130:859–65.
85. Paris F, Fuks Z, Kang A, Capodieci P, Juan G, Ehleiter D, et al. Endothelial apoptosis as the primary lesion initiating intestinal radiation damage in mice. Science. 2001;293:293–7.
86. Pegram M, Ngo D. Application and potential limitations of animal models utilized in the development of trastuzumab (Herceptin): a case study. Adv Drug Deliv Rev. 2006;58(5–6):723–34.
87. Peterson D, Keefe D, Sonis S. New frontiers in mucositis. ASCO Educational Book, ASCO; 2012.
88. Plosker G, Keam S. Trastuzumab: a review of its use in the management of HER2-positive metastatic and early stage breast cancer. Drugs. 2006;66:449–75.
89. Potten C, Booth C, Pritchard D. The intestinal epithelial stem cell: the mucosal governor. Int J Exp Pathol. 1997;78:219–43.
90. Potten C, Wilson J, Booth C. Regulation and significance of apoptosis in the stem cells of the gastrointestinal epithelium. Stem Cells. 1997;15:82–93.
91. Prisciandaro L, Geier M, Butler R, Cummins A, Howarth G. Probiotic factors partially improve parameters of 5-fluorouracil-induced intestinal mucositis in rats. Cancer Biol Ther. 2011;11:671–7.
92. Quarmby S, Hunter R, Kumar S. Irradiation induced expression of CD31, ICAM-1 and VCAM-1 in human microvascular endothelial cells. Anticancer Res. 2000;20:3375.
93. Ridwelski K, Gebauer T, Fahlke J, Kroning H, Kettner E, Meyer F, et al. Combination chemotherapy with docetaxel and cisplatin for locally advanced and metastatic gastric cancer. Ann Oncol. 2001;12:47–51.
94. Roche A, Bognel J, Bognel C, Bernier J. Correlation between the histological changes and glucose intestinal absorption following a single dose of 5-fluorouracil. Digestion. 1970;3:195–212.
95. Roseth A. Determination of faecal calprotectin, a novel marker of organic gastrointestinal disorders. Dig Liver Dis. 2003;35:607–9.
96. Rothenberg M, Eckardt J, Kuhn J, Burris H, Nelson J, Hilsenbeck S, et al. Phase II trial of irinotecan in patients with progressive or rapidly recurrent colorectal cancer. J Clin Oncol. 1996;14:1128–35.
97. Ruescher T, Sodeifi A, Scrivani S, Kaban L, Sonis S. The impact of mucositis on alpha-hemolytic streptococcal infection in patients undergoing autologous bone marrow transplantation for hematologic malignancies. Cancer. 1998;82:2275–81.
98. Ruiz M, Salvador J, Bayo J, Lomas M, Moreno A, Valero M, et al. Phase-II study of weekly schedule of trastuzumab, paclitaxel, and carboplatin followed by a week off every 28 days for HER2+ metastatic breast cancer. Cancer Chemother Pharmacol. 2008;62:1085–90.
99. Sawaki M, Ito Y, Tada K, Mizunuma N, Takahashi S, Horikoshi N, et al. Efficacy and safety of trastuzumab as a single agent in heavily pretreated patients with HER-2/neu-overexpressing metastatic breast cancer. Tumori. 2004;90(1):40–3.

100. Scrimger R, Kanji A, Parliament M, Warkentin H, Field C, Jha N, et al. Correlation between saliva production and quality of life measurements in head and neck cancer patients treated with intensity-modulated radiotherapy. Am J Clin Oncol. 2007;30:271–7.
101. Shao Z, Tang Z, Yan C, Jiang Y, Ma R, Liu Z, et al. Effects of intensity-modulated radiotherapy on human oral microflora. J Radiat Res. 2011;52:834–9.
102. Small M, Cavanagh R, Gottlieb L, Colon P, Zamcheck N. The effect of aminopterin on the absorption of xylose from the rat small intestine. Am J Dig Dis. 1959;4:700–5.
103. Soga Y, Maeda Y, Ishimaru F, Tanimoto M, Maeda H, Nishimura F, et al. Bacterial substitution of coagulase-negative staphylococci for streptococci on the oral mucosa after hematopoietic cell transplantation. Support Care Cancer. 2011;19:995–1000.
104. Sonis A, Sonis S. Oral complications of cancer therapy in pediatric patients. J Pedodot. 1979;3:12–6.
105. Sonis S. The biologic role for nuclear factor-kappaB in disease and its potential involvement in mucosal injury associated with anti-neoplastic therapy. Crit Rev Oral Biol Med. 2002;13:380–9.
106. Sonis S. A biological approach to mucositis. J Support Oncol. 2004;2:21–32.
107. Sonis S. The pathobiology of mucositis. Nat Rev Cancer. 2004;4:277–84.
108. Sonis S, et al. Prevention of chemotherapy-induced ulcerative mucositis by transforming growth factor Beta-3. Cancer Res. 1994;54:1135–8.
109. Sonis S, Costa J, Evitts S, Lindquist L, Nicholson M. Effect of epidermal growth factor on ulcerative mucositis in hamsters that receive chemotherapy. Oral Surg Oral Med Oral Pathol. 1992;74:749–55.
110. Sonis S, Costa J, Evitts S, Lindquist L, Nicolson M. Effect of epidermal growth factor on ulcerative mucositis in hamsters that receive cancer chemotherapy. Oral Surg Oral Med Oral Pathol. 1992;74:749–55.
111. Sonis S, Elting L, Keefe D, Peterson D, Schubert M. Perspectives on cancer therapy-induced mucosal injury: pathogenesis, measurement, epidemiology, and consequences for patients. Cancer. 2004;100:1995–2025.
112. Sonis S, Linquist L, VanVugt A, Stewart A, Stam K, Qu G, et al. Prevention of chemotherapy-induced ulcerative mucositis by transforming growth factor-beta3. Cancer Res. 1994;54:1135–8.
113. Sonis S, O'Donnell K, Popat R, Bragdon C, Phelan S, Cocks D, et al. The relationship between mucosal cyclooxygenase-2 (COX-2) expression and experimental radiation-induced mucositis. Oral Oncol. 2004;40:170–6.
114. Sonis S, Peterson R, Edwards L, Lucey C, Wang L, Mason L, et al. Defining mechanisms of action of interleukin-11 on the progression of radiation-induced oral mucositis in hamsters. Oral Oncol. 2000;36:373–81.
115. Sonis S, Scherer J, Phelan S, Lucey C, Barron J, O'Donnell K, et al. The gene expression sequence of radiated mucosa in an animal mucositis model. Cell Prolif. 2002;35:s92–102.
116. Sonis S, Shklar T, Jenson J, Florine D. An animal model for mucositis induced by cancer chemotherapy. Oral Surg Oral Med Oral Pathol. 1990;69:437–43.
117. Sonis S, Tracey C, Shklar G, Jenson J, Florine D. An animal model for mucositis induced by cancer chemotherapy. Oral Surg Oral Med Oral Pathol. 1990;69:437–43.
118. Sonis S, Treister N, Chawla S, Demetri G, Haluska F. Preliminary characterization of oral lesions associated with inhibitors of mammalian target of rapamycin in cancer patients. Cancer. 2010;116:210–5.
119. Squier CA, Hill MW. Oral mucosa. In: Ten Cate AR, editor. Oral histology development, structure and function. St Louis: Mosby-Year Book Inc; 1994. p. 389–431.
120. Stringer A, Gibson R, Bowen J, Ashton K, Logan R, Al-Dasooqi N, et al. Irinotecan-induced mucositis manifesting as diarrhoea corresponds with an amended intestinal flora and mucin profile. Int J Exp Pathol. 2009;90:489–99.
121. Stringer A, Gibson R, Bowen J, Keefe D. Chemotherapy-induced changes to microflora: evidence and implications of change. Curr Drug Metab. 2009;10:79–83.

122. Stringer A, Gibson R, Logan R, Bowen J, Yeoh A, Burns J, et al. Chemotherapy-induced diarrhea is associated with changes in the luminal environment in the DA rat. Exp Biol Med. 2007;232:96–106.
123. Stringer A, Gibson R, Logan R, Bowen J, Yeoh A, Keefe D. Faecal microflora and β-glucuronidase expression are altered in an irinotecan-induced diarrhoea model in rats. Cancer Biol Ther. 2008;7:1919–25.
124. Stringer A, Gibson R, Logan R, Bowen J, Yeoh A, Keefe D. Gastrointestinal microflora and mucins play a role in the development of 5-Fluorouracil-induced gastrointestinal mucositis in rats. Exp Biol Med. 2009;234:430–41.
125. Takasuna K, Hirohashi M, Kato M, et al. Involvement of Beta-Glucuronidase in intestinal microflora in the intestinal toxicity of antitumor camptothecin derivative irinotecan hydrochloride (CPT-11) in rats. Cancer Res. 1996;56:3752–7.
126. Toi M, Iwata H, Fujiwara Y, Ito Y, Nakamura S, Tokuda Y, et al. Lapatinib monotherapy in patients with relapsed, advanced, or metastatic breast cancer: efficacy, safety, and biomarker results from Japanese patients phase II studies. Br J Cancer. 2009;101(10):1676–82.
127. Tortora GJ, Grabowski SR. Principles of anatomy and physiology. New York: HarperCollins; 1993.
128. van der Velden W, Donnelly J, Blijlevens N. Lymphocyte subsets, granulocyte-colony-stimulating factor responsiveness and post-stem cell transplantation infections: mucositis is the underestimated confounder? Cytotherapy. 2012;14:381–3.
129. van Vliet M, Harmsen H, de Bont E, Tissing W. The role of intestinal microbiota in the development and severity of chemotherapy-induced mucositis. PLoS Pathog. 2010;6(5):e1000879.
130. Verburg M, Renes I, Meijer H, Tamininau J, Buller H, Einerhand A, et al. Selective sparing of goblet cells and paneth cells in the intestine of methotrexate-treated rats. Am J Physiol. 2000;279:G1037–47.
131. Wang R, Iganej S, Cosmatos H, Latino F, Kagan R. Pilot trial of concurrent low-dose weekly paclitaxel and radiotherapy in patients with advanced squamous cell carcinoma of the head and neck (SCCHN) [abstract 1685]. Prog/Proc Am Soc Clin Oncol. 2000;19:426a.
132. Weichselbaum R, Hallahan D, Fuks Z, Kufe D. Radiation induction of immediate early genes: effectors of the radiation-stress response. Int J Radiat Oncol Biol Phys. 1994;30:229.
133. Wisplinghoff H, Seifert H, Wenzel R, Edmond M. Current trends in the epidemiology of nosocomial bloodstream infections in patients with hematological malignancies and solid neoplasms in hospitals in the United States. Clin Infect Dis. 2003;36:1103–10.
134. Wright NA. Stem cell repertoire in the intestine. In: Potten CS, editor. Stem cells. London: Academic; 1997. p. 315–30.
135. Wu X, Chen P, Sonis S, Lingen M, Berger A, Toback F. A novel peptide to treat oral mucositis blocks endothelial and epithelial cell apoptosis. Int J Radiat Oncol Biol Phys. 2012;83:e409–15.
136. Wurth M, Musacchila X. Mechlorethamine effects on intestinal absorption in vitro and on cell proliferation. Am J Physiol. 1973;225:73–80.
137. Yeoh A, Bowen J, Gibson R, Keefe D. Nuclear factor kappaB (NFkappaB) and cyclooxygenase-2 (Cox-2) expression in the irradiated colorectum is associated with subsequent histopathological changes. Int J Radiat Oncol Biol Phys. 2005;63:1295–303.
138. Zhao J, Huang L, Belmar N, Buelow R, Fong T. Oral RDP58 allows CPT-11 dose intensification for enhanced tumor response by decreasing gastrointestinal toxicity. Clin Cancer Res. 2004;10:2851–9.

Dermatitis and Alopecia

Yevgeniy Balagula and Mario E. Lacouture

Abstract A fundamental shift in chemotherapeutic approach has occurred over the last several decades. A new class of so-called targeted agents, designed to target aberrant molecular pathways responsible for promoting carcinogenesis, was added to the arsenal of cytotoxic drugs that target rapidly proliferating tumor cells. These advances improved patient care and decreased morbidity and mortality. While systemic adverse events such as bone marrow suppression and gastrointestinal toxicities have diminished with the use of targeted agents, the spectrum of dermatologic toxicities have expanded. Mechanisms underlying the development of such a variety of cutaneous adverse events remain to be elucidated, but a significant amount of data shedding light on potential pathophysiology has accumulated. This chapter will review these data.

Remarkable progress in anticancer therapeutics throughout the second half of the twentieth century, and particularly the last decade, has improved patient survival and care. The ability of conventional cytotoxic chemotherapy agents and radiotherapy to target neoplastic cells has been partially dependent on their rapid proliferation, one of the hallmarks of cancer [1]. However, the nonspecific nature of these therapies results in significant toxicities of hematopoietic and dermatologic systems, prototypical tissues with rapidly proliferating cells. Novel targeted cancer therapies, inhibiting specific molecular pathways responsible for tumorigenesis, have reduced the degree of systemic, nonspecific adverse events such as myelosuppression [2]. Nevertheless, since cell surface receptors and intracellular enzymes targeted by these agents are abundantly expressed in normal skin and adnexal structures, many of these have been similarly fraught with a wide spectrum of dermatologic

---

Y. Balagula • M.E. Lacouture (✉)
Memorial Sloan Kettering Cancer Center, New York, NY, USA
e-mail: lacoutum@mskcc.org

S.T. Sonis and D.M. Keefe (eds.), *Pathobiology of Cancer Regimen-Related Toxicities*,
DOI 10.1007/978-1-4614-5438-0_8,

toxicities. This chapter will review the current understanding of pathogenesis and potential mechanisms responsible for these toxicities, with the emphasis on the available preclinical and clinical data.

Skin Manifestations

Papulopustular (Acneiform) Rash Associated with Epidermal Growth Factor Receptor Inhibitors

A papulopustular (acneiform) rash is the most common dermatologic toxicity of epidermal growth factor receptor inhibitors (EGFRI). The rash can affect up to 100% of patients but is seen to a variable degree depending on the EGFRI type [3]. This is a "class-effect" toxicity and individual lesion morphology and distribution is similar across different EGFRIs, including monoclonal antibodies (cetuximab and panitumumab) and intracellular tyrosine kinase inhibitors (erlotinib and gefitinib) [3]. Characterized by folliculocentric erythematous papules and/or pustules, it most commonly affects areas rich in pilosebaceous glands such as the scalp, face, and the upper trunk, less commonly involving lower trunk, buttocks, and extremities (Fig. 8.1) [4, 5]. The onset of the rash can range but is typically seen within 2 weeks of therapy [4]. The rash is dose-dependent and tends to have a higher incidence and severity with monoclonal antibodies than kinase inhibitors [6].

It is currently believed that the two main events contributing to the development of the rash are direct EGFR inhibition and consequent induction of an inflammatory

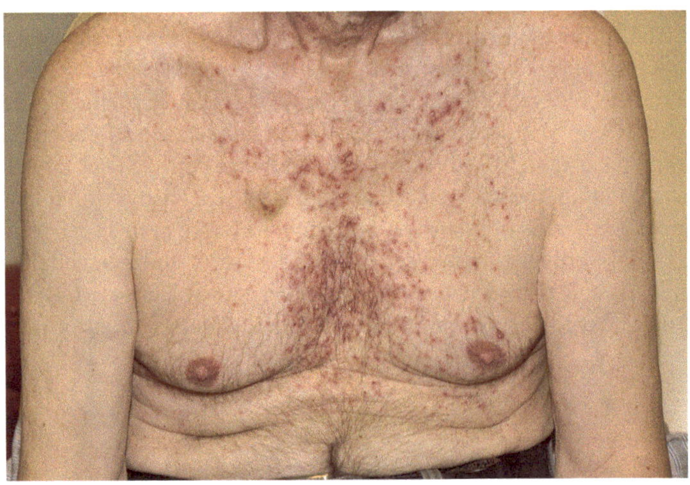

Fig. 8.1 Acneiform (papulopustular) rash affecting seborrheic-rich areas of the chest

response. In skin, EGFR is predominately localized to undifferentiated, actively proliferating basal and suprabasal keratinocytes, outer layers of the hair follicle, pilosebaceous, and eccrine sweat glands [7, 8]. The cell surface is coupled to various intracellular signaling pathways, the activation of which subsequently alters DNA expression and transcription. Through pathways such as Ras-Raf-MEK-ERK mitogen-activated protein kinase (MAPK), phosphatidylinositol 3-kinase (PI3K)-Akt, stress-activated protein kinase, and Jak/Stat, the EGFR regulates keratinocyte proliferation, differentiation, migration, and survival, orchestrating the formation of a protective epidermal barrier [9]. Therefore, it is not unexpected that interference with EGFR activation, which normally initiates upward keratinocyte migration and induces terminal differentiation [10], results in characteristic clinical changes. EGFR inhibition is associated with multiple downstream cellular effects, one of which is induction of apoptosis [11]. In fact, the level of apoptosis is amplified by almost 50% between days 4 and 12 in keratinocytes exposed to monoclonal antibody targeting the EGFR [11], which coincides with the clinical onset of the rash. Increased cell attachment impairing normal migration is reflected by increased expression of adhesion molecules such desmoglein 2 [12]. In human skin biopsies, the impact on keratinocyte proliferation is reflected by decreased expression of a proliferation marker KI67 and increased expression of the negative growth regulator p27 [13, 14]. Early differentiation is reflected by premature expression of terminal differentiation markers keratin 1 and STAT3 in the basal keratinocyte layer, which normally are only identified in suprabasal keratinocytes [13, 14]. Interestingly, a clinically less severe rash associated with lapatinib, a dual inhibitor of EGFR (HER1) and EGFR-2 (HER2) corresponds to a more "benign" immunohistochemical milieu. Decreased expression of p27 and enhanced expression of pAKT, a signal transduction marker, was seen in skin of patients treated with lapatinib as compared to cetuximab, panitumumab, and erlotinib [15], providing immunohistochemical–clinical correlation.

The inhibition of EGFR and its corresponding downstream signaling cascades appears to be the primary and the fundamental step responsible for the characteristic papulopustular eruption [16]. This concept is reinforced by mouse models, which initially demonstrate significant alterations in hair follicles and pilosebaceous units with inflammatory changes as a secondary event [17, 18]. The level of phosphorylated EGFR is markedly downregulated in basal and suprabasal keratinocytes and the outer layer of hair follicle in biopsies of the eruption [19]. Furthermore, decreased level of phosphorylated receptor is observed in unaffected skin and this reduction is prominent in hair follicles as compared to interfollicular epidermis [19]. These changes highlight the critical role of EGFR in rash pathogenesis and may explain folliculocentric manifestation of the rash. The importance of EGFR in maintaining epidermal homeostasis is reflected in histological changes with the most frequently affected structures including hair follicles (61%), epidermis (52%), and sebaceous glands (45%) and less involvement of eccrine sweat glands (22%) [20]. Epidermal atrophy, keratinocyte dysmaturation, and dyskeratosis are also seen [15]. Sebaceous glands are hypoplastic and contain poorly differentiated sebocytes and keratinocytes. These striking alterations of pilosebaceous units are observed in normal skin in the absence of inflammatory infiltrate, again implicating receptor inhibition as the primary event in rash pathogenesis [20]. The predilection for sebaceous gland involvement also explains the typical distribution of the papulopustular rash. However, these changes are in contrast to morphologic alterations observed in acne vulgaris, where sebaceous gland hyperplasia, lipid accumulation, and comedo formation are observed [20], which emphasizes the microscopic and mechanistic differences despite their similar clinical appearance.

Synthesis and release of inflammatory chemokines and cytokines including monocyte chemoattractant protein 1 (CCL-2), RANTES (CCL-5), T-lymphocyte-recruiting CCL-18 and CCL-3 are stimulated by inhibition of EGFR on keratinocytes. Subsequent recruitment of immune cells further amplifies the inflammatory response [16]. Infiltration of CD45RO positive T lymphocytes is an early event [5]. Leukocytic infiltration can be seen in lesional interfollicular dermis, hair follicles, and sebaceous glands. Additionally, Langerhans cells, macrophages, and Tumor Necrosis Factor Related Apoptosis Inducing Ligand (TRAIL) positive cells with dendritic morphology are seen [20]. Further ongoing and sustained follicular damage may lead to suppurative inflammation, characterized by either neutrophil-rich superficial perifolliculitis or suppurative folliculitis with hyperkeratosis, keratotic plugging, and rupture of follicular infundibulum [5, 19]. Alterations in pilosebaceous units are seen in both affected and unaffected skin, while inflammation accompanies these changes only within the area of the eruption [20]. This supports a view that inflammation may be induced by altered growth and differentiation of follicular epithelium [20]. Yet others propose EGFRI-induced inflammation to be an early event that induces follicular acanthosis and plugging that subsequently results in pustular inflammation. Inflammatory changes along with abnormal keratinocyte differentiation may be driven by upregulation of IL-1 and TNF-alpha detected in perilesional epidermis [19]. Interestingly, treatment of mice with TNF-alpha inhibitor or interleukin-1 antagonist significantly reduced characteristic EGFRI-induced skin changes [18].

Bacterial infection does not directly contribute to the development of the rash and early inflammation appears to be sterile [5]. However, disruption of epidermal protective barrier and follicular structures results in increased susceptibility to secondary bacterial superinfection, most commonly with *Staphylococcus aureus*, and in seborrheic-rich areas [21]. It has also been hypothesized that *Staphylococcus aureus* infection may have a greater contribution to the development of a late phase of papulopustular rash, with median onset of 200 days [22]. Another retrospective analysis demonstrated a significantly increased density of Demodex folliculorum in patients' lesions, highlighting the impaired antimicrobial defense mechanisms [23]. Besides the physical alterations in the epidermis and hair follicles, EGFR inhibition may impair antimicrobial-induced keratinocyte migration and proliferation [24, 25], further increasing predisposition to cutaneous infections.

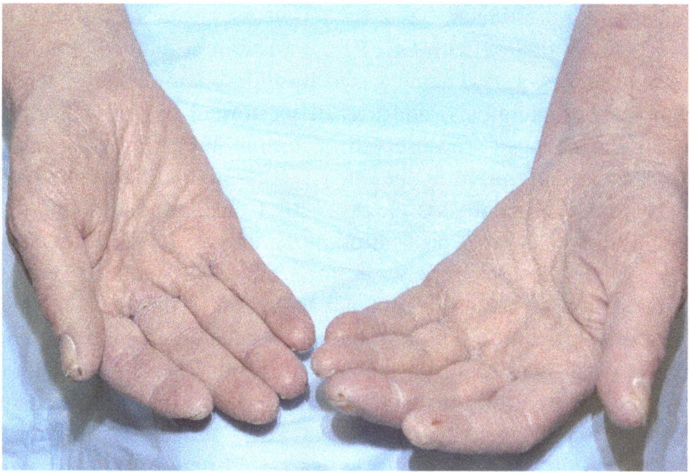

Fig. 8.2 Dry skin (xerosis), with fissures affecting fingertips, associated with pain and limiting activities of daily living

EGFRI-Induced Xerosis and Pruritus

Skin dryness (xerosis) is a frequently observed toxicity in patients receiving EGFRI, which can affect up to 35% of patients [6]. This adverse event is subacute/chronic in nature, manifesting after several weeks of therapy and typically subsequent to the appearance of the rash [4, 6]. Up to 100% of patients treated with EGFRI for greater than 6 months develop xerosis [26]. The involvement is not limited to the areas of the eruption and extends to affect larger body surface areas [27]. Associated with pruritus and characterized by fine superficial diffuse scaling, it can progress to severe xerosis, ichthyosis-like changes, and asteatotic dermatitis [28]. Further disruption of the epidermis leads to physical breakdown of skin and painful fissures, which tend to form on the distal finger tufts, dorsal surface of interphalangeal joints, and nail folds (Fig. 8.2) [4, 5]. However, any region can be affected, including palms and trunk, with formation of deep and long fissures [22].

The formation of a cornified layer, also referred to as stratum corneum, is the end-result of a tightly regulated process of differentiation and migration of keratinocytes from the basal layer. The cornified layer is composed of terminally differentiated, anucleate keratinocytes (corneocytes), which are shed to the environment at a rate similar to proliferation of undifferentiated cells in the basal layer. Corneocytes, in turn, mostly consist of keratin filaments that are ensheathed in various structural proteins, including filaggrin, loricrin, and involucrin, and are surrounded by insoluble lipids [29]. Filaggrin, which is degraded into hydrophilic amino acids, plays a critical role in the water-retention function of the stratum corneum [29]. EGFRI-induced xerosis stems from the disruption of epidermal homeostasis necessary for precise differentiation and upward keratinocyte migration. Normal stratum corneum has a basket-weave appearance. In contrast, skin biopsies in patients treated with EGFRI demonstrate a thinned stratum corneum with abnormal architecture, characterized by tightly and closely packed corneocytes (compact orthokeratosis) and loss of its normal basket-weave appearance [13, 30]. Similar changes are observed in normal appearing skin unaffected by papulopustular rash, highlighting the impact on terminal differentiation that is not limited to regions of clinically visible rash [20]. Considering that it normally takes approximately 2 weeks for a post-mitotic keratinocyte to reach the stratum corneum may explain the typical delay in onset of EGFRI-induced xerosis [10]. A recent investigation revealed a significant reduction in thickness of the filaggrin layer in stratum corneum in both lesional and perilesional skin of patients treated with EGFRI, which was more pronounced in lesions [19]. Thus, in addition to abnormal keratinocyte life-cycle, the loss of filaggrin with its water-retention function contributes to the development of xerosis [19]. Parakeratosis (abnormal retention of nuclei in corneocytes) in affected skin [13] may explain the tendency of skin to scale and abnormally shed.

Pruritus may be due to underlying xerosis, similar to other conditions in which there is a disruption of an epidermal barrier (e.g., eczema), or rash-associated inflammatory changes [16]. Normally, in non-EGFRI related skin inflammation, pruritus is often mediated by histamine release from dermal mast cells [31]. Substance P is a potent mast cell activator that may mediate the release of pruritogenic substance such as histamine, through binding to the neurokinin-1 receptor. Significant peri-adnexal accumulation of mast cells has been demonstrated in lesional skin of patients treated with erlotinib, thought to be mediated by EGFRI-induced keratinocyte release of stem cell factor [32]. It was also recently reported that aprepitant, a neurokinin-1–receptor antagonist has resulted in marked relief of erlotinib-induced pruritus in several patients [33, 34]. Therefore neurokinin-1, blocked by aprepitant, may be the principal driving event in EGFRI-induced pruritus [35]. However, in a setting of EGFRI therapy, pruritus is often difficult to control and besides promoting sedation, the typical anti-histamine agents do not provide consistent and reliable relief. The failure of EGFRI-associated pruritus to fully respond to anti-histamine agents suggests that additional factors may play a role.

Hand Foot Skin Reaction Associated with Multikinase Inhibitors

Hand foot skin reaction (HFSR) is a distinct adverse event associated with multikinase inhibitors (MKIs) such as sorafenib, sunitinib, and pazopanib. As the name implies, the manifestation is predominately limited to palmar and plantar surfaces. The typical onset of HFSR is observed within 6 weeks of therapy and often during the first 2 weeks of treatment [36]. The onset of clinical signs is often preceded by dysesthesias and paresthesias [37]. The initial manifestation is variable and patients may present with localized erythematous lesions with or without superficial desquamation, erythematous plaques with discrete and tense bullae, yellow hyperkeratotic

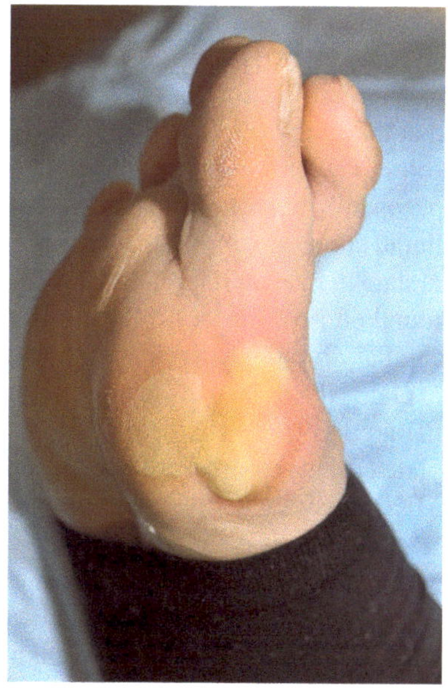

Fig. 8.3 Hand foot skin reaction affecting the foot, associated with pain and limiting activities of daily living

lesions, which may be surrounded by edematous or erythematous borders, or only with bullae (Fig. 8.3) [37, 38]. Hyperkeratosis is a late manifestation, once the blisters have subsided [37]. The typical sites include areas exposed to frequent and repeated friction and trauma, such as distal tips of digits, thenar, and hypothenar eminence, metacarpo- or metatarsophalangeal, and interphalangeal joints, and heels. Lateral aspects of digits, periungual tissues, and the dorsal surface of hands and feet may also be affected [37, 39]. The presentation is usually bilateral and tends to me more severe on plantar surfaces [40]. All-grade and high-grade (grade 3) HFSR is observed in 34% and 9% of patients treated with sorafenib, respectively [41], but can affect up to 60% [37]. The incidence of all-grade and high-grade HFSR due to sunitinib is 19% and 6%, respectively [42]. Pazopanib, however, is associated with a markedly lower incidence of all-grade (4.5%) and high-grade (1.8%) HFSR [43].

The underlying mechanisms have not been elucidated but multiple studies have shown a clinicohistopathologic correlation. The predominant histologic alteration in HFSR is keratinocyte damage, ranging from dyskeratosis, keratinocyte vacuolar degeneration, to band-like necrosis [39, 40, 44]. Perivascular lymphocytic infiltrate is observed consistently and superficial telangiectasias with dilated dermal vessels can be seen [39, 40, 44], which may explain clinically observed erythema. Keratinocyte necrosis is an early event and confluent necrosis with subsequent discohesiveness, intraepidermal separation, and serous fluid accumulation, can lead to clinically observed bullous lesions [39, 40, 44]. Proliferation rate of keratinocytes is significantly accelerated with resultant epidermal acanthosis (thickened spinous layer), hyperkeratosis (thickened stratum corneum), and parakeratosis (retention of pyknotic nuclei), which reflects accelerated cornification [39]. Hyperkeratosis and parakeratosis contribute to thickened (hyperkeratotic) lesions and scaling, respectively.

The predilection for palmar-plantar involvement is unknown. It was hypothesized that the drug may be actively excreted via eccrine sweat glands found in palms and soles [45]. However, this has not yet been demonstrated for sorafenib, sunitinib, or pazopanib, and one study, albeit small, failed to detect sorafenib in sweat by using liquid chromatography-tandem spectroscopy [46]. Alternatively, the drug may accumulate in the capillary network of large and wide dermal papillae, characteristically found on acral surfaces. Repetitive trauma and friction to these areas may result in capillary rupture and release of the drug [38]. However, this hypothesis would fail to explain HFSR changes in non-pressure bearing areas, such as finger webs and perianal area [39]. Lastly, palmoplantar involvement may be due to high expression of platelet derived growth factor receptor (PDGFR) and c-kit, the target receptors of MKIs, in sweat duct epithelium of eccrine sweat glands [47, 48].

Similarly, it is unclear how the drugs or their metabolites cause damage. Dose-escalation studies demonstrate dose-dependency of HFSR, suggesting a direct toxic effect [49], which may be mediated through receptor inhibition. Nevertheless, the culprit receptors are currently unknown. Sorafenib, sunitinib, and pazopanib overlap in blocking PDGFR, vascular endothelial growth factor receptor (VEGFR) and c-kit. Weak inhibition of Flt-3 and Ret by pazopanib is also shared by sorafenib and sunitinib [43]. However, Flt-3 and Ret expression has not been demonstrated in cutaneous structures and are not likely involved [40]. The detection of eccrine gland syringometaplasia [39] and cystic degeneration [40] has been described, implicating PDGFR and c-kit expressed on eccrine glands as the culprit. However, these changes are observed late into therapy (>30 days) [40], are not consistently seen [39, 44] and would not explain keratinocyte necrosis. Keratinocytes represent another direct target, but they do not express VEGFR and immunostaining for PDGFR and c-kit failed to reveal any differences between normal skin and HFSR lesions [39, 50]. Moreover, imatinib, which inhibits PDGFR and c-kit, is not commonly associated with HFSR [51]. Endothelial cells with VEGFR and the surrounding pericytes and fibroblasts, expressing PDGFR, represent important targets that may play a critical role in HFSR pathogenesis [40]. It is unlikely that individual inhibition of PDGFR or VEGFR alone is responsible for HFSR since similar to imatinib, bevacizumab, which targets VEGFR, is not typically associated with HFSR [52, 53]. However, simultaneous inhibition of these receptors may be the cause. PDGFR-expressing pericytes and fibroblasts play a role in cellular differentiation and vascular survival [54, 55]. Targeting of endothelial cells and pericytes by dual inhibition of VEGFR-2 and PDGFR-β, enhances tumor vessel regression due to interference with pericyte-mediated endothelial cell survival [56]. PDGF has also been shown to mediate wound healing [57]. Sunitinib can induce significant capillary rarefaction (reduction in microvascular density) in patients' digits [58]. Additionally, vascular ectasia [40], endothelial cell modifications, and distention/apoptosis in dermal capillaries are observed histologically [50]. Thus, it has been hypothesized that dual inhibition of these receptors may interfere with the ability of fibroblasts and endothelial cells to undergo continuous repair and diminish vascular reserve, predisposing areas subject to repeated friction, pressure, and daily subclinical trauma to develop changes of HFSR [37, 40]. Involvement of elbows [40], tips of the fingers and thumbs due to handheld device use [59], or a stump from a previous below the knee amputation [45] supports this hypothesis. The importance of dual inhibition of VEGFR and PDGFR is also suggested by observation that bevacizumab results in increased severity and frequency of HFSR when added to sorafenib [60]. Furthermore, novel agents targeting PDGFR and VEGFR, including axitinib (AG-013736) [61], cediranib (AZD2171) [62], vandetanib (ZD6474) [63], and telatinib (BAY-9352) [64], have all been associated with HFSR. Nevertheless, further investigations are needed to elucidate the precise pathogenesis.

Hand Foot Syndrome

Hand foot syndrome (HFS), also known as acral erythema and palmar-plantar erythrodysesthesia, is a reaction that has been classically associated with conventional cytotoxic chemotherapy agents. It is most commonly seen with capecitabine (28–74%), 5-FU (34%), cytarabine (19.7%), and pegylated liposomal doxorubicin (3.2–48%) [38]. While HFS shares several clinical and histopathologic characteristics with HFSR, it is believed to represent a distinct adverse event [37, 39, 40]. Similarities include palmoplantar involvement and a prodrome of dysesthesias and tingling sensations. There may be formation of well-demarcated, painful erythematous, and edematous plaques. Erythema is symmetric but can be more pronounced over the pads of the distal phalanges. Alternatively, erythema may diffusely involve the entire palmar or plantar surfaces. Patients may also present with superficial fine desquamation without erythema or alternating erythematous bands over the joint surfaces. With increased severity, erythematous plaques may blister and desquamate and rare full-thickness necrosis can occur [65–67]. Hyperkeratosis is less common and mostly occurs during the chronic phases of HFS and is more diffuse as compared to distinct hyperkeratotic lesions of HFSR [37, 66]. Typically, the palms are more severely affected. Occasionally, there may be extension to dorsal aspects of feet and hands or periungual tissues [65]. The onset of HFS varies from 24 h to 10 months following the initiation of the offending drug [65].

Histopathologic features are similar to HFSR and are characterized by dyskeratotic keratinocytes at various stages of necrosis, vacuolar degeneration, ectatic vessels, and a mild perivascular lymphocytic infiltrate [38, 66]. Severe cases, clinically characterized by peeling and intense erythema, demonstrate extensive necrosis and epidermal detachment [67]. Hyperkeratosis and parakeratosis are commonly seen, but the degree of epidermal proliferation and acanthosis is more pronounced in HFSR to multikinase inhibitors [38]. Sweat duct necrosis and eccrine squamous syringometaplasia can be observed as well [38, 67].

The predilection for palms and soles may be explained by high density of eccrine sweat glands on these surfaces. Pegylated doxorubicin and epirubicin have been detected by in vivo fluorescence deep in the palmar and plantar sweat ducts and surrounding their external openings in the upper skin layers, suggesting that the drug is excreted to the skin surface by sweat and subsequently penetrates the stratum corneum and deeper layers [68, 69]. The drugs are predominately detected on the palms and soles as compared to forehead, axilla, and forearms. Thick stratum corneum found on palmoplantar surfaces is believed to be an additional factor in HFS development, serving as a reservoir for the offending drug [69]. Alternatively, areas such as palms and soles or those subject to increased friction may achieve higher concentration of pegylated doxorubicin secondary to increased blood flow in rich dermal capillary network [70]. Damage to these capillaries may result in drug extravasation and tissue damage [38].

Direct cytotoxicity against rapidly proliferating keratinocytes is believed to be the main driving factor in HFS pathogenesis [38, 66]. The cytotoxic effect is supported by increased incidence and severity of HFS with increased doses of the drug and marked improvement with dose reduction [66]. Furthermore, findings of vacuolar degeneration and dyskeratosis are consistent with direct cytotoxic injury to the epidermis [38]. In the case of capecitabine, it has been hypothesized that the relative paucity of dihydropyrimidine dehydrogenase in palmar skin, which normally catabolizes 5-FU produced from ingested capecitabine, may result in a localized buildup of 5-FU [71]. However, this hypothesis is not supported by the lack of clinically observed increased incidence of HFS in patients treated with tegafur (5-FU pro-drug) or with 5-FU in combination with uracil and eniluracil, respectively, which are selective dihydropyrimidine dehydrogenase inhibitors [72]. It is possible that inflammation mediated by COX-2 overexpression may contribute to the development of HFS and treatment with celecoxib, a selective COX-2 inhibitor, has been shown to significantly reduce the incidence of HFS in patients treated with capecitabine [73]. An alternative hypothesis of host-vs.-altered host reaction was also proposed for 5-FU-induced HFS [74]. This was based on findings of apoptotic keratinocytes with satellitosis of lymphocytes and the ability of 5-FU to alter cell surface glycoproteins, which can elicit an acute immune graft vs. host disease (GVHD)-like response [74]. Interestingly, in the absence of extracutaneous involvement, acute GVHD clinically mimics HFS and in early stages is identical histologically [66].

Radiation Dermatitis

Radiation-induced cutaneous changes can be categorized as acute, consequential, or late. Acute alterations are observed during treatment or within weeks following radiotherapy. Non-healing, persistent acute changes are classified as consequential effects, and late or chronic reactions can manifest months to years after completion of radiotherapy [75]. The development of consequential late effects is related to severity of acute reactions [76]. Depending on severity, acute radiation dermatitis may range from erythema and dry desquamation to tender and edematous erythema, moist desquamation, ulceration, hemorrhage, and necrosis. Consequential late effects are characterized by chronic ulcerations, fibrosis, or

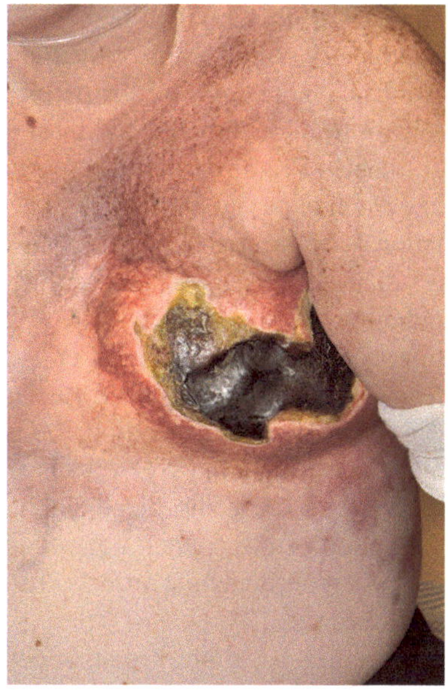

Fig. 8.4 Radiation dermatitis grade 4

even necrosis of subcutaneous structures such as bone. The clinical spectrum of chronic radiation dermatitis includes an often transient edematous peau d'orange appearance, poikilodermatous changes with pigmentary abnormalities, atrophy and telangiectasias, loss of hair and skin appendages, absent sweating, fibrosis, ulceration, skin atrophy and necrosis (Fig. 8.4) [77].

As ionizing radiation permeates through cutaneous structures it can either excite electrons to higher energy levels or eject them from their corresponding atoms, so-called ionization. These electrons together with radiation set in motion an ionizing cascade, generating free radicals, which subsequently induce direct DNA damage with resultant cell death and alterations of proteins, lipids, carbohydrates and complex molecules [78, 79]. The importance of DNA damage is highlighted in patients with genetic defects in DNA repair, such as ataxia-telangiectasia, who are radiosensitive and develop severe radiation toxicity [80]. With each additional exposure to radiation, there is accumulation of direct tissue injury and induction of an inflammatory cell infiltrate [78]. The damage is not limited to rapidly proliferating basal epidermal cells but also affects fibroblasts, endothelial, Langerhans and germinal cells of sebaceous glands and hair follicles, limiting the pool for adequate replacement of functional cells [77, 81]; wound healing is also impaired [82]. The injury stemming from radiotherapy has been referred to as a "complex wound", characterized by ongoing damage to already perturbed tissues with the simultaneous compensatory reparative process [78].

One of the earliest sublethal radiation-induced changes in endothelial cells include altered expression of adhesion molecules, prostacyclin, thromboxane, and nitric oxide, that contribute to an overall pro-coagulant environment [78]. The initial phase of erythema, often invisible clinically and detected with reflectance spectrophotometry within hours of treatment [83], may be due to radiation-induced vasodilation and increased vascular permeability, mediated by histamine and prostaglandin release from endothelial and mast cells [78]. In contrast, the characteristic clinical erythema of acute dermatitis observed after 1–2 weeks of radiotherapy, is likely due to inflammatory changes from direct endothelial and other cell destruction, upregulation of cytokines (IL-1, IL-6, and TNF-alpha), and an inflammatory infiltrate, with resultant increased vascular permeability, thrombosis, and vasodilation [78]. With radiation doses exceeding 40 Gy, erythema may be followed by moist desquamation, typically manifesting after 4–5 weeks of therapy [84]. Skin denudation is the result of misbalance between the rate of cellular destruction and regeneration, which can occur at doses exceeding 10 Gy/week [78]. This can progress to skin necrosis and ulcer formation [81]. Xerosis and hair loss that may accompany acute dermatitis is due to sebaceous gland and hair follicle damage, respectively [77]. While the disruption of the epithelial basement membrane is not necessary for the development of consequential effects, its breakdown is associated with a significantly increased risk [78].

Changes of chronic radiation dermatitis stem from a complex interplay between microvascular injury, inflammation, fibroblast dysfunction, collagen deposition, and wound healing [78, 79]. Vasculature and connective tissues are the predominantly affected structures [85]. Late radiation effects may be triggered and facilitated by the early radiation-induced pro-coagulant state [78]. Similarly, early leukocyte influx secondary to increased expression of endothelial cell adhesion molecules contributes to the development of parenchymal atrophy, fibrosis, and necrosis [86]. Transforming growth factor (TGF)-beta 1 is also one of the critical mediators of initiation, development and persistence of chronic radiation dermatitis and fibrosis [87]. Elevated levels of TGF-beta are found in fibrotic tissues of patients who have received radiotherapy, contributing to chemotaxis of fibroblasts and collagen synthesis [85]. Subcutaneous fibrosis may result in tissue retraction and limit mobility [77]. Additionally, irradiated fibroblasts and cells participating in tissue repair are functionally impaired, which contributes to long lasting abnormal tissue remodeling [78]. Synthesis and accumulation of abnormal collagen can lead to persistent, non-healing ulcers and skin breakdown [79]. Skin atrophy as a result of dermal fibroblast diminution and collagen reabsorption contributes to skin fragility [88]. Small arteries and arterioles demonstrate eccentric myointimal proliferation that diminishes or prevents blood flow with subsequent ischemia that potentiates the development of ulcers and necrosis [84]. Vascular sclerosis with luminal narrowing is also a significant component of late radiation injury [78]. The importance of microvasculature in mediating late effects is further supported by increased incidence of late toxicities in patients with hypertension and diabetes mellitus [89]. Dilated vessels (telangiectasias) readily visible through thinned epidermis and atrophic dermis may stem from endothelial cell loss or destruction of smooth muscle cells with subsequent loss of capillary pressure control [78].

Hair Growth Alterations

Chemotherapy-Induced Alopecia

Chemotherapy-induced alopecia (CIA) from conventional cytotoxic agents is a prevalent and significant toxicity associated with multiple agents. Taxanes (paclitaxel), topoisomerase inhibitors (doxorubicin), alkylators (cyclophosphamide), and antimetabolites (5-fluorouracil) can lead to CIA in >80%, 60–100%, >60%, and 10–50% of cases, respectively [90].

A hair follicle cycle consists of three phases, of which anagen is characterized by rapidly proliferating hair matrix cells in the bulb that form the hair shaft. Lasting from 2 to 6 years, it is followed by a short (2–3 weeks), apoptosis-driven regression phase (catagen). The final resting phase (telogen; 3–4 months) leads to loss of fully keratinized hair [91]. Due to their non-selectivity, cytotoxic agents are also toxic to normally rapidly proliferating and mitotically active hair matrix cells. Therefore, chemotherapy predominately targets hair follicles in anagen phase and has been traditionally referred to as "dystrophic" anagen effluvium [90, 92]. It results in decreased proliferation of matrix keratinocytes, partial loss of inner and outer root sheath layers, shrinkage of dermal papilla, premature apoptosis-driven catagen, and dystrophy of the hair follicle [93]. Morphologically, there is diminution of hair bulb diameter, thinning and tapering of the shaft, or marked atrophy of the hair root [94, 95]. Massive apoptosis induced by chemotherapy-induced DNA damage in hair matrix keratinocytes has been demonstrated in vitro, in animal models, and human hair follicles in vivo [93, 96, 97]. Investigations in animal models demonstrated p53 and its target genes to be a critical mediator of apoptosis and CIA [98]. Alternatively, p53-independent induction of apoptosis is observed as well [99]. Higher concentrations of cytostatic agents lead to necrosis as has been shown in in vitro human hair follicle model [93]. The toxic effects on the inner root sheath cells may compromise its attachment to the surrounding structures, and hair is also weakened at points of constriction and thinning. Thus, hair loss may be spontaneous or facilitated by mild force such as hair combing. Alternately, weakened hair shafts can break with minimal pressure [94, 100]. Since up to 90% of hair follicles in human scalp are in anagen, which tends to be long in duration, the head it is the most common location of CIA. However, eyebrows, eyelashes, axillary and pubic hair, which have a lower percentage of anagen hairs, may be affected as well [90]. Because of acute disruption of such a large number of anagen follicles, anagen effluvium has an early onset, typically within 1–3 weeks and is most prominent at 1–2 months following initiation of chemotherapy [101]. Although anagen effluvium due to chemotherapy has been traditionally described as diffuse, some studies demonstrated patterned hair loss, with predilection to specific scalp areas and sparing the hair line, suggesting that multiple factors, such as hormonal status, gender, and hair follicle location contribute to the overall shedding pattern [92].

CIA can also occur due to telogen effluvium. Direct toxicity to hair follicles, particularly those that are in late anagen phase characterized by spontaneous diminution of mitotic rate, facilitates their transition into telogen phase [102]. Since hair loss occurs at the end of the normal telogen phase, it predominately manifests 3–4 months following the drug exposure. Rarely affecting greater than 50% of scalp hair, it is significantly less severe, characterized by thinning rather than pronounced shedding [100]. Telogen effluvium is most commonly seen with 5-fluorouracil and methotrexate [100]. Furthermore, a recent investigation in breast cancer patients receiving cyclophosphamide, methotrexate, and 5-fluorouracil demonstrated that shed telogen hairs significantly exceeded the number of lost fractured and anagen hairs during treatment, accounting for a mean of 81%, 12%, and 6%, respectively (p-value = 0.0002). More importantly, telogen hairs were morphologically abnormal, characterized by tapering of the proximal root shaft, which varied in both diameter and the length, and diminution of the bulb [103]. Shaft tapering was also observed in proximally fractured hair. The authors postulated that in addition to apoptosis-driven tapering of anagen hair shafts, there was a subsequent premature induction of telogen and proposed a term "atrophic telogen effluvium" to reflect these two predominant features [103]. These findings are in contrast to earlier results that failed to show any alterations of hair cycle [94, 95] that led to the traditional term of chemotherapy-induced anagen effluvium, attributed to loss of fractured hair and the assumption that follicles remain in anagen. Of note, "atrophic telogen effluvium" should be distinguished from the classical telogen effluvium of morphologically normal hair, which was also observed in this study.

Typically, there is spontaneous hair regrowth, observed after 3–6 months following the completion of a telogen phase. However, alterations in hair texture and color can occur in approximately 65% of patients [92, 104, 105]. In addition to keratinocytes, melanocytes in the hair follicle pigmentary unit are also significantly affected [93]. Disruption of melanin production and transfer associated with melanin clumping, ectopic melanin location, and irregular banding pattern of hair shafts characterize hair follicle dystrophy following chemotherapy [93], demonstrating the impact of chemotherapy on normal pigmentation. Data from animal models highlight the two distinct pathways of follicular response to chemotherapy-induced damage. Exposure to lower doses of chemotherapy results in "dystrophic anagen" pathway, which after causing hair shedding, is followed by generation of a defective hair shaft with abnormal pigmentation. Subsequent to catagen-telogen transition, normal anagen hairs develop during a delayed"secondary recovery" phase. Alternatively, high-doses of chemotherapy rapidly induce a so-called "dystrophic catagen" with resultant hair loss, followed by a very short dystrophic telogen phase. The end-result is a "secondary recovery" with premature and faster transition to normal anagen phase that leads to faster regrowth of normal hair [106].

Rarely, persistent CIA, defined as lack of or incomplete regrowth of thin hair 6 months following the completion of treatment, may ensue [105]. Typically seen in a setting of ablative chemotherapy regimens for bone marrow transplantation, it is most commonly associated with busulfan, but cyclophosphamide, thiotepa, melphalan, etoposide, carboplatin, docetaxel, and paclitaxel have been implicated as well. The histopathologic hallmark in the majority of cases is a significant reduction of hair follicles with absence or minimal perifollicular fibrosis and inflammation [105]. The mechanisms of permanent CIA are unknown, but may be due to follicular stem cell damage located in the hair bulge, or separation of the hair matrix cells from the underlying dermal papillae [107], which plays a critical role in regulating follicular growth cycle [91]. Additionally, it may result from disruption of signaling pathways to the secondary hair germ at the base of the follicle [105], which contains actively cycling stem and progenitor cells [108].

Conclusion

Major cancer treatment modalities have an enormous impact on numerous cutaneous and adnexal structures with clinically significant sequelae. Furthermore, when used in combination, the degree of toxicity may be potentiated. While momentous strides have been made in elucidating the underlying pathophysiology, much remains unknown. Improved understanding is critical in our ability to designing effective, evidence and mechanism-based therapeutic approaches and should be the emphasis of future research efforts.

References

1. Hanahan D, Weinberg RA. The hallmarks of cancer. Cell. 2000;100:57–70.
2. Vokes EE, Chu E. Anti-EGFR therapies: clinical experience in colorectal, lung, and head and neck cancers. Oncology (Williston Park). 2006;20:15–25.
3. Perez-Soler R, Delord JP, Halpern A, et al. HER1/EGFR inhibitor-associated rash: future directions for management and investigation outcomes from the HER1/EGFR inhibitor rash management forum. Oncologist. 2005;10:345–56.
4. Roe E, Garcia Muret MP, Marcuello E, Capdevila J, Pallares C, Alomar A. Description and management of cutaneous side effects during cetuximab or erlotinib treatments: a prospective study of 30 patients. J Am Acad Dermatol. 2006;55:429–37.
5. Busam KJ, Capodieci P, Motzer R, Kiehn T, Phelan D, Halpern AC. Cutaneous side-effects in cancer patients treated with the antiepidermal growth factor receptor antibody C225. Br J Dermatol. 2001;144:1169–76.
6. Lynch Jr TJ, Kim ES, Eaby B, Garey J, West DP, Lacouture ME. Epidermal growth factor receptor inhibitor-associated cutaneous toxicities: an evolving paradigm in clinical management. Oncologist. 2007;12:610–21.
7. Nanney LB, Stoscheck CM, King Jr LE, Underwood RA, Holbrook KA. Immunolocalization of epidermal growth factor receptors in normal developing human skin. J Invest Dermatol. 1990;94:742–8.
8. Nanney LB, Magid M, Stoscheck CM, King Jr LE. Comparison of epidermal growth factor binding and receptor distribution in normal human epidermis and epidermal appendages. J Invest Dermatol. 1984;83:385–93.
9. Mendelsohn J, Baselga J. Status of epidermal growth factor receptor antagonists in the biology and treatment of cancer. J Clin Oncol. 2003;21:2787–99.
10. Fuchs E, Raghavan S. Getting under the skin of epidermal morphogenesis. Nat Rev Genet. 2002;3:199–209.

11. Rodeck U, Jost M, Kari C, et al. EGF-R dependent regulation of keratinocyte survival. J Cell Sci. 1997;110(Pt 2):113–21.
12. Lorch JH, Klessner J, Park JK, et al. Epidermal growth factor receptor inhibition promotes desmosome assembly and strengthens intercellular adhesion in squamous cell carcinoma cells. J Biol Chem. 2004;279:37191–200.
13. Albanell J, Rojo F, Averbuch S, et al. Pharmacodynamic studies of the epidermal growth factor receptor inhibitor ZD1839 in skin from cancer patients: histopathologic and molecular consequences of receptor inhibition. J Clin Oncol. 2002;20:110–24.
14. Baselga J, Rischin D, Ranson M, et al. Phase I safety, pharmacokinetic, and pharmacodynamic trial of ZD1839, a selective oral epidermal growth factor receptor tyrosine kinase inhibitor, in patients with five selected solid tumor types. J Clin Oncol. 2002;20:4292–302.
15. Nardone B, Nicholson K, Newman M, et al. Histopathologic and immunohistochemical characterization of rash to human epidermal growth factor receptor 1 (HER1) and HER1/2 inhibitors in cancer patients. Clin Cancer Res. 2010;16:4452–60.
16. Lacouture ME. Mechanisms of cutaneous toxicities to EGFR inhibitors. Nat Rev Cancer. 2006;6:803–12.
17. Murillas R, Larcher F, Conti CJ, Santos M, Ullrich A, Jorcano JL. Expression of a dominant negative mutant of epidermal growth factor receptor in the epidermis of transgenic mice elicits striking alterations in hair follicle development and skin structure. EMBO J. 1995; 14:5216–23.
18. Surguladze D, Deevi D, Claros N, et al. Tumor necrosis factor-alpha and interleukin-1 antagonists alleviate inflammatory skin changes associated with epidermal growth factor receptor antibody therapy in mice. Cancer Res. 2009;69:5643–7.
19. Han SS, Lee M, Park GH, et al. Investigation of papulopustular eruptions caused by cetuximab treatment shows altered differentiation markers and increases in inflammatory cytokines. Br J Dermatol. 2009;162:371–9.
20. Guttman-Yassky E, Mita A, De Jonge M, et al. Characterisation of the cutaneous pathology in non-small cell lung cancer (NSCLC) patients treated with the EGFR tyrosine kinase inhibitor erlotinib. Eur J Cancer. 2010;46:2010–9.
21. Eilers Jr RE, Gandhi M, Patel JD, et al. Dermatologic infections in cancer patients treated with epidermal growth factor receptor inhibitor therapy. J Natl Cancer Inst. 2010;102:47–53.
22. Amitay-Laish I, David M, Stemmer SM. Staphylococcus coagulase-positive skin inflammation associated with epidermal growth factor receptor-targeted therapy: an early and a late phase of papulopustular eruptions. Oncologist. 2010;15:1002–8.
23. Gerber PA, Kukova G, Buhren BA, Homey B. Density of Demodex folliculorum in patients receiving epidermal growth factor receptor inhibitors. Dermatology. 2011;222:144–7.
24. Pastore S, Mascia F, Gulinelli S, et al. Stimulation of purinergic receptors modulates chemokine expression in human keratinocytes. J Invest Dermatol. 2007;127:660–7.
25. Niyonsaba F, Ushio H, Nakano N, et al. Antimicrobial peptides human beta-defensins stimulate epidermal keratinocyte migration, proliferation and production of proinflammatory cytokines and chemokines. J Invest Dermatol. 2007;127:594–604.
26. Osio A, Mateus C, Soria JC, et al. Cutaneous side-effects in patients on long-term treatment with epidermal growth factor receptor inhibitors. Br J Dermatol. 2009;161:515–21.
27. Hu JC, Sadeghi P, Pinter-Brown LC, Yashar S, Chiu MW. Cutaneous side effects of epidermal growth factor receptor inhibitors: clinical presentation, pathogenesis, and management. J Am Acad Dermatol. 2007;56:317–26.
28. Segaert S, Van Cutsem E. Clinical signs, pathophysiology and management of skin toxicity during therapy with epidermal growth factor receptor inhibitors. Ann Oncol. 2005; 16:1425–33.
29. Candi E, Schmidt R, Melino G. The cornified envelope: a model of cell death in the skin. Nat Rev Mol Cell Biol. 2005;6:328–40.
30. Malik SN, Siu LL, Rowinsky EK, et al. Pharmacodynamic evaluation of the epidermal growth factor receptor inhibitor OSI-774 in human epidermis of cancer patients. Clin Cancer Res. 2003;9:2478–86.

31. Ikoma A, Steinhoff M, Stander S, Yosipovitch G, Schmelz M. The neurobiology of itch. Nat Rev Neurosci. 2006;7:535–47.
32. Gerber PA, Buhren BA, Cevikbas F, Bolke E, Steinhoff M, Homey B. Preliminary evidence for a role of mast cells in epidermal growth factor receptor inhibitor-induced pruritus. J Am Acad Dermatol. 2010;63:163–5.
33. Vincenzi B, Tonini G, Santini D. Aprepitant for erlotinib-induced pruritus. N Engl J Med. 2010;363:397–8.
34. Leveque D. Aprepitant for erlotinib-induced pruritus. N Engl J Med. 2010;363:1680–1; author reply 1681.
35. Gerber PA, Buhren BA, Homey B. More on aprepitant for erlotinib-induced pruritus. N Engl J Med. 2011;364:486–7.
36. Porta C, Paglino C, Imarisio I, Bonomi L. Uncovering Pandora's vase: the growing problem of new toxicities from novel anticancer agents. The case of sorafenib and sunitinib. Clin Exp Med. 2007;7:127–34.
37. Autier J, Escudier B, Wechsler J, Spatz A, Robert C. Prospective study of the cutaneous adverse effects of sorafenib, a novel multikinase inhibitor. Arch Dermatol. 2008;144: 886–92.
38. Lipworth AD, Robert C, Zhu AX. Hand-foot syndrome (hand-foot skin reaction, palmar-plantar erythrodysesthesia): focus on sorafenib and sunitinib. Oncology. 2009;77:257–71.
39. Yang CH, Lin WC, Chuang CK, et al. Hand-foot skin reaction in patients treated with sorafenib: a clinicopathological study of cutaneous manifestations due to multitargeted kinase inhibitor therapy. Br J Dermatol. 2008;158:592–6.
40. Lacouture ME, Reilly LM, Gerami P, Guitart J. Hand foot skin reaction in cancer patients treated with the multikinase inhibitors sorafenib and sunitinib. Ann Oncol. 2008;19: 1955–61.
41. Chu D, Lacouture ME, Fillos T, Wu S. Risk of hand-foot skin reaction with sorafenib: a systematic review and meta-analysis. Acta Oncol. 2008;47:176–86.
42. Chu D, Lacouture ME, Weiner E, Wu S. Risk of hand-foot skin reaction with the multitargeted kinase inhibitor sunitinib in patients with renal cell and non-renal cell carcinoma: a meta-analysis. Clin Genitourin Cancer. 2009;7:11–9.
43. Balagula Y, Wu S, Su X, Feldman DR, Lacouture ME. The risk of hand foot skin reaction to pazopanib, a novel multikinase inhibitor: a systematic review of literature and meta-analysis. Invest New Drugs. 2011;30:1773–81.
44. Lee WJ, Lee JL, Chang SE, et al. Cutaneous adverse effects in patients treated with the multitargeted kinase inhibitors sorafenib and sunitinib. Br J Dermatol. 2009;161:1045–51.
45. Lai SE, Kuzel T, Lacouture ME. Hand-foot and stump syndrome to sorafenib. J Clin Oncol. 2007;25:341–3.
46. Jain L, Gardner ER, Figg WD, Chernick MS, Kong HH. Lack of association between excretion of sorafenib in sweat and hand-foot skin reaction. Pharmacotherapy. 2010;30:52–6.
47. Lammie A, Drobnjak M, Gerald W, Saad A, Cote R, Cordon-Cardo C. Expression of c-kit and kit ligand proteins in normal human tissues. J Histochem Cytochem. 1994;42:1417–25.
48. Ponten F, Ren Z, Nister M, Westermark B, Ponten J. Epithelial-stromal interactions in basal cell cancer: the PDGF system. J Invest Dermatol. 1994;102:304–9.
49. Strumberg D, Awada A, Hirte H, et al. Pooled safety analysis of BAY 43-9006 (sorafenib) monotherapy in patients with advanced solid tumours: is rash associated with treatment outcome? Eur J Cancer. 2006;42:548–56.
50. Faivre S, Delbaldo C, Vera K, et al. Safety, pharmacokinetic, and antitumor activity of SU11248, a novel oral multitarget tyrosine kinase inhibitor, in patients with cancer. J Clin Oncol. 2006;24:25–35.
51. Breccia M, Carmosino I, Russo E, Morano SG, Latagliata R, Alimena G. Early and tardive skin adverse events in chronic myeloid leukaemia patients treated with imatinib. Eur J Haematol. 2005;74:121–3.
52. Schenone S, Bondavalli F, Botta M. Antiangiogenic agents: an update on small molecule VEGFR inhibitors. Curr Med Chem. 2007;14:2495–516.

53. Munehiro A, Yoneda K, Nakai K, et al. Bevacizumab-induced hand-foot syndrome: circumscribed type. Br J Dermatol. 2010;162:1411–3.
54. Song S, Ewald AJ, Stallcup W, Werb Z, Bergers G. PDGFRbeta+ perivascular progenitor cells in tumours regulate pericyte differentiation and vascular survival. Nat Cell Biol. 2005;7:870–9.
55. Heldin CH, Backstrom G, Ostman A, et al. Binding of different dimeric forms of PDGF to human fibroblasts: evidence for two separate receptor types. EMBO J. 1988;7:1387–93.
56. Erber R, Thurnher A, Katsen AD, et al. Combined inhibition of VEGF and PDGF signaling enforces tumor vessel regression by interfering with pericyte-mediated endothelial cell survival mechanisms. FASEB J. 2004;18:338–40.
57. Judith R, Nithya M, Rose C, Mandal AB. Application of a PDGF-containing novel gel for cutaneous wound healing. Life Sci. 2010;87:1–8.
58. van der Veldt AA, de Boer MP, Boven E, et al. Reduction in skin microvascular density and changes in vessel morphology in patients treated with sunitinib. Anticancer Drugs. 2010;21:439–46.
59. Boone SL, Jameson G, Von Hoff D, Lacouture ME. Blackberry-induced hand-foot skin reaction to sunitinib. Invest New Drugs. 2009;27:389–90.
60. Azad NS, Posadas EM, Kwitkowski VE, et al. Combination targeted therapy with sorafenib and bevacizumab results in enhanced toxicity and antitumor activity. J Clin Oncol. 2008;26:3709–14.
61. Mukohara T, Nakajima H, Mukai H, et al. Effect of axitinib (AG-013736) on fatigue, thyroid-stimulating hormone, and biomarkers: a phase I study in Japanese patients. Cancer Sci. 2010;101:963–8.
62. Drevs J, Siegert P, Medinger M, et al. Phase I clinical study of AZD2171, an oral vascular endothelial growth factor signaling inhibitor, in patients with advanced solid tumors. J Clin Oncol. 2007;25:3045–54.
63. Annunziata CM, Walker AJ, Minasian L, et al. Vandetanib, designed to inhibit VEGFR2 and EGFR signaling, had no clinical activity as monotherapy for recurrent ovarian cancer and no detectable modulation of VEGFR2. Clin Cancer Res. 2010;16:664–72.
64. Eskens FA, Steeghs N, Verweij J, et al. Phase I dose escalation study of telatinib, a tyrosine kinase inhibitor of vascular endothelial growth factor receptor 2 and 3, platelet-derived growth factor receptor beta, and c-Kit, in patients with advanced or metastatic solid tumors. J Clin Oncol. 2009;27:4169–76.
65. Baack BR, Burgdorf WH. Chemotherapy-induced acral erythema. J Am Acad Dermatol. 1991;24:457–61.
66. Nagore E, Insa A, Sanmartin O. Antineoplastic therapy-induced palmar plantar erythrodysesthesia ('hand-foot') syndrome. Incidence, recognition and management. Am J Clin Dermatol. 2000;1:225–34.
67. Hueso L, Sanmartin O, Nagore E, et al. [Chemotherapy-induced acral erythema: a clinical and histopathologic study of 44 cases]. Actas Dermosifiliogr. 2008;99:281–90.
68. Jacobi U, Waibler E, Schulze P, et al. Release of doxorubicin in sweat: first step to induce the palmar-plantar erythrodysesthesia syndrome? Ann Oncol. 2005;16:1210–1.
69. Martschick A, Sehouli J, Patzelt A, et al. The pathogenetic mechanism of anthracycline-induced palmar-plantar erythrodysesthesia. Anticancer Res. 2009;29:2307–13.
70. Lotem M, Hubert A, Lyass O, et al. Skin toxic effects of polyethylene glycol-coated liposomal doxorubicin. Arch Dermatol. 2000;136:1475–80.
71. Ferrero JM, Lassalle S, Mari M, et al. Hand-foot syndrome (HFS) in patients receiving capecitabine: a pharmacological explanation. J Clin Oncol. 2006 ASCO Annual Meeting Proceedings Part I. 2006;24(18S)2019.
72. Saif MW, Elfiky A, Diasio R. Hand-foot syndrome variant in a dihydropyrimidine dehydrogenase-deficient patient treated with capecitabine. Clin Colorectal Cancer. 2006;6:219–23.

73. Zhang RX, Wu XJ, Lu SX, Pan ZZ, Wan DS, Chen G. The effect of COX-2 inhibitor on capecitabine-induced hand-foot syndrome in patients with stage II/III colorectal cancer: a phase II randomized prospective study. J Cancer Res Clin Oncol. 2011;137:953–7.
74. Beard JS, Smith KJ, Skelton HG. Combination chemotherapy with 5-fluorouracil, folinic acid, and alpha-interferon producing histologic features of graft-versus-host disease. J Am Acad Dermatol. 1993;29:325–30.
75. Stone HB, Coleman CN, Anscher MS, McBride WH. Effects of radiation on normal tissue: consequences and mechanisms. Lancet Oncol. 2003;4:529–36.
76. Dorr W, Hendry JH. Consequential late effects in normal tissues. Radiother Oncol. 2001;61:223–31.
77. Hymes SR, Strom EA, Fife C. Radiation dermatitis: clinical presentation, pathophysiology, and treatment. J Am Acad Dermatol. 2006;54:28–46.
78. Denham JW, Hauer-Jensen M. The radiotherapeutic injury—a complex 'wound'. Radiother Oncol. 2002;63:129–45.
79. Dormand EL, Banwell PE, Goodacre TE. Radiotherapy and wound healing. Int Wound J. 2005;2:112–27.
80. Cox R, Masson WK, Weichselbaum RR, Nove J, Little JB. The repair of potentially lethal damage in x-irradiated cultures of normal and ataxia telangiectasia human fibroblasts. Int J Radiat Biol Relat Stud Phys Chem Med. 1981;39:357–65.
81. Archambeau JO, Pezner R, Wasserman T. Pathophysiology of irradiated skin and breast. Int J Radiat Oncol Biol Phys. 1995;31:1171–85.
82. Bernstein EF, Harisiadis L, Salomon GD, et al. Healing impairment of open wounds by skin irradiation. J Dermatol Surg Oncol. 1994;20:757–60.
83. Simonen P, Hamilton C, Ferguson S, et al. Do inflammatory processes contribute to radiation induced erythema observed in the skin of humans? Radiother Oncol. 1998;46:73–82.
84. Mendelsohn FA, Divino CM, Reis ED, Kerstein MD. Wound care after radiation therapy. Adv Skin Wound Care. 2002;15:216–24.
85. Canney PA, Dean S. Transforming growth factor beta: a promotor of late connective tissue injury following radiotherapy? Br J Radiol. 1990;63:620–3.
86. Quarmby S, Kumar P, Kumar S. Radiation-induced normal tissue injury: role of adhesion molecules in leukocyte-endothelial cell interactions. Int J Cancer. 1999;82:385–95.
87. Martin M, Lefaix J, Delanian S. TGF-beta1 and radiation fibrosis: a master switch and a specific therapeutic target? Int J Radiat Oncol Biol Phys. 2000;47:277–90.
88. Harper JL, Franklin LE, Jenrette JM, Aguero EG. Skin toxicity during breast irradiation: pathophysiology and management. South Med J. 2004;97:989–93.
89. Chon BH, Loeffler JS. The effect of nonmalignant systemic disease on tolerance to radiation therapy. Oncologist. 2002;7:136–43.
90. Trueb RM. Chemotherapy-induced alopecia. Semin Cutan Med Surg. 2009;28:11–4.
91. Jankovic SM, Jankovic SV. The control of hair growth. Dermatol Online J. 1998;4:2.
92. Yun SJ, Kim SJ. Hair loss pattern due to chemotherapy-induced anagen effluvium: a cross-sectional observation. Dermatology. 2007;215:36–40.
93. Bodo E, Tobin DJ, Kamenisch Y, et al. Dissecting the impact of chemotherapy on the human hair follicle: a pragmatic in vitro assay for studying the pathogenesis and potential management of hair follicle dystrophy. Am J Pathol. 2007;171:1153–67.
94. Crounse RG, Van Scott EJ. Changes in scalp hair roots as a measure of toxicity from cancer chemotherapeutic drugs. J Invest Dermatol. 1960;35:83–90.
95. Kligman AM. Pathologic dynamics of human hair loss I. Telogen effuvium. Arch Dermatol. 1961;83:175–98.
96. Schilli MB, Paus R, Menrad A. Reduction of intrafollicular apoptosis in chemotherapy-induced alopecia by topical calcitriol-analogs. J Invest Dermatol. 1998;111:598–604.
97. Goldberg MT, Tackaberry LE, Hardy MH, Noseworthy JH. Nuclear aberrations in hair follicle cells of patients receiving cyclophosphamide. A possible in vivo assay for human exposure to genotoxic agents. Arch Toxicol. 1990;64:116–21.

98. Botchkarev VA. Molecular mechanisms of chemotherapy-induced hair loss. J Investig Dermatol Symp Proc. 2003;8:72–5.
99. Perego P, Corna E, De Cesare M, et al. Role of apoptosis and apoptosis-related genes in cellular response and antitumor efficacy of anthracyclines. Curr Med Chem. 2001;8:31–7.
100. Olsen A. Chemotherapy-induced alopecia: overview and methodology for characterizing hair changes and regrowth. In: Olver IN, editor. The MASCC textbook of cancer supportive care and survivorship. New York: Springer; 2011.
101. Wang J, Lu Z, Au JL. Protection against chemotherapy-induced alopecia. Pharm Res. 2006;23:2505–14.
102. Trueb RM. Chemotherapy-induced anagen effluvium: diffuse or patterned? Dermatology. 2007;215:1–2.
103. Bleiker TO, Nicolaou N, Traulsen J, Hutchinson PE. 'Atrophic telogen effluvium' from cytotoxic drugs and a randomized controlled trial to investigate the possible protective effect of pretreatment with a topical vitamin D analogue in humans. Br J Dermatol. 2005;153: 103–12.
104. Trueb RM. Chemotherapy-induced alopecia. Curr Opin Support Palliat Care. 2010;4:281–4.
105. Tallon B, Blanchard E, Goldberg LJ. Permanent chemotherapy-induced alopecia: case report and review of the literature. J Am Acad Dermatol. 2010;63:333–6.
106. Hendrix S, Handjiski B, Peters EM, Paus R. A guide to assessing damage response pathways of the hair follicle: lessons from cyclophosphamide-induced alopecia in mice. J Invest Dermatol. 2005;125:42–51.
107. Tran D, Sinclair RD, Schwarer AP, Chow CW. Permanent alopecia following chemotherapy and bone marrow transplantation. Australas J Dermatol. 2000;41:106–8.
108. Ito M, Kizawa K, Hamada K, Cotsarelis G. Hair follicle stem cells in the lower bulge form the secondary germ, a biochemically distinct but functionally equivalent progenitor cell population, at the termination of catagen. Differentiation. 2004;72:548–57.

Fibrosis

Lynne A. Murray and Matthew A. Sleeman

Abstract Tissue remodeling is a common pathology in many diseases. During the reparative processes of normal wound healing an increase in extracellular matrix (ECM) proteins provides barrier protection and ultimately restoration of the normal tissue architecture. However, during repeated injury or an initial abnormal response to the trauma, a loss of tissue elasticity and function ensues resulting in a chronic fibrotic response that compromises organ function. Fibrosis is a common occurrence during many types of cancer-related therapies. It can ensue as a direct consequence of the treatment such as systemic chemotherapy or locally after site-directed radiation. Fibrotic remodeling can also occur indirectly to cancer treatments, for example in the case of chemotherapy inducing oral mucositis, which then transitions into fibrosis. This Chapter will describe the cellular pathways associated with the development and maintenance of fibrosis and how they arise in response to cancer treatments.

The lung is particularly sensitive to the development of fibrosis and will be the main focus of this Chapter. The lung has a high turnover of matrix and this balance of generation and destruction is tightly regulated by a balance of pro-fibrotic and proteolytic mediators. Once this balance is altered, detrimental ECM deposition occurs. Moreover, the lung contains a large pool of resident matrix producing cells, fibroblasts and myofibroblasts, which, when activated, can generate ECM and pro-fibrotic and pro-inflammatory mediators rapidly, adding to the fibrotic remodeling environment. This excess matrix deposition stiffens the lung tissue which ultimately impacts adequate gas exchange. Due to various cancer treatments promoting lung fibrosis, current therapeutic management is to stop cancer treatment to limit detrimental effects on lung function as there is currently no approved anti-fibrotic therapeutics in the United States. Pirfenidone, a small molecule inhibitor of TGFβ synthesis, has shown promising results in patients with idiopathic pulmonary fibrosis

L.A. Murray (✉) • M.A. Sleeman
Respiratory, Inflammation and Autoimmunity,
MedImmune Limited, Cambridge CB21 6GH, UK
e-mail: murrayl@medimmune.com

S.T. Sonis and D.M. Keefe (eds.), *Pathobiology of Cancer Regimen-Related Toxicities*,
DOI 10.1007/978-1-4614-5438-0_9, © Springer Science+Business Media New York 2013

(IPF) [1]. However, the clinical benefit was only a reduction in the rate of lung function decline [1]. Therefore understanding the mechanisms that underlie fibrosis may allow for anti-fibrotic drugs to be given alongside cancer treatments to ensure completion of the chemotherapy or radiotherapy course.

Interestingly, some of the key pro-fibrotic mechanisms are phenomena described in tumorigenesis. Thus targeting these pathways may not only limit the development of aberrant fibrosis, but also target underlying tumor establishment and development. In addition, certain therapeutic, such as antimetabolites, alkylating agents, and corticosteroids, agents exert anti-fibrotic effects due to inhibiting normal re-epithelialization and fibroproliferation, which can inhibit normal wound healing (reviewed in [2]). However, this Chapter will focus on the various cellular pathways associated with promoting fibrosis and how they may be targeted clinically.

Mechanisms of Fibrosis: Wound Healing Gone Awry

Fibrosis is the deposition of excess extracellular matrix, which in the case of dermal wound healing serves to seal off any exposed underlying tissue by restoring barrier function, thus minimizing further injury and allowing for wound sterilization. Although the matrix material, including fibronectin and collagen, rapidly provides protection against further injury, the collagen is often dense and poorly organized and is eventually remodeled; resulting in a more normal tissue structure and appearance. In normal wound healing, myofibroblasts, fibroblast-like cells with markers of smooth muscle cells, that typically reside below the epithelium, sequentially perpetuate and then dampen inflammation via the secretion of chemokines, cytokines, arachidonic acid metabolites, and protease inhibitors [3]. When activated, they express cell surface adhesion molecules and extracellular matrixes (ECM) allowing specific interactions with inflammatory cells, including lymphocytes, mast cells, and neutrophils. At sites of normal wound healing, once sufficient ECM has been deposited and remodeled, fibroblasts and myofibroblasts undergo apoptosis [4, 5]. This serves to limit the excessive deposition of ECM and also dampen the pro-inflammatory and pro-fibrotic milieu. However, for reasons still unclear, myofibroblasts persist in fibrotic diseases, such as idiopathic pulmonary fibrosis and cancer therapy-induced comorbidities.

Fibroblasts and Myofibroblasts

Fibroblasts play a myriad of important roles in normal tissue function. These cells are typically spindle-shaped cells with an oval flat nucleus found in the interstitial spaces of organs. They reside in highly complex multicellular environments, usually closely apposed to the epithelium or endothelium. Fibroblasts are metabolically active cells, capable of synthesizing, secreting and degrading ECM components, including collagens, proteoglycans, tenascin, laminin, and fibronectin. These cells continually synthesize ECM proteins although the amount they secrete is tightly regulated. For example, up to 90% of all procollagen molecules are degraded intracellularly prior to secretion, depending on tissue and age. Further, fibroblasts generate matrix metalloproteinases (MMPs) and their inhibitors, tissue inhibitor of metalloproteinases (TIMPs) thus controlling tissue architecture and matrix turnover rates. As well as being the primary source of ECM proteins, which, in addition to providing a scaffold for cells, play key roles in determining cell phenotype and function, fibroblasts contribute to injury responses in both the initiation and resolution phases. Fibroblasts are activated by numerous signals such as mechanical forces imposed during tissue function, ECM interactions, and local cell death.

MMPs exert proteolytic activities on various proteins including many ECM components and are thus central to ECM formation [6]. They have been shown to be elevated in asthma and COPD (reviewed in [6]); as well as in IPF where MMP1, 2, and 9 were co-localized to the epithelium surrounding fibrotic lesions, whereas increased TIMP2 was also observed suggesting that the MMP activity may be inhibited and the fibrotic region not degraded [7]. Another function of MMPs is to activate growth factors and chemokines, thus potentially promoting the fibrotic and inflammatory milieu [8–10].

In the lung, fibroblasts are found in the greatest number in the subepithelial layer of the conducting airways and the interstitium of the lung parenchyma. Here they are in a prime location to interact with the epithelial and endothelial cells. These interactions are likely important in disease settings and will be further discussed in this Chapter. Furthermore, a large number of mediators, produced by many different cell types, and proteases of the coagulation cascade are known to promote fibroblast proliferation, collagen synthesis, migration, and differentiation.

Myofibroblasts express α-smooth muscle actin (α-SMA) and have contractile and secretory properties that are central to controlling tissue architecture [11]. They express a panel of markers that have been correlated with site of origin. For example, myofibroblasts found in the peripheral and subpleural regions of fibrosis express α-SMA, vimentin, and desmin, whereas cells found in other regions of the lung do not express desmin [12]. In vitro, fibroblast-to-myofibroblast trans-differentiation can be induced by TGFβ1 and it has been hypothesized that TGFβ1 found locally at sites of fibrosis trans-differentiates resident fibroblasts into myofibroblasts [13, 14].

Altered Cellular Phenotypes in Fibrosis

There are numerous cell phenotype differences observed during fibrosis (outlined in Fig. 1). As the key hallmark of fibrosis is aberrant matrix deposition, the fibroblast and myofibroblast have been extensively researched. It has been proposed that the contractile properties of myofibroblasts are central to wound healing by limiting the amount of exposed wound area [15]. However, the sustained abnormal presence of contractile myofibroblasts in tissue may cause a retraction of normally elastic tissue resulting in loss of function. Fibroblasts, when activated, produce various ECM components and can also differentiate into myofibroblasts, which are more

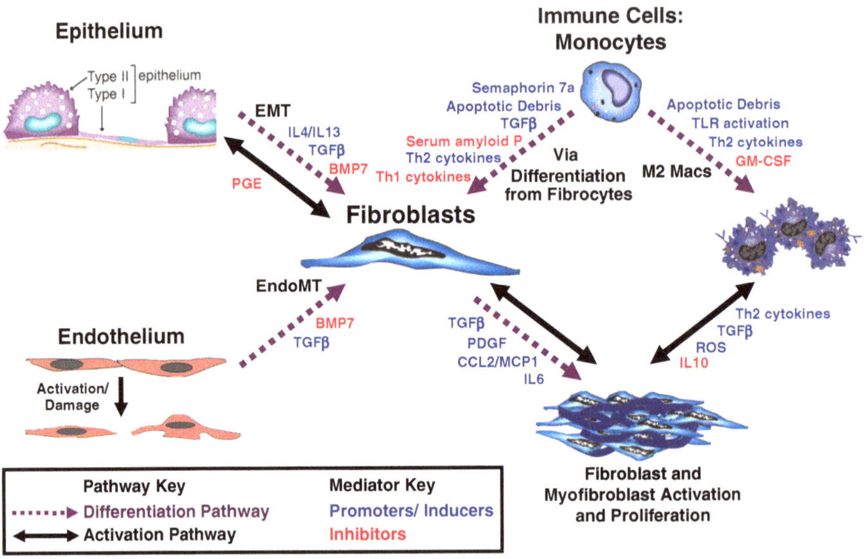

Fig. 1 Key cell types associated with fibrosis and the differentiation pathways and mediators that promote and inhibit these pathways. Fibrosis is characterized by excess extracellular matrix deposition from collagen-producing cells including fibroblasts and myofibroblasts. These cells are resident within tissues, generating matrix to support tissue function and homeostasis. However, during fibrosis, fibroblasts become activated, proliferate at greater rates than unactivated cells and are resistant to apoptosis. There is also an increased number of myofibroblasts, which are normally at low levels in non-fibrotic tissue. Monocytes activate fibroblasts through either differentiating to fibrocytes or alternatively activated M2 macrophages. Both of these monocyte-derived cells are found at elevated numbers in the circulation of patients with chronic lung fibrosis and at sites of tissue remodeling. Both epithelial cell and endothelial cell injury is observed at sites of fibrotic tissue remodeling. Moreover, the transition of epithelial cells to mesenchymal cells and endothelial cells to mesenchymal cells has been hypothesized as sources for increased mesenchymal cell numbers in fibrosis. Multiple mediators have been shown to promote fibrosis and to direct cells to a pro-fibrotic or anti-fibrotic phenotype, with the cellular environment also helping to direct cell phenotype. Many of the mediators shown to induce pro-fibrotic responses are pleiotropic, directing more than cellular response on multiple cell types

contractile and more readily synthesize ECM, resulting in a greater loss of tissue elasticity.

Another key cell type demonstrated to promote fibrosis is the epithelial cell. These cells stimulated with the prototypic growth factor TGF-β result in a transition of cell phenotype away from the resident epithelial cell and towards a motile, ECM-producing mesenchymal cell. This process is referred to as epithelial to mesenchymal transition (EMT) and is understood to contribute to lung, liver, and kidney fibrosis [16–18].

In chronic lung diseases, the dual expression of epithelial and mesenchymal markers in the same cells has lead investigators to postulate EMT as a mechanism resulting in more ECM-generating cells in the lung [19, 20]. The differentiation of airway epithelial cells has been previously described, for example type I pneumocytes transitioning into goblet cells [17, 21–23]. However, the switching of an epithelial cell into a phenotype that moves beyond the original cell's embryonic lineage, that is, into mesenchymal cells, has only recently been hypothesized as a driving factor in fibrosis [16, 24, 25].

Information of phenotypic differences in fibroblasts found in fibrotic regions vs. those located in normal tissue has been garnered from in vitro studies using fibroblasts isolated from fibrotic lung or skin diseases. Fibroblasts isolated from these environments are phenotypically different to non-fibrotic fibroblasts [26–28]. Furthermore, fibroblasts from a pro-fibrotic environment exhibit both altered responsiveness to growth factors and enhanced chemokine receptor expression, which has also been observed in murine models of remodeling [29]. Taken together, these studies suggest a distinct heterogeneity in fibroblast function and phenotype in fibrotic tissue.

As remodeling of the lung is associated with the accumulation of fibroblasts and myofibroblasts, understanding the derivation of these cells is critical to our understanding of disease processes. Current thinking suggests that there may be multiple pathways through which fibroblasts and myofibroblasts are derived [3]. These include proliferation of resident fibroblasts. Also, plasticity and transitioning of cells such as epithelial or endothelial cells into a mesenchymal phenotype has been reported at sites of active fibrosis. Further, emerging data highlight a significant role of circulating, bone marrow-derived cells called fibrocytes at remodeling lung architecture [30]. The same magnitude of involvement of fibrocytes in experimental models of renal fibrosis has not been observed [31]. Therefore the relative contributions of different cell types may be dependent on both the nature of the trigger and the organ site of fibrosis.

Epithelial to Mesenchymal Transition

The dynamic process by which epithelial cells undergo phenotypic transition to fully differentiated motile mesenchymal cells, such as fibroblasts and myofibroblasts, is called EMT. This process occurs normally during early fetal development where there appears to be a seamless plasticity between epithelial and mesenchymal cells [32]. Furthermore, this phenomenon is well accepted in cancer as a key mechanism that supports tumor metastases [33]. However, the process of EMT in development and cancer differ greatly in that, unlike developmental EMT, the tumorigenic EMT process is poorly regulated [34].

During EMT, the downregulation of epithelial and tight junction proteins is associated with a concomitant increase in mesenchymal cell markers [25, 35]. Epithelial cells exposed to TGFβ1, alone or in combination with other growth factors such as epidermal growth factor (EGF), begin the process of EMT by increasing the expression of MMPs that promote basement membrane degradation and cell detachment [33, 36]. The cells also undergo cytoskeletal changes, as well as altered expression of surface molecules which allow for the migration and transition of these cells to a mesenchymal phenotype [36, 37]. The majority of the work evaluating EMT has been performed in vitro; however the full extent of this pathogenic pathway in vivo is currently being evaluated. In animal models of kidney fibrosis, it has been estimated that up to 20% of the fibroblasts found in the fibrotic lesions were derived from the epithelium through EMT [16, 37–39]. Therefore, if an anti-fibrotic therapy has a mechanism which inhibits EMT, another clinical benefit may be a reduction in tumor metastases.

A transition of endothelial cells to mesenchymal cells (EndoMT) has been described in a murine model of cardiac fibrosis [40]. Further, this process may have a role in systemic sclerosis and the downstream fibrotic pathologies observed in the lung, heart, and kidney, as endothelial alterations are observed [41]. In lung fibrosis models, bleomycin promotes alterations of the endothelium, increasing cell surface adhesion molecules and thus promoting inflammation [42]. Murine models of systemic sclerosis via the systemic delivery of bleomycin, either by osmotic pump delivery of bleomycin or repeated subcutaneous injections, results in lung fibrosis that originates from the vascular region [43–46]. Moreover, bleomycin can induce direct toxicity to endothelial cells in vitro [47]. EndoMT has not been assessed in these models. However, endothelial changes in response to chemotherapeutics have been observed [48, 49], therefore this may be one pro-fibrotic mechanism induced by chemotherapy.

Linking Inflammation and Apoptosis to Fibrosis

Monocyte-Derived Cells: Alternatively Activated Macrophages

Although the key hallmark of fibrosis is excessive ECM deposition, there is often altered leukocyte infiltration and phenotype associated with fibrotic lesions. Monocyte/macrophages exhibit great degrees of plasticity depending on the tissue location, as well as any underlying changes in the inflammatory milieu due to disease states [50, 51]. In fibrotic tissue, tissue resident macrophages demonstrate enhanced pro-fibrotic mediator production, indicative of an alternatively activated or M2 macrophage phenotype. M2 macrophages express elevated levels of scavenger receptors such as macrophage scavenger receptor (MSR) and mannose receptor C (MRC/CD206) [52, 53]. In the lungs of patients with chronic lung disease such as IPF and COPD, the M2 macrophage phenotype predominates [54–56]. In healthy tissue, M2 macrophages act as sentinels of the immune system, phagocytosing debris, and pathogens in a non-phlogistic mechanism, in that downstream inflammation is limited and the inflammatory process is attenuated [57]. However under fibrotic conditions, M2 macrophages are defective in phagocytosis and do not dampen the inflammatory response [52]. These macrophages are capable of synthesizing pro-fibrotic mediators, which supports their role in wound healing, but this cell type is inefficient at supporting host defence [52]. Further, in granulomatous conditions, such as schistomiasis liver disease, M2 macrophages generate matrix to encapsulate the pathogen, and inhibition of M2 macrophage development is lethal in murine models of granuloma liver disease. Along with chronic immunosuppression associated with cancer treatment, a predominance of M2 macrophages at sites of fibrosis may also enhance the susceptibility of patients to mucosally initiated infections.

Tumor-associated macrophages (TAMs) have similar markers to M2 macrophages [58], and also produce pro-fibrotic mediators such as TGFβ and can inhibit M1 macrophage function [59, 60]. However, unlike M2 macrophages, TAMs express IFNγ-inducible genes, suggesting that these cells are a unique subset to classical M1 and M2 macrophages [59]. Solid tumors contain levels of chemokines that are potent monocytes recruiting and activating chemokines such as CCL2/MCP1 and CCL5/RANTES [61–64]. Experimental mouse models of cancer have demonstrated both positive and negative effects of TAMs; with TAMs either promoting tumor growth or enhancing tumor clearance depending on the extent of TAM infiltration [65]. It would be interesting to determine the phenotype of TAMs in tumors being treated with chemotherapy and/or radiotherapy where the surrounding tissue exhibits fibrotic remodeling. This may elucidate any link for this cell type at promoting therapy-related fibrosis and highlight alternative approaches in tumor therapy management.

Monocyte-Derived Cells: Fibrocytes

Another monocyte-derived cell type associated with fibrosis is the fibrocyte. These cells are a population of circulating bone marrow-derived cells that express surface markers for leukocytes and mesenchymal cells and are capable of differentiating into myofibroblasts [66]. They have been observed at sites of active fibrosis and increased circulating numbers correlate with mortality in IPF [67]. They are induced by pro-fibrotic mediators such as TGFβ1 and Th2 cytokines [68]; and express a variety of markers including leukocyte markers (CD45, CD34), mesenchymal markers (collagen I, fibronectin), and chemokine receptors (CCR3, CCR5, CCR7, and CXCR4) [69]. Human and mouse studies have demonstrated that fibrocytes from peripheral blood migrate to skin wound chambers [69–71] and bronchial mucosa after antigen challenge [72].

Furthermore, these cells have been reported in disease states with fibrotic pathologies including hypertrophic scars, asthma, and IPF [72–75]. Fibrocytes are pleiotropic and may contribute to fibrogenesis by directly producing collagen, as well as inflammatory cytokines, hematopoietic growth factors, and chemokines [74–78]. While it was originally thought that fibrocytes promote fibrosis through production of ECM components, it is becoming increasingly hypothesized that their primary role in tissue remodeling may be through secretion of soluble factors.

Assessing circulating primary cells from IPF patients, we have determined an elevation of CD163+CD14+ cells and M2-associated soluble mediators in the circulation which was more pronounced in progressive IPF patients, suggestive of an overall elevated M2 background in these patients [79]. Due to the similarities in histopathology of bleomycin-induced or radiation-induced lung fibrosis, it can be hypothesized that these patients will also have increased circulating fibrocytes and M2-prone monocyte/macrophages. We have also shown that peripheral blood monocytes from patients with scleroderma-related lung disease display a pro-fibrotic phenotype, characterized by increased CD163 expression and CCL18 expression [80].

Studies of bleomycin-induced fibrosis have assessed either M2 macrophage or fibrocyte number [79, 81]. It is increasingly recognized that there is some overlap between these cell subsets in terms of both function and markers [81]. However, although both fibrocytes and M2 macrophages can be derived from monocytes, they are not completely redundant in function. Recently, using a lung-specific TGFβ1 over-expression model of lung fibrosis, we determined that depletion of lung monocyte/macrophages using liposomal clodronate reduced collagen accumulation, but this had no effect on the TGFβ-induced fibrocyte recruitment [82]. Therefore, for novel therapeutic approaches such as cell depletion or specific targeting, the M2 macrophage may be a more compelling target for chronic lung remodeling.

Treatment-Related Fibrosis

Chemotherapy-Induced Fibrosis

Bleomycin is a common chemotherapeutic agent which alkylates DNA, thus interfering with cell replication. It is commonly used for squamous cell carcinomas and testicular cancer. However, one of the major side effects of bleomycin treatment is bleomycin-induced pneumonitis (BIP) and progression to lung fibrosis [83, 84]. BIP is believed to start gradually during the course of treatment but can continue unabated up to 6 months post-therapy [85]. Pneumonitis is typically seen in approximately 46% of patients being treated with bleomycin [86]. The clinical presentation and diagnosis of BIP is challenging, particularly as it resembles other conditions such as pulmonary metastases or pneumonia. Even the use of radiography, CT scanning or transbronchial biopsy can prove inconclusive. However, the most common lesions associated with BIP are metaplasia of the bronchial epithelium, inflammation in the alveolar spaces with oedema, collagen deposition, and evidence of fibrosis. As bleomycin appears to have a similar histological phenotype in rodents, the exact etiology of BIP in patients is largely based on our understanding of rodent models [86]. Moreover, the effects on promoting lung fibrosis are so robust that bleomycin is also used as the gold-standard preclinical model of assessing lung fibrosis in rodents [87]. One single direct intratracheal administration of bleomycin to rodents is sufficient to mediate significant lung remodeling that has features observed in the lungs of patients with the fatal chronic lung disease, IPF. Repeated systemic bleomycin administration to mice results in systemic fibrosis, with the most pronounced effects in the lung. It has been hypothesized that the reason for the localization of fibrosis to the lung is a pulmonary deficiency of the enzyme bleomycin hydrolase, which breaks down bleomycin. Absence of this enzyme in the lung will allow pulmonary accumulation of the chemotherapeutic, thus potentiating the damaging effect(s) of the agent.

In vivo preclinical bleomycin-induced lung fibrosis has been reduced with various therapeutic compounds, such as inhibiting pro-inflammatory cell recruitment and activation with anti-chemokine antibodies [75–78], inhibiting pro-fibrotic growth factor activation [88] or attenuating growth factor receptor signaling [89–91]. Clinically, patients that are receiving the drug for chemotherapy are closely monitored for lung function alterations to determine if the drug is promoting lung fibrosis [92]. If the patient's lung function declines, bleomycin therapy is either stopped or changed to an alternative chemotherapeutic. In experimental models of bleomycin-induced lung fibrosis, the remodeling can spontaneously resolve [93]. Whether this resolution occurs in patients after cessation of bleomycin therapy is currently unknown.

One of the initial responses to intratracheal bleomycin is epithelial cell damage and apoptosis [94]. The mechanism of action of many chemotherapeutics is to also promote cellular apoptosis. Recent evidence has demonstrated the pro-inflammatory and pro-fibrotic effects of apoptotic and necrotic debris. Intratracheal administration of apoptotic cells is sufficient to cause fibrosis in the murine lung [95]. Moreover, these DAMPs (damage-associated molecular patterns) utilize receptors and recognition pathways on innate immune cells used for host defence such as the toll-like receptors (TLRs) and carbohydrate receptors [96, 97]. There is a significant degree of cell death in the murine model of lung-specific $TGF\beta_1$ over-expression, a model used to elucidate lung fibrosis mechanisms and explore the anti-fibrotic potential of therapeutic agents [98, 99]. Here, inhibition of apoptosis in this mouse model inhibits collagen deposition and fibrocyte accumulation [100–102].

Another class of chemotherapeutic drugs that has been associated with inducing lung fibrosis as a side effect is EGF receptor inhibitors including gefitinib and erlotinib. These and drugs are used for the treatment of various tumors including head and neck cancer, and lung cancer. However, the risk benefit ratio is such that these drugs are administered, but with close monitoring of lung function [103, 104]. The mechanism by which inhibition of EGFR promotes lung injury and interstitial lung disease (ILD) is unknown and also controversial, as gefitinib has been shown to reduce lung fibrosis in experimental models of disease [105, 106]. However, increased lung injury and an alteration in Th1-associated mediators have been hypothesized as a downstream pro-fibrotic consequence to EGFR inhibition [107, 108].

Radiation-Induced Fibrosis

Another key tool in the oncologists' armamentarium to treat cancer is the use of radiotherapy, whether it is used preoperatively to slow tumor growth and reduce volume, or postoperatively to help reduce the viability of remaining tumor cells that could not be removed during debulking of the tumor. External beam radiation is the most commonly administered radiotherapy and this can result in damage to the normal, healthy tissue that the beam passes through prior to or after reaching the target tumor. However, the beam of radiation delivered through the skin often results in unwanted acute and chronic tissue damage in the associated surrounding tissues. The acute damage results in inflammation and erythema, that may promote desquamation of the epithelial layer, but these early changes are often reversible and correlate with the dose of radiation [109]. These changes can, however, interfere with any normal wound healing processes. Cell sloughing occurs due to the growth-arresting effects of radiation. Then once the dermis has turned over, approximately 2–3 months [110, 111], after radiation has first been administered, the new skin layer at the site of where the radiation was administered is thinner and more susceptible to injury and tearing.

The initial changes observed at the dermal level are also extended more than 6 months after completion of the course of radiotherapy. There are histological changes in cell appearance and matrix deposition, with alterations in skin thickness and elasticity due to aberrations in fibrotic remodeling [109, 112]. Blood vessels supplying the skin also become fibrotic, causing local tissue hypoxia, which exacerbates the altered phenotype of resident cell types including fibroblasts [113].

Although preoperative radiotherapy may decrease tumor volume size and is thought to promote tumor containment prior to surgery, the radiation may cause healthy tissue to become fibrotic and become more involved in the normal tumor margins, thus requiring greater tissue removal [114]. Postoperative radiation has fewer side effects [109]; however, as described above, radiation can alter wound healing. Therefore, in order to administer the radiotherapy whilst it can still target tumorigenesis, there is a small window in which radiation can be effectively used therapeutically [115–118]. Therefore, if there was a therapeutic which could be co-administered with radiation to minimize the non-tumor responses, this would allow for a more effective course of therapy.

Oral Mucositis and Subsequent Fibrosis

A common side effect of chemotherapy and radiotherapy is oral mucositis, with mucositis occurring to some degree in more than one-third of patients receiving anti-neoplastic drug therapy [119]. This is an important side effect of cancer treatment as the downstream fibrotic reparative processes following oral mucositis can cause significant loss of tissue function, requiring parenteral feeding, and have significant effects on morbidity. Oral mucositis is caused directly by the cytotoxic effect of chemotherapeutic agents and indirectly by sustained neutropenia. In the case of bone marrow transplantation, mucositis is a serious transplant-related morbidity. Moderate to severe mucositis occurs in virtually all patients who receive radiation therapy for tumors of the head and neck. It typically begins at cumulative exposures of 15 Gy and then worsens when total doses exceed 60 Gy [119–122]. The ensuing injury significantly impairs quality of life and can hamper the scheduled course of therapy, thus reducing efficacy of treatment thereby affecting long-term survival [121]. Further, in patients with immunosuppression including granulocytopenia, the ulcerations that accompany mucositis are common portals of entry for indigenous oral bacteria, often leading to sepsis or bacteremia [121].

Oral mucositis is scored using the established WHO criteria [123]. Using this scale, a score of ≥ 3 represents severe mucositis that affects a patient's quality of life, with 70% of patients requiring feeding tubes to maintain adequate nutrition [123]. In approximately 35% of patients with grade 3–4 mucositis, subsequent rounds of radiotherapy may be delayed or prevented, leading to less successful clinical therapy [123]. Lower grades of mucositis may also significantly affect patient quality of life. Grade 1 is associated with pain that is sufficiently severe to require opioid analgesia, whereas grade 2 patients have difficulty in swallowing due to ulcerations in the mouth and throat, which, if severe, may necessitate total parenteral nutrition and rehydration. More than 90% of patients develop grade 4 mucositis as manifested by severe erosive mucosal damage and the requirement for parenteral narcotics [123].

The pathogenesis of mucositis is multifactorial, involving the interaction of oral mucosal epithelial cells, endothelial cells, connective tissue, and the submucosal infiltrate [124]. At the cellular level there are four interdependent phases: (1) an inflammatory/vascular phase, (2) an epithelial phase, (3) an ulcerative/bacteriological phase, and finally (4) a healing/resolution phase [125]. In the oral mucosa, directed radiation therapy induces apoptosis of the epithelial barrier which initiates an inflammatory cascade, characterized by NFκB activation [126]. The generation of pro-inflammatory cytokines such as tumor necrosis factor (TNF)-α and interleukin (IL)-1β [124], and pro-fibrotic mediators such as transforming growth factor (TGF)-β [124], result in the breakdown of the oral mucosa, causing formation of ulcerative lesions. Early expression of mitogen-activated protein kinase (MAPK) is observed within 8 h post-radiation [127]. MAPK may function with NF-κB as a coactivator of a large number of genes involved in mucosal injury [128]. The activation of NF-κB due to radiation results in up-regulation of a variety of genes leading to increased production of pro-inflammatory cytokines including TNF-α, IL-1β and IL-6 [129]. Targeted radiation has been shown to increase cyclooxygenase-2 expression in submucosal tissues, and this parallels the development of ulcerative mucositis [130]. The ensuing fibrotic response serves to provide adequate local wound healing, however this can be over-exuberant and limit tissue elasticity and function.

Radiation directly damages cellular DNA, which results in cellular apoptosis and also inhibits homeostatic cell renewal. The debris generated during apoptosis promotes inflammation via numerous cascades including Toll-like receptor signaling and NFκB activation. Non-phlogistic clearance of these pro-inflammatory signals by macrophages acts to switch off the immune response, thus limiting damage to host tissue [131]. Inadequate clearance by macrophages has been shown to contribute directly to a number of acute and chronic diseases, such as atherosclerosis, autoimmunity, and fibrosis [132]. This may be due to the over-riding mechanism in which clearing apoptotic cells and debris dampens inflammation resulting in resolution [133]. Further, inhibiting apoptosis has been shown to reduce oral mucositis [134].

Apart from direct epithelial and endothelial apoptosis to the chemotherapeutics and radiation, leukocyte activation can also enhance lung damage and downstream fibrosis. The lung is susceptible to the development of injury following systemic exposure or cellular trauma due in part to the small size of the microcapillary network. Leukocytes often slow their transit time and have to deform in order to pass through the vasculature. Also, if these cells are activated and expressing elevated levels of adhesion molecules, there is an increase in the likelihood of adhesion to the damaged endothelium, resulting in local leukocyte activation.

The standard therapy for mucositis is predominantly palliative, including application of topical analgesics such as lidocaine and/or systemic administration of opiates and antibiotics. However treating the underlying fibrotic responses to prevent the excessive deposition of collagen yet not attenuate normal wound healing responses is an attractive therapeutic strategy. Also, some of the pathophysiologies associated with tumorigenesis such as EMT and enhanced cell survival, which have been discussed in this Chapter, are also associated with fibrosis. Therefore, clinical compounds that can limit fibrosis may also have some beneficial antitumor effects. Chemotherapy and radiotherapy provide the early signals that trigger fibrosis. One way to limit lung fibrosis is to stop treatment. However, interfering with the course of therapy may allow the cancer to further develop.

Using an animal model of radiation-induced mucositis and fibrosis as a surrogate for head and neck cancer radiotherapy, it has been shown that site-directed radiation to the hamster cheek pouch results in cellular apoptosis, mucositis, and subsequently fibrosis (Fig. 2). In this model, administering a serum protein, serum amyloid P (SAP, PTX-2), which reduces lung models of fibrosis [79, 82], inhibits both the mucositic response and the fibrotic pathology [135]. PTX-2 binds to externalized chromatin and opsonizes dying cells [136, 137]. The alternate binding face of PTX-2 is recognized by the Fcγ receptor on phagocytes and allows clearance of cellular debris [138]. Therefore it is hypothesized that PTX-2 clears apoptotic and/or necrotic debris from sites of inflammation thus preventing the development of detrimental tissue fibrosis.

Therapeutically Targeting Fibrosis During Cancer Treatment

Various potential therapeutics have been assessed clinically for oral mucositis including recombinant IL-11 [139], granulocyte–macrophage colony-stimulating factor (GM-CSF) [140, 141], G-CSF [142], and TGFβ3 [143]. Currently, the only approved treatment for mucositis is palifermin (Kepivance, Biovitrum, Stockholm, Sweden), which is approved for the treatment of oral mucositis in patients undergoing bone marrow ablation for transplant. Palifermin is a recombinant keratinocyte growth factor (KGF) and shares a common underlying mechanism with the other experimental agents tested clinically, in that it promotes epithelial cell proliferation. This therefore provides a denser epithelial barrier, to protect the vulnerable mucosal tissue.

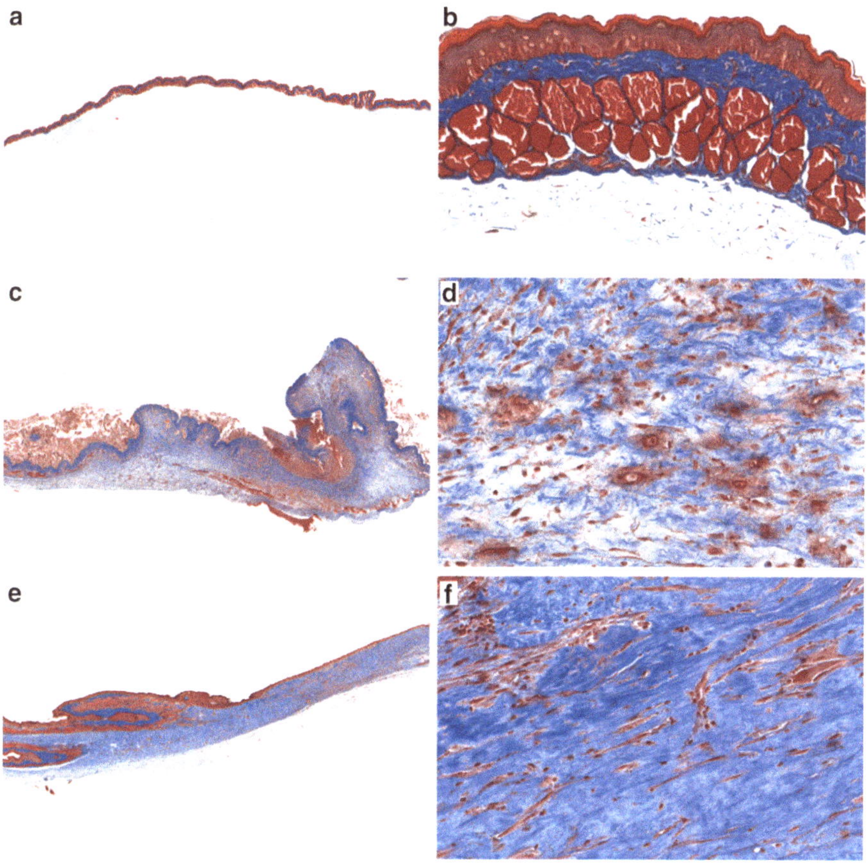

Fig. 2 Radiation-induced fibrosis in a hamster cheek pouch model of mucositis. Representative histopathology of a hamster cheek pouch that has been exposed to a single dose of radiation (40 Gy) at Day 0 and histologically analyzed at Day 16 (**c** and **d**) or Day 28 (**e** and **f**). *Top panels* (**a** and **b**) are non-irradiated hamster cheek pouch controls. Collagen deposition is visualized using Masson's Trichrome staining, with collagen appearing *blue/magenta* in color

During a small trial assessing the clinical efficacy of recombinant human IL11 (rhIL11) in patients receiving cyclophosphamide, cytoxan, and allogenic stem cells for hematological malignancies, the trial was stopped due to excessive early mortality and significant toxicity, making it impossible to determine whether rhIL-11 given in this schedule can reduce the rate of graft vs. host disease (GVHD) [139]. More promising results were observed in a trial assessing recombinant GM-CSF mouth rinses [140]. In that study, the duration of severe mucositis appeared to be reduced, with 60% of the GM-CSF mouth rinse patients having severe mucositis for less than 9 days, but only 28% of the controls had severe mucositis for less than 9 days. There was also a reduction in the number of patients that had severe mucositis for more than 20 days with GM-CSF compared to control (10% vs. 34%). In another study using topical oral r-metHuG-CSF (filgrastim; rG-CSF) in high-grade lymphoma patients, there was a trend towards a reduction in the severity of methotrexate-induced mucositis compared to placebo [142]. Interestingly, the number of days in hospital was reduced significantly in the G-CSF group. In conclusion, topical oral G-CSF mouth rinses may be beneficial to reduce oral mucositis [142]. TGFβ3 mouthwashes have been reported to be well tolerated in breast cancer patients undergoing chemotherapy; however, the efficacy has not been reported [143]. All of these published trials have presented encouraging data, however they only consist of small numbers of patients on specific chemotherapeutic regimens; therefore more controlled, double-blind clinical trials are now required [140].

In conclusion, understanding common mechanisms promoting fibrosis may result in a paradigm shift in the approach to drug development. The current focus in drug discovery is targeting the disease when exploring novel therapeutic candidates. However, taking a pathway approach to tailoring therapies, where underlying pathologies become the initial focus, may allow for repositioning of therapeutics already approved for non-lung indications. Reducing fibrosis that occurs following chemotherapy and/or radiotherapy may enhance subsequent clinical efficacy of the on-going treatment regimen, by increasing the likelihood of completing the prescribed course of treatment. Also, determining if subsets of cancer patients that have increased therapy-related fibrosis, such as head and neck cancer patients, have more of an TAM/M2 macrophage-driven disease, and treating these patients with an M2-specific therapeutic may allow for a greater probability of clinical success. EMT is another common mechanism between cancer and fibrosis and understanding the significance of this pathology in promoting fibrosis may allow for tailoring of EMT-directed cancer therapies to inhibit both primary tumorigenesis and prevent fibrosis that may be induced from concomitant radiotherapy. There are numerous novel therapies being examined clinically for IPF and determining the impact that these drugs have on preventing and potentially reversing lung fibrosis may elucidate novel therapies that can be co-administered with cancer treatments.

References

1. Noble PW, et al. Pirfenidone in patients with idiopathic pulmonary fibrosis (CAPACITY): two randomised trials. Lancet. 2011;377(9779):1760–9.
2. Payne WG, et al. Wound healing in patients with cancer. Eplasty. 2008;8:e9.
3. Hinz B, et al. The myofibroblast: one function, multiple origins. Am J Pathol. 2007;170(6): 1807–16.
4. Desmouliere A, et al. Apoptosis mediates the decrease in cellularity during the transition between granulation tissue and scar. Am J Pathol. 1995;146(1):56–66.
5. Darby I, Skalli O, Gabbiani G. Alpha-smooth muscle actin is transiently expressed by myofibroblasts during experimental wound healing. Lab Invest. 1990;63(1):21–9.
6. Demedts IK, et al. Matrix metalloproteinases in asthma and COPD. Curr Opin Pharmacol. 2005;5(3):257–63.
7. Fukuda Y, et al. Localization of matrix metalloproteinases-1, -2, and -9 and tissue inhibitor of metalloproteinase-2 in interstitial lung diseases. Lab Invest. 1998;78(6):687–98.

8. Li Q, et al. Matrilysin shedding of syndecan-1 regulates chemokine mobilization and transepithelial efflux of neutrophils in acute lung injury. Cell. 2002;111(5):635–46.
9. Ludbrook SB, et al. The integrin alphavbeta3 is a receptor for the latency-associated peptides of transforming growth factors beta1 and beta3. Biochem J. 2003;369(Pt 2):311–8.
10. Sheppard D. Roles of alphav integrins in vascular biology and pulmonary pathology. Curr Opin Cell Biol. 2004;16(5):552–7.
11. Ohta K, et al. Immunohistochemical identification and characterization of smooth muscle-like cells in idiopathic pulmonary fibrosis. Am J Respir Crit Care Med. 1995;152(5 Pt 1): 1659–65.
12. Zhang K, et al. Myofibroblasts and their role in lung collagen gene expression during pulmonary fibrosis. A combined immunohistochemical and in situ hybridization study. Am J Pathol. 1994;145(1):114–25.
13. Chambers RC, et al. Global expression profiling of fibroblast responses to transforming growth factor-beta1 reveals the induction of inhibitor of differentiation-1 and provides evidence of smooth muscle cell phenotypic switching. Am J Pathol. 2003;162(2):533–46.
14. Desmouliere A, et al. Transforming growth factor-beta 1 induces alpha-smooth muscle actin expression in granulation tissue myofibroblasts and in quiescent and growing cultured fibroblasts. J Cell Biol. 1993;122(1):103–11.
15. Gabbiani G, Badonnel MC. Contractile events during inflammation. Agents Actions. 1976;6(1–3):277–80.
16. Kalluri R, Neilson EG. Epithelial-mesenchymal transition and its implications for fibrosis. J Clin Invest. 2003;112(12):1776–84.
17. Kim KK, et al. Alveolar epithelial cell mesenchymal transition develops in vivo during pulmonary fibrosis and is regulated by the extracellular matrix. Proc Natl Acad Sci U S A. 2006;103(35):13180–5.
18. Zeisberg M, et al. Fibroblasts derive from hepatocytes in liver fibrosis via epithelial to mesenchymal transition. J Biol Chem. 2007;282(32):23337–47.
19. Zavadil J, Bottinger EP. TGF-beta and epithelial-to-mesenchymal transitions. Oncogene. 2005;24(37):5764–74.
20. Zavadil J, et al. Integration of TGF-beta/Smad and Jagged1/Notch signalling in epithelial-to-mesenchymal transition. EMBO J. 2004;23(5):1155–65.
21. Borok Z, et al. Modulation of t1alpha expression with alveolar epithelial cell phenotype in vitro. Am J Physiol. 1998;275(1 Pt 1):L155–64.
22. Danto SI, et al. Reversible transdifferentiation of alveolar epithelial cells. Am J Respir Cell Mol Biol. 1995;12(5):497–502.
23. Torday JS, Torres E, Rehan VK. The role of fibroblast transdifferentiation in lung epithelial cell proliferation, differentiation, and repair in vitro. Pediatr Pathol Mol Med. 2003;22(3): 189–207.
24. Selgas R, Jimenez-Heffernan J, Lopez-Cabrera M. On the epithelial-mesenchymal transition of mesothelial cells. Kidney Int. 2004;66(2):866–7; author reply 867.
25. Valcourt U, et al. TGF-beta and the Smad signaling pathway support transcriptomic reprogramming during epithelial-mesenchymal cell transition. Mol Biol Cell. 2005;16(4): 1987–2002.
26. Moodley YP, et al. Fibroblasts isolated from normal lungs and those with idiopathic pulmonary fibrosis differ in interleukin-6/gp130-mediated cell signaling and proliferation. Am J Pathol. 2003;163(1):345–54.
27. Scaffidi AK, et al. Oncostatin M stimulates proliferation, induces collagen production and inhibits apoptosis of human lung fibroblasts. Br J Pharmacol. 2002;136(5):793–801.
28. Keerthisingam CB, et al. Cyclooxygenase-2 deficiency results in a loss of the anti-proliferative response to transforming growth factor-beta in human fibrotic lung fibroblasts and promotes bleomycin-induced pulmonary fibrosis in mice. Am J Pathol. 2001;158(4):1411–22.
29. Hogaboam CM, et al. Differential monocyte chemoattractant protein-1 and chemokine receptor 2 expression by murine lung fibroblasts derived from Th1- and Th2-type pulmonary granuloma models. J Immunol. 1999;163(4):2193–201.

30. Quan TE, Cowper SE, Bucala R. The role of circulating fibrocytes in fibrosis. Curr Rheumatol Rep. 2006;8(2):145–50.
31. Castano AP, et al. Serum amyloid P inhibits fibrosis through Fc gamma R-dependent monocyte-macrophage regulation in vivo. Sci Transl Med. 2009;1(5):5–13.
32. Thiery JP. Epithelial-mesenchymal transitions in development and pathologies. Curr Opin Cell Biol. 2003;15(6):740–6.
33. Kalluri R, Weinberg RA. The basics of epithelial-mesenchymal transition. J Clin Invest. 2009;119(6):1420–8.
34. Dasari V, et al. Epithelial-mesenchymal transition in lung cancer: is tobacco the "smoking gun"? Am J Respir Cell Mol Biol. 2006;35(1):3–9.
35. Willis BC, Borok Z. TGF-{beta}-induced EMT: mechanisms and implications for fibrotic lung disease. Am J Physiol Lung Cell Mol Physiol. 2007;293:525–34.
36. Grunert S, Jechlinger M, Beug H. Diverse cellular and molecular mechanisms contribute to epithelial plasticity and metastasis. Nat Rev Mol Cell Biol. 2003;4(8):657–65.
37. Iwano M, et al. Evidence that fibroblasts derive from epithelium during tissue fibrosis. J Clin Invest. 2002;110(3):341–50.
38. Okada H, et al. Early role of Fsp1 in epithelial-mesenchymal transformation. Am J Physiol. 1997;273(4 Pt 2):F563–74.
39. Saika S, et al. Smad3 signaling is required for epithelial-mesenchymal transition of lens epithelium after injury. Am J Pathol. 2004;164(2):651–63.
40. Zeisberg EM, et al. Endothelial-to-mesenchymal transition contributes to cardiac fibrosis. Nat Med. 2007;13(8):952–61.
41. Varga J, Abraham D. Systemic sclerosis: a prototypic multisystem fibrotic disorder. J Clin Invest. 2007;117(3):557–67.
42. Azuma A, et al. Role of E-selectin in bleomycin induced lung fibrosis in mice. Thorax. 2000;55(2):147–52.
43. Lakos G, et al. Targeted disruption of TGF-beta/Smad3 signaling modulates skin fibrosis in a mouse model of scleroderma. Am J Pathol. 2004;165(1):203–17.
44. Yamamoto T, et al. Anti-sclerotic effect of transforming growth factor-beta antibody in a mouse model of bleomycin-induced scleroderma. Clin Immunol. 1999;92(1):6–13.
45. Yamamoto T, et al. Animal model of sclerotic skin I: local injections of bleomycin induce sclerotic skin mimicking scleroderma. J Invest Dermatol. 1999;112(4):456–62.
46. Murray LA, et al. BMP-7 does not protect against bleomycin-induced lung or skin fibrosis. PLoS One. 2008;3(12):e4039.
47. Martin 2nd WJ, Kachel DL. Bleomycin-induced pulmonary endothelial cell injury: evidence for the role of iron-catalyzed toxic oxygen-derived species. J Lab Clin Med. 1987;110(2):153–8.
48. Lazo JS. Endothelial injury caused by antineoplastic agents. Biochem Pharmacol. 1986;35(12):1919–23.
49. Mailloux A, et al. Anticancer drugs induce necrosis of human endothelial cells involving both oncosis and apoptosis. Eur J Cell Biol. 2001;80(6):442–9.
50. Geissmann F, et al. Development of monocytes, macrophages, and dendritic cells. Science. 2012;327(5966):656–61.
51. Martinez FO, Helming L, Gordon S. Alternative activation of macrophages: an immunologic functional perspective. Annu Rev Immunol. 2009;27:451–83.
52. Gordon S. Alternative activation of macrophages. Nat Rev Immunol. 2003;3(1):23–35.
53. Tiemessen MM, et al. CD4+CD25+Foxp3+ regulatory T cells induce alternative activation of human monocytes/macrophages. Proc Natl Acad Sci U S A. 2007;104(49):19446–51.
54. Hancock A, et al. Production of interleukin 13 by alveolar macrophages from normal and fibrotic lung. Am J Respir Cell Mol Biol. 1998;18(1):60–5.
55. Zhou Y, et al. Alterations in adenosine metabolism and signaling in patients with chronic obstructive pulmonary disease and idiopathic pulmonary fibrosis. PLoS One. 2010;5(2):e9224.
56. Barnes PJ. Alveolar macrophages as orchestrators of COPD. COPD. 2004;1(1):59–70.

57. Savill J. Apoptosis in resolution of inflammation. J Leukoc Biol. 1997;61(4):375–80.
58. Mantovani A, et al. Macrophage polarization: tumor-associated macrophages as a paradigm for polarized M2 mononuclear phagocytes. Trends Immunol. 2002;23(11):549–55.
59. Biswas SK, et al. A distinct and unique transcriptional program expressed by tumor-associated macrophages (defective NF-kappaB and enhanced IRF-3/STAT1 activation). Blood. 2006;107(5):2112–22.
60. Saccani A, et al. p50 nuclear factor-kappaB overexpression in tumor-associated macrophages inhibits M1 inflammatory responses and antitumor resistance. Cancer Res. 2006;66(23):11432–40.
61. Mantovani A, et al. The origin and function of tumor-associated macrophages. Immunol Today. 1992;13(7):265–70.
62. Gu L, Tseng SC, Rollins BJ. Monocyte chemoattractant protein-1. Chem Immunol. 1999;72:7–29.
63. Bottazzi B, et al. Regulation of the macrophage content of neoplasms by chemoattractants. Science. 1983;220(4593):210–2.
64. Azenshtein E, et al. The CC chemokine RANTES in breast carcinoma progression: regulation of expression and potential mechanisms of promalignant activity. Cancer Res. 2002;62(4):1093–102.
65. Nesbit M, et al. Low-level monocyte chemoattractant protein-1 stimulation of monocytes leads to tumor formation in nontumorigenic melanoma cells. J Immunol. 2001;166(11):6483–90.
66. Lama VN, Phan SH. The extrapulmonary origin of fibroblasts: stem/progenitor cells and beyond. Proc Am Thorac Soc. 2006;3(4):373–6.
67. Moeller A, et al. Circulating fibrocytes are an indicator of poor prognosis in idiopathic pulmonary fibrosis. Am J Respir Crit Care Med. 2009;179(7):588–94.
68. Shao DD, et al. Pivotal advance: Th-1 cytokines inhibit, and Th-2 cytokines promote fibrocyte differentiation. J Leukoc Biol. 2008;83(6):1323–33.
69. Abe R, et al. Peripheral blood fibrocytes: differentiation pathway and migration to wound sites. J Immunol. 2001;166(12):7556–62.
70. Bucala R, et al. Circulating fibrocytes define a new leukocyte subpopulation that mediates tissue repair. Mol Med. 1994;1(1):71–81.
71. Chesney J, et al. Regulated production of type I collagen and inflammatory cytokines by peripheral blood fibrocytes. J Immunol. 1998;160(1):419–25.
72. Schmidt M, et al. Identification of circulating fibrocytes as precursors of bronchial myofibroblasts in asthma. J Immunol. 2003;171(1):380–9.
73. Abe S, et al. Cells derived from the circulation contribute to the repair of lung injury. Am J Respir Crit Care Med. 2004;170(11):1158–63.
74. Moore BB, et al. CCR2-mediated recruitment of fibrocytes to the alveolar space after fibrotic injury. Am J Pathol. 2005;166(3):675–84.
75. Moore BB, et al. The role of CCL12 in the recruitment of fibrocytes and lung fibrosis. Am J Respir Cell Mol Biol. 2006;35(2):175–81.
76. Hashimoto N, et al. Bone marrow-derived progenitor cells in pulmonary fibrosis. J Clin Invest. 2004;113(2):243–52.
77. Phillips RJ, et al. Circulating fibrocytes traffic to the lungs in response to CXCL12 and mediate fibrosis. J Clin Invest. 2004;114(3):438–46.
78. Murray LA, et al. Hyper-responsiveness of IPF/UIP fibroblasts: interplay between TGFbeta1, IL-13 and CCL2. Int J Biochem Cell Biol. 2008;40(10):2174–82.
79. Murray LA, et al. Serum amyloid P therapeutically attenuates murine bleomycin-induced pulmonary fibrosis via its effects on macrophages. PLoS One. 2010;5(3):e9683.
80. Mathai SK, et al. Circulating monocytes from systemic sclerosis patients with interstitial lung disease show an enhanced profibrotic phenotype. Lab Invest. 2010;90(6):812–23.
81. Pilling D, et al. Reduction of bleomycin-induced pulmonary fibrosis by serum amyloid P. J Immunol. 2007;179(6):4035–44.

82. Murray LA, et al. TGF-beta driven lung fibrosis is macrophage dependent and blocked by serum amyloid P. Int J Biochem Cell Biol. 2011;43(1):154–62.
83. Evans WE, et al. Clinical pharmacology of bleomycin and cisplatin. Head Neck Surg. 1981;4(2):98–110.
84. Krakoff IH, et al. Clinical pharmacologic and therapeutic studies of bleomycin given by continuous infusion. Cancer. 1977;40(5):2027–37.
85. White DA, Stover DE. Severe bleomycin-induced pneumonitis. Clinical features and response to corticosteroids. Chest. 1984;86(5):723–8.
86. Sleijfer S. Bleomycin-induced pneumonitis. Chest. 2001;120(2):617–24.
87. Moeller A, et al. The bleomycin animal model: a useful tool to investigate treatment options for idiopathic pulmonary fibrosis? Int J Biochem Cell Biol. 2008;40(3):362–82.
88. Horan GS, et al. Partial inhibition of integrin alpha(v)beta6 prevents pulmonary fibrosis without exacerbating inflammation. Am J Respir Crit Care Med. 2008;177(1):56–65.
89. Higashiyama H, et al. Inhibition of activin receptor-like kinase 5 attenuates bleomycin-induced pulmonary fibrosis. Exp Mol Pathol. 2007;83(1):39–46.
90. Aono Y, et al. Imatinib as a novel antifibrotic agent in bleomycin-induced pulmonary fibrosis in mice. Am J Respir Crit Care Med. 2005;171(11):1279–85.
91. Daniels CE, et al. Imatinib mesylate inhibits the profibrogenic activity of TGF-beta and prevents bleomycin-mediated lung fibrosis. J Clin Invest. 2004;114(9):1308–16.
92. Comis RL. Bleomycin pulmonary toxicity: current status and future directions. Semin Oncol. 1992;19(2 Suppl 5):64–70.
93. Chung MP, et al. Role of repeated lung injury and genetic background in bleomycin-induced fibrosis. Am J Respir Cell Mol Biol. 2003;29(3 Pt 1):375–80.
94. Hay J, Shahzeidi S, Laurent G. Mechanisms of bleomycin-induced lung damage. Arch Toxicol. 1991;65(2):81–94.
95. Wang L, et al. Induction of secondary apoptosis, inflammation, and lung fibrosis after intratracheal instillation of apoptotic cells in rats. Am J Physiol Lung Cell Mol Physiol. 2006;290(4):L695–702.
96. Zhang Q, et al. Circulating mitochondrial DAMPs cause inflammatory responses to injury. Nature. 2010;464(7285):104–7.
97. Murray LA, et al. Deleterious role of TLR3 during hyperoxia-induced acute lung injury. Am J Respir Crit Care Med. 2008;178(12):1227–37.
98. Kang HR, et al. Transforming growth factor (TGF)-beta1 stimulates pulmonary fibrosis and inflammation via a Bax-dependent, bid-activated pathway that involves matrix metalloproteinase-12. J Biol Chem. 2007;282(10):7723–32.
99. Kang HR, et al. Semaphorin 7A plays a critical role in TGF-beta1-induced pulmonary fibrosis. J Exp Med. 2007;204(5):1083–93.
100. Kuwano K, et al. Attenuation of bleomycin-induced pneumopathy in mice by a caspase inhibitor. Am J Physiol Lung Cell Mol Physiol. 2001;280(2):L316–25.
101. Lee CG, et al. Early growth response gene 1-mediated apoptosis is essential for transforming growth factor beta1-induced pulmonary fibrosis. J Exp Med. 2004;200(3):377–89.
102. Peng X, et al. Local apoptosis promotes collagen production by monocyte-derived cells in transforming growth factor beta1-induced lung fibrosis. Fibrogenesis Tissue Repair. 2011;4(1):12.
103. Danson S, et al. Interstitial lung disease in lung cancer: separating disease progression from treatment effects. Drug Saf. 2005;28(2):103–13.
104. Tsuboi M, Le Chevalier T. Interstitial lung disease in patients with non-small-cell lung cancer treated with epidermal growth factor receptor inhibitors. Med Oncol. 2006;23(2):161–70.
105. Ishii Y, Fujimoto S, Fukuda T. Gefitinib prevents bleomycin-induced lung fibrosis in mice. Am J Respir Crit Care Med. 2006;174(5):550–6.
106. Wang C, et al. Effects of gefitinib on radiation-induced lung injury in mice. J Nippon Med Sch. 2008;75(2):96–105.
107. Inomata S, et al. Acute lung injury as an adverse event of gefitinib. Anticancer Drugs. 2004;15(5):461–7.

108. Hoshino T, et al. Role of proinflammatory cytokines IL-18 and IL-1beta in bleomycin-induced lung injury in humans and mice. Am J Respir Cell Mol Biol. 2009;41(6):661–70.
109. Bernstein EF, et al. Biology of chronic radiation effect on tissues and wound healing. Clin Plast Surg. 1993;20(3):435–53.
110. Hopewell JW. The skin: its structure and response to ionizing radiation. Int J Radiat Biol. 1990;57(4):751–73.
111. Halprin KM. Epidermal "turnover time"—a re-examination. Br J Dermatol. 1972;86(1):14–9.
112. Rudolph R, Arganese T, Woodward M. The ultrastructure and etiology of chronic radiotherapy damage in human skin. Ann Plast Surg. 1982;9(4):282–92.
113. Moore MJ. The effect of radiation on connective tissue. Otolaryngol Clin North Am. 1984;17(2):389–99.
114. Wang Z, Qiu W, Mendenhall WM. Influence of radiation therapy on reconstructive flaps after radical resection of head and neck cancer. Int J Oral Maxillofac Surg. 2003;32(1):35–8.
115. Mendelsohn FA, et al. Wound care after radiation therapy. Adv Skin Wound Care. 2002;15(5):216–24.
116. Hom DB, Adams GL, Monyak D. Irradiated soft tissue and its management. Otolaryngol Clin North Am. 1995;28(5):1003–19.
117. Vikram B. Importance of the time interval between surgery and postoperative radiation therapy in the combined management of head & neck cancer. Int J Radiat Oncol Biol Phys. 1979;5(10):1837–40.
118. Isaacs Jr JH, et al. Postoperative radiation of open head and neck wounds. Laryngoscope. 1987;97(3 Pt 1):267–70.
119. Plevova P. Prevention and treatment of chemotherapy- and radiotherapy-induced oral mucositis: a review. Oral Oncol. 1999;35(5):453–70.
120. Knox JJ, Puodziunas AL, Feld R. Chemotherapy-induced oral mucositis. Prevention and management. Drugs Aging. 2000;17(4):257–67.
121. Peterson DE. Research advances in oral mucositis. Curr Opin Oncol. 1999;11(4):261–6.
122. Sonis ST, et al. Oral mucositis and the clinical and economic outcomes of hematopoietic stem-cell transplantation. J Clin Oncol. 2001;19(8):2201–5.
123. Sonis ST, et al. Perspectives on cancer therapy-induced mucosal injury: pathogenesis, measurement, epidemiology, and consequences for patients. Cancer. 2004;100 Suppl 9: 1995–2025.
124. Sonis ST, et al. Defining mechanisms of action of interleukin-11 on the progression of radiation-induced oral mucositis in hamsters. Oral Oncol. 2000;36(4):373–81.
125. Sonis ST. Mucositis as a biological process: a new hypothesis for the development of chemotherapy-induced stomatotoxicity. Oral Oncol. 1998;34(1):39–43.
126. Lee SW, et al. Effect of epidermal growth factor against radiotherapy-induced oral mucositis in rats. Int J Radiat Oncol Biol Phys. 2007;67(4):1172–8.
127. Sonis ST, et al. The gene expression sequence of radiated mucosa in an animal mucositis model. Cell Prolif. 2002;35 Suppl 1:93–102.
128. Shi Q, et al. Cooperation between transcription factor AP-1 and NF-kappaB in the induction of interleukin-8 in human pancreatic adenocarcinoma cells by hypoxia. J Interferon Cytokine Res. 1999;19(12):1363–71.
129. Logan RM, et al. The role of pro-inflammatory cytokines in cancer treatment-induced alimentary tract mucositis: pathobiology, animal models and cytotoxic drugs. Cancer Treat Rev. 2007;33(5):448–60.
130. Sonis ST, et al. The relationship between mucosal cyclooxygenase-2 (COX-2) expression and experimental radiation-induced mucositis. Oral Oncol. 2004;40(2):170–6.
131. Gregory CD, Devitt A. The macrophage and the apoptotic cell: an innate immune interaction viewed simplistically? Immunology. 2004;113(1):1–14.
132. Weigert A, Jennewein C, Brune B. The liaison between apoptotic cells and macrophages—the end programs the beginning. Biol Chem. 2009;390(5–6):379–90.

133. Serhan CN, Savill J. Resolution of inflammation: the beginning programs the end. Nat Immunol. 2005;6(12):1191–7.
134. Hwang D, et al. Effects of ceramide inhibition on experimental radiation-induced oral mucositis. Oral Surg Oral Med Oral Pathol Oral Radiol Endod. 2005;100(3):321–9.
135. Murray LA, et al. Serum amyloid P ameliorates radiation-induced oral mucositis and fibrosis. Fibrogenesis Tissue Repair. 2010;3:11.
136. Hintner H, et al. Amyloid P component binds to keratin bodies in human skin and to isolated keratin filament aggregates in vitro. J Invest Dermatol. 1988;91(1):22–8.
137. Breathnach SM, et al. Serum amyloid P component binds to cell nuclei in vitro and to in vivo deposits of extracellular chromatin in systemic lupus erythematosus. J Exp Med. 1989;170(4):1433–8.
138. Mold C, Gresham HD, Du Clos TW. Serum amyloid P component and C-reactive protein mediate phagocytosis through murine Fc gamma Rs. J Immunol. 2001;166(2):1200–5.
139. Antin JH, et al. A phase I/II double-blind, placebo-controlled study of recombinant human interleukin-11 for mucositis and acute GVHD prevention in allogeneic stem cell transplantation. Bone Marrow Transplant. 2002;29(5):373–7.
140. Bez C, et al. GM-CSF mouthrinses in the treatment of severe oral mucositis: a pilot study. Oral Surg Oral Med Oral Pathol Oral Radiol Endod. 1999;88(3):311–5.
141. Chi KH, et al. Effect of granulocyte-macrophage colony-stimulating factor on oral mucositis in head and neck cancer patients after cisplatin, fluorouracil, and leucovorin chemotherapy. J Clin Oncol. 1995;13(10):2620–8.
142. Karthaus M, et al. Effect of topical oral G-CSF on oral mucositis: a randomised placebo-controlled trial. Bone Marrow Transplant. 1998;22(8):781–5.
143. Wymenga AN, et al. Phase I study of transforming growth factor-beta3 mouthwashes for prevention of chemotherapy-induced mucositis. Clin Cancer Res. 1999;5(6):1363–8.

Myelosuppression

David C. Dale

Abstract Myelosuppression is the principal dose-limiting toxicity of cancer chemotherapy. Neutropenia, febrile neutropenia, and infections are the most important consequences. Cancer chemotherapy also causes anemia, thrombocytopenia, and immunodeficiency, but these are usually less severe complications. The risk and severity of myelosuppression is dependent on several factors including the tumor type (location, grade, and metastases), comorbidities, and the treatment regimen. Based on randomized clinical trials and broader clinical experience, several organizations have developed and regularly update treatment recommendations for optimal regimens and guidelines regarding strategies to avoid adverse effects including the risk of myelosuppression. Myeloid growth factors are now widely used to prevent chemotherapy-associated febrile neutropenia and facilitate the treatment of cancer.

Introduction

Modern chemotherapy originated with the German chemical industry and war gas programs of World War I [1, 2]. These gases were alkylating agents, and they caused death due to their gastrointestinal and hematopoiesis toxicities. Because nitrogen mustard caused marrow and lymphoid hypoplasia, it was first used to treat lymphocytic leukemia and Hodgkin's disease in 1943 [3]. After World War II, folic acid antagonists were introduced for the treatment of childhood lymphoblastic leukemia, again with dose-limiting myelotoxicity [4, 5]. The advent of combination chemotherapy for Hodgkin's disease in the 1960s was pivotal for the development of modern cancer therapy, but the myelotoxic drugs in combination had even more

D.C. Dale, M.D. (✉)
Department of Medicine, University of Washington, Seattle, WA, USA
e-mail: dcdale@uw.edu

S.T. Sonis and D.M. Keefe (eds.), *Pathobiology of Cancer Regimen-Related Toxicities*,
DOI 10.1007/978-1-4614-5438-0_10,

complex effects on hematopoiesis than single drugs, and these effects limited the application of drugs in combinations as potentially curative therapies [2, 6]. Myelosuppression still is the major dose limiting toxicity for most cancer chemotherapy regimens [7].

This chapter provides a foundation for understanding and managing the myelotoxicity associated with cancer chemotherapy. Readers are referred to specific information sources—drug formularies, cancer treatment guidelines, and reviews of specific agents for detailed therapeutic recommendations [8, 9].

The Hematopoietic System

All hematopoietic cells are derived from pluripotent stem cells with the property to differentiate along several specific pathways to form blood cells and also the capacity to self-renew [10]. In general, the hematopoietic progenitor cells in the bone marrow have a relatively high proliferative rate, making them particularly vulnerable to agents affecting cell division. However, the earliest progenitors, the cells classified as stem cells, have lower proliferative indices than their progeny, thereby protecting the stem cells, or some of them, from drugs affecting proliferating cells, at least when administered at usual therapeutic doses. Thus, marrow recovery after each course of chemotherapy depends on the presence of adequate numbers of residual stem and progenitor cells capable of proliferating, a marrow environment which will support their proliferation and survival, and the availability of growth factors to stimulate marrow recovery. Within the hematopoietic lineages, myeloid cells, particularly progenitor cells of the neutrophil lineage, are the most proliferative population, and therefore the most susceptible to the toxicities of cancer chemotherapy.

To understand the clinical consequences of myelotoxicity, it is important to appreciate the kinetics of the system—the supply of progenitors, the rates of new cell formation, and the dynamics of their deployment. In addition the physiological regulation of the hematopoietic system by cytokines and growth factors is vitally important. The principal targets for the myelotoxic effects of chemotherapy are shown in Table 1.

An interesting feature of the hematopoietic system is its capacity to expand locally and at distant sites through release of stem cells and other early progenitors from the bone marrow into the blood [11]. The mobilization of progenitor cell by this mechanism occurs after chemotherapy, cytokine administration, and other mobilizing agents [12, 13]. The released cells travel through the blood to favorable sites for their survival and proliferation in the bone marrow, spleen, and occasionally other tissues. There they may serve to progressively expand the body's mass of hematopoietic tissues. The stem and progenitor cells released into the blood after chemotherapy and other agents can also be "harvested" for hematopoietic transplantation. Clinical studies have shown that sequential doses of chemotherapy drive down the number of early progenitors that can be mobilized in the blood in a

Table 1 Hematological effects of chemotherapy

Leukocytes	Cells primarily affected	Effects
Neutrophils	Promyelocytes Myelocytes	Cause moderate to severe neutropenia after 4–6 days with gradual recovery normally by 14–21 days
Monocytes	Promonocytes	Monocytopenia Tissue monocytes and macrophages not affected
Lymphocytes	T and B cells and precursors	Transient immunosuppression
Erythrocytes	Proerythrocytes	Transient reticulocytopenia Mild to moderate anemia
Platelets	Megakaryocytes	Transient mild thrombocytopenia

stepwise fashion [14, 15]. The cumulative effects of chemotherapy may be related to the progressive loss of the early progenitor cells that are necessary for hematopoietic recovery.

Myelopoiesis

Neutrophils

Neutrophils are derived from hematopoietic stem cells through an orderly process of cell division and differentiation. The earliest recognizable cells of the lineage are promyelocytes; cells with cytoplasmic granules containing in myeloperoxidase, proteases, and several other enzymes. These cells are in active DNA synthesis and vulnerable to most myelotoxic agents. Their progeny become myelocytes, the most numerous proliferating neutrophil precursors, and therefore the population of cells most severely affected by myelotoxic agents. Metamyelocytes, bands, and mature marrow neutrophils are the succeeding categories of precursors for blood neutrophils; all cells in this category are nondividing cells. Conceptually, it is the loss of cells from the myelocyte compartment which has the greatest impact on the depth of the nadir of blood neutrophils after administration of chemotherapy and it is the timing for recovery and maturation of cells from this population which largely determines the duration of neutropenia.

It is estimated that about 10–14 days are required for cells of the neutrophil lineage to transition from early myeloid progenitors to mature neutrophils in the blood [15, 16]. Normally about 65% of promyelocytes are in cell cycle and DNA synthesis [16]. With stress or in response to myeloid growth factors the proportion of the progenitors in cycle increases; but many factors determine the exact proportion of cells in cycle and the proliferative rate for various cell populations. These include the specific mechanism of injury to the hematopoietic cells, the health status of the capillaries and microenvironment in the marrow, and the capacity of the patient to generate cytokines and growth factors.

Beyond the promyelocyte–myelocyte stage, developing cells of the neutrophil lineage enter a non-proliferative, maturational stage. The time that cells spend in this maturational stage, the marrow transit time, can be measured by injection of tritiated thymidine or infusion of deuterium-labeled glucose to label a population of myelocytes and then determining the time required for labeled cell to enter the blood. Many studies have shown that the marrow transit time for neutrophils is normally about 6 days [16–18]. Infections or administration of growth factors shorten the marrow transit time. For example, administration of daily granulocyte colony-stimulating factor (G-CSF) reduces the marrow transit time to about 3 days [19] and granulocyte macrophage colony-stimulating factor (GM-CSF) reduces the transit time to about 4.5 days [20].

Maintenance of the supply of neutrophils in the blood requires constant production by the marrow. This is necessary because the half-life of neutrophils is very short, the blood half-life is only about 6–10 h and the tissue life span is probably even shorter [21]. For this reason, the supply must be replaced continually, and any interruption in production soon leads to a fall in blood counts. The short life span and high turnover rate necessitate a high proliferative rate for the neutrophil precursor pool and the very high proportion of marrow cells being neutrophil progenitors. Normally about half of the hematopoietic cells in the bone marrow are neutrophils and their progenitors.

During the maturation stage the developing neutrophil's nuclear chromatin condenses and the cell develops the densely staining, lobular nucleus which characterizes the mature blood neutrophil. During this period the cell also acquires secondary and tertiary granules, develops a generous supply of cytoplasmic glycogen and adds the cell surface receptors to regulate the trafficking of the cells. The maturing neutrophils in bone marrow serve as a reserve population, cells which can be mobilized into the blood in response to infections and inflammatory signals. The size of this population, the marrow neutrophil reserves, can be determined with careful bone marrow sectioning and ferrokinetic studies, but these methods are very tedious [18]. The reserves can be estimated more simply by administration of agents such as endotoxin, corticosteroids, or G-CSF [22, 23]. Normally these agents will double or triple the circulating neutrophil count within a few hours. Myelosuppressive agents reduce this response; as the counts fall the reserves diminish and then recover with marrow recovery [24].

Myelotoxic drugs affect progenitors through the myelocyte stage but generally do not directly affect the completion of development of neutrophils in the post-mitotic pool in the bone marrow. For this reason, there is usually a gradual fall of blood neutrophil counts after chemotherapy, with counts decreasing by about 30–50% each day until the absolute neutrophil count (ANC) reaches its nadir at about 6 days after a bolus of chemotherapy [25, 26]. If the ANC falls more rapidly, the chemotherapy agents are probably triggering apoptosis of maturing neutrophils or altering their capacity to traffic from the marrow to the blood.

Myelosuppressive chemotherapy has substantial secondary effects on myelopoiesis largely attributed to the natural release of the cytokines affecting neutrophil production. Readily observed effects are the shift to more bands and metamyelocytes in the blood and toxic granulation of blood neutrophils, due to enhanced staining of the neutrophil's primary granules. Most studies have shown that blood neutrophils from patients recovering from chemotherapy have transiently reduced functions including chemotaxis, chemiluminescence, and killing of bacteria [27–31].

Monocytes

Monocytes, tissue macrophages, and dendritic cells are derived from hematopoietic stem cells along a parallel pathway to neutrophils. Monocytes enter the blood earlier than neutrophils, and there is no reserve supply of these cells in the marrow. After most chemotherapy agents, monocyte counts are suppressed similar to neutrophil counts because of interrupted production. Probably because monocytes have a much shorter maturation period in the marrow, blood monocyte counts usually recover 1 or 3 days before neutrophils. Functions of monocytes and dendritic cells after chemotherapy are reported to be normal [32, 33]. The rise in monocyte counts after chemotherapy usually heralds the recovery of neutrophils and a failure of monocyte recovery signals protracted neutropenia [34].

Lymphopoiesis

Lymphocyte production and trafficking is a complex process involving the generation of cells in the thymus, spleen, lymph nodes, and bone marrow and the subsequent trafficking of the cells through the blood, lymphatic channels, and marrow. Although chemotherapy induces lymphocytopenia by interrupting cell proliferation, the reserve of nondividing cells is large enough to permit rapid recovery of blood counts in most circumstances. Chemotherapy, however, affects all lymphatic tissues and induces transient immunodeficiency, the severity increasing with the intensity and repetitiveness of the treatments. Chemotherapy-related immunodeficiency is an important factor leading to the susceptibility to infections, particularly with intensive regimens such as intensive treatments combined with hematopoietic transplantation [35].

Erythropoiesis

Erythroid cells are derived from the common hematopoietic stem cells and chemotherapy agents suppress the proliferation and production of red blood cells. Erythroid cell turnover, however, is far slower than neutrophil turnover because red blood cells have a longer life span, normally about 100 days [36]. For this reason a brief interruption of erythroid cell production is tolerated more readily than an interruption of production of neutrophils and monocytes. Although most cancer patients develop anemia during chemotherapy, the anemia is usually multifactorial. It is the consequence of cancer-related inflammation suppressing erythropoietin production, change in iron intake and utilization as well as anorexia, and blood loss for testing and for other reasons [37].

Thrombopoiesis

Platelets are also derived from the common hematopoietic stem cells [38]. Platelet production depends upon the formation and maturation of megakaryocytes and fragmentation of their cytoplasm to form individual platelets. Thrombopoiesis is regulated by thrombopoietin (TPO), a glycoprotein primarily produced in the liver. Platelet formation normally takes 7–10 days, the time from the proliferative stimulus until the mature platelet enters the blood. Platelet turnover rates are intermediate between those for neutrophils and red blood cells; the platelet life span is about 7–10 days. Interruption of platelet production by most chemotherapy agents causes thrombocytopenia, but it is less severe and less significant clinically than chemotherapy-induced neutropenia.

Cancer Therapies and Myelotoxicity

The pharmacology and myelosuppressive effects of anticancer drugs are quite complex. The complexity is increased by the use of combinations of drugs and multiple chemical drugs combined with biological agents. The following section briefly describes the effects of the principal categories of anticancer therapies on hematopoiesis. Table 2 outlines mechanisms and adverse effects, including myelosuppression for some specific agents.

Cycle-Phase Specific Agents

Cycle-phase specific agents include: antimetabolites, e.g., docetaxel, paclitaxel, etoposide, hydroxylurea, and irinotecan; folic acid antagonists, e.g., methotrexate; vinca alkaloids, e.g., vincristine and vinblastine; and some additional agents, e.g., bleomycin. The antimetabolites interfere with normal nucleic acid biosynthesis by competing in the normal pathways for incorporation of purines and pyrimidines or their precursors into DNA. They may also be incorporated into nucleic acids in place of the normal nucleotides. In brief, docetaxel and paclitaxel cause assembly of nonfunctional microtubules. Etoposide damages DNA and induces G 2-phase arrest. Irinotecan and its analogs induce lethal double strand breaks in DNA. The folic acid antagonists prevent the enzyme dihydrofolinic acid reductase from the necessary conversion of folic acid to tetrahydrofolinic acid. The vinca alkaloids and hydroxylurea

Table 2 Myelosuppressive drugs

	Mechanism	Adverse effects
Cycle-specific agents		
Methotrexate	Inhibits dihydrofolatereductase, depletes tetrahydrofolate coenzymes, blocks DNA synthesis	Myelosuppression Neutropenia Thrombocytopenia Mucositis Diarrhea Bleeding
Purine analogs	Block purine synthesis	Myelosuppression Neutropenia Thrombocytopenia Hepatotoxicity Neuropathy
Cytarabine	Analog of cytidine, inhibits DNA polymerase, terminates DNA strand elongation	Myelosuppression Neutropenia Thrombocytopenia Neurological Gastrointestinal Hepatic toxicities
Gemcitabine	Similar to cytarabine	Similar to cytarabine
Vinca alkaloids Vincristine Vinblastine	Binds to tubulin, inhibits formation of mitotic spindle	Neurotoxicity Neuropathy Myelosuppression
Topoisomerase I inhibitors Irinotecan Topotecan	Binds to topoisomerase I, prevents resealing of DNA single-strand breaks, activates apoptosis	Myelosuppression Neutropenia Mucositis Diarrhea
Topoisomerase II inhibitors Anthracyclines Doxorubicin Idarubicin Mitoxantrone	Interact with topoisomerase II, preventing resealing of broken DNA strands, activates apoptosis	Myelosuppression Neutropenia Nausea, vomiting Alopecia Cardiotoxicity
Cycle nonspecific agents		
Alkylating agents Cyclophosphamide Procarbazine Chlorambucil Melphalan	Generate carbonium, intermediate attacking specific sites on DNA	Myelosuppression Neutropenia Gastrointestinal, lung, cardiac toxicities Secondary malignancies
Other agents		
Bleomycin	Causes single and double-stranded breaks in DNA	Add to myelosuppression in combined chemotherapy regimens
Thalidomide	Antiangiogenic, immune modulation, inhibitor of growth factor and cytokine release	Neuropathy Fatigue Sedation Teratogenic

(continued)

Table 2 (continued)

	Mechanism	Adverse effects
Lenalidomide	Same as thalidomide	More myelosuppression than thalidomide
Retinoids ATRA—all-trans retinoic acid	Binds to a nuclear receptor formed by RARα and retinoid receptor X	Hyperleukocytosis with fever Pulmonary infiltrates, effusions Confusion Dry eyes, cheilitis Hepatic dysfunction
Molecular-targeted agents BCR-ABL tyrosine kinase inhibitors Imatinib Nilotinib Dasatinib	Inhibit BCR-ABL kinase, cKIT kinase and platelet growth factor kinase	GI toxicity Neutropenia Anemia
Bortezomib	Proteasome inhibitor, causes apoptosis of myeloma cells	Thrombocytopenia Neuropathy
Monoclonal antibodies		
Rituximab	Binds to CD20 on B lymphocytes, mechanism incompletely understood	Local reactions, anaphylaxis Late-onset neutropenia Immunosuppression
Bevacizumab	Antiangiogenic antibody, targets vascular endothelial growth factor	Neutropenia with combination chemotherapy
Trastuzumab	Antibody targeting HER2/neu receptor	Increases neutropenia in combined chemo-therapy regimens
Drugs under investigation		
Poly ADP ribose inhibitors	Inhibitor of PARP I—DNA repair enzyme	Increases neutropenia with combination chemotherapy

prevent cell division by interrupting normal microtubule assembly. The net effect of these agents is to block the pathway of DNA synthesis in the target cancer cells and also in all other proliferative cells, including proliferating hematopoietic cells. This interruption of proliferation usually causes gradual, dose-dependent onset of neutropenia, monocytopenia, lymphocytopenia, thrombocytopenia, and anemia.

Cycle-Phase Nonspecific Agents

Alkylating agents include a variety of drugs influencing cell division by substituting an alkyl group for hydrogen in many organic compounds. This results in covalent linkages of various groups causing depurination, base-pair miscoding, cross-linking of DNA strands and interference with DNA replication and RNA transcription. The alkylating agent group is large and includes several subgroups of compounds: nitrogen mustards (e.g., cyclophosphamide, chlorambucil), and melphalan; ethylenimines (e.g., thiotepa); alkylsulfonates (e.g., busulfan); triazenes (e.g., dacarbazine); and nitrosureas (e.g., carmusatine, streptozocin). Because of their more complex mechanisms of action, they are more likely to alter the functions of post-mitotic neutrophils, monocytes, and lymphocytes and cause more rapid or longer immunosuppression.

Other Agents

This category includes drugs affecting cell differentiation and survival through enzyme inhibition or triggering of apoptosis. Imatinib mesylate and its analogs are competitive inhibitors of Bcr-Abl tyrosine kinase; all are used in the treatment of chronic myeloid leukemia and several other types of tumors. These agents induce apoptosis of cells expressing this kinase and also cause myelosuppression, particularly neutropenia [39–41]. Trastuzumab is a recombinant monoclonal antibody binding to HER2, a transmembrane receptor protein over expressed on some cancer cells. This antibody also has myelosuppressive activity as used with combination chemotherapy [42–45]. Similarly rituximab is a monoclonal antibody directed against CD20 expressed selectively on B lymphocytes, but it can also cause late onset of neutropenia [46]. As "targeted therapies" for cancer therapy emerge, careful studies have generally shown diverse unexpected toxicities including myelosuppression, particularly when these agents are combined with other myelosuppressive agents. The precise mechanisms for the added toxicity of monoclonal antibodies and other newer agents are currently not well understood.

Combination Chemotherapy

Most malignancies, both hematological malignancies and solid tumors, are currently treated with combination chemotherapy or a combination of surgery, chemotherapy, and radiation therapy. Since the 1960s the evolution of treatment regimens has been both scientific and empiric. For most malignancies, years of observation are required to establish long-term benefits and adverse effects of a specific treatment plan.

For most chemotherapy regimens the risk and severity of myelosuppression, severe neutropenia and infections is established in clinical trials [47] (Table 3).

Data from these trials is then used by expert panels and organizations such as the National Comprehensive Cancer Network (NCCN) to develop treatment guidelines and specific recommendations for managing malignancies and the attendant risks and benefits of the tumor and stage-specific strategies, including the risk of

Table 3 Risk of febrile neutropenia with some common chemotherapy regimens

Cancer type	Regimen	Risk of febrile neutropenia
Breast	AC →docetaxel	5–25
	Docetaxel→ AC	40
	TAC	22–25
	AC	10–20
	CMF	0–32
Small cell lung cancer	ACE	24–57
	ICE	24
	Etoposide/carboplatin	10–20
	Paclitaxel/carboplatin	9
Non-small cell lung cancer	Docetaxel/carboplatin	26
	Etoposide/cisplatin	54
	Paclitaxel/cisplatin	5–11
	Gemcitabine/cisplatin	1–7
Non-Hodgkins lymphoma	CHOP-21	17–50
	R-CHOP-21	19
Colorectal cancer	5FU/leucovorin	1–15
	FOLFIRI	3–14
	FOLFOX	0–8

Data from Aapro et al. [47]
The abbreviations used are those in common usage in each indication: *5FU* 5-fluorouracil; *AC* doxorubicin/cyclophosphamide; *ACE* doxorubicin/cyclophosphamide/etoposide; *CHOP21* cyclophosphamide/doxorubicin/vincristine/prednisone; *CMF* cyclophosphamide/methotrexate/fluorouracil; *FOLFIRI* 5-FU/folinic acid/irinotecan; *FOLFOX* 5-FU/folinic acid/oxaliplatin; *ICE* ifosfamide/carboplatin/etoposide; *R-CHOP-21* rituximab/CHOP

myelosuppression, bleeding, and anemia. In addition, the NCCN publishes guidelines for the use of myeloid growth factors for the management of chemotherapy-induced neutropenia. These guidelines stratify the risk of severe myelosuppression and febrile neutropenia as high, intermediate, and low risk with supporting documentation. Readers are referred to continually updated resources of the American Society of Clinical Oncology, the American Cancer Society, and the European Society for Medical Oncology (ESMO) for current treatment recommendations.

Combination chemotherapy is the most effective current approach to the management of many forms of cancer, but knowing the benefits and risk for individual patients is often difficult, particularly if the patient has an unusually severe or aggressive cancer, has risk factors predisposing to greater toxicity, or the treatment requires relatively high doses of agents which are myelosuppressive. Efforts are underway to individualize treatment based on the demographic and genetic profiles of the patient, the tumor type and stage, and the toxicities of regimen [48–50]. Because of the number of variables involved, monitoring of patients receiving myelosuppressive therapies is currently the cornerstone of management. Monitoring includes regular contact with patient—via telephone and computer-based monitoring systems as well as traditional office and laboratory visits. Better understanding of the risk of myelosuppression depends on the analysis of carefully maintained records of treatments and clinical events for large groups of unselected patients, as well as detailed analysis of data from randomized trials.

Monitoring Patients for the Myelosuppressive Effects of Cancer

Monitoring involves observations of signs, symptoms, laboratory, and imaging results as well as cancer and non-cancer-related medications, dietary information, and sense of illness or wellness of the patient. In the early stages of development of new anticancer agents, serial complete blood cell counts (CBCs) and, in some cases, serial bone marrow examinations are performed to assess a new drug's myelosuppressive effects. However, for more complete and comprehensive understanding of myelosuppressive activity of chemotherapy agents and combinations of agents, it would be helpful if there were even more observations and more systematically collected data in these investigations. For example, the bone marrow is the target tissue for myelosuppression, but bone marrow data is not available and often the number of blood counts performed is limited. The occurrence of fever or febrile neutropenia is regularly used to infer the severity of myelotoxicity.

For combination chemotherapy, the risk of myelosuppression is usually characterized as "high risk," "intermediate risk," or "low risk" based on the estimated risk of the regimen causing neutropenia and/or febrile neutropenia. For regimens with low or intermediate risk, monitoring usually involves obtaining CBCs before their first and every subsequent course of chemotherapy. For intermediate and high risk regimens, CBCs generally are done more frequently, e.g., before each cycle of treatment and at the predicted neutrophil nadir. Clinical monitoring includes responsibilities of the patient and home care givers as well as nurses, physicians, and laboratory specialists. Prompt reporting of symptoms—fever, worsening anorexia, nausea, and vomiting—and tissue-specific symptoms—cough and shortness of breath, diarrhea, abdominal pain, and oral or skin lesions—are vitally important. Treatment guidelines generally recommend white blood cell and neutrophil counts to be in the normal range before administration of the first and subsequent courses of chemotherapy.

Innovations for Management of Myelotoxicity

For many years myelotoxicity of cancer chemotherapy was simply managed by limiting doses and extending intervals between administrations of these agents. After establishing patterns of toxicity in preclinical models and dose-response studies, randomized and non-randomized clinical studies provided the basis for drug approvals and use. The duration of neutropenia and time to recovery of blood neutrophil counts to the normal range was the usual basis for the administration of chemotherapy at 3–4 week intervals, the time interval usually required for recovery of myelopoiesis.

Table 4 FDA approved indications for hematopoietic growth factors to reduce myelosuppression

Name	Abbreviation	Use
Granulocyte colony-stimulating factor	G-CSF	Accelerate neutrophil recovery in patients receiving chemotherapy
Granulocyte-macrophage colony-stimulating factor	GM-CSF	Reduce risk of death due to infection in patients >55 years old undergoing induction chemotherapy for acute myeloid leukemia
Epoetin alpha/darbepoetin	EPO	Chemotherapy-induced anemia[a]
Oprelvekin	IL-II	Prevent severe thrombocytopenia and need for platelet transfusions after myelosuppressive chemotherapy

[a]Several restrictions are applicable—see text

Innovations in cancer management, particularly the possibility of administering chemotherapy on schedule and at shorter intervals than was previously deemed possible, came with the development of the hematopoietic growth factors, most importantly the development of G-CSF (Table 4).

Granulocyte Colony-Stimulating Factor

G-CSF is the glycoprotein that regulates the production and deployment of neutrophils. It is produced in the body by many types of cells. Infections, endotoxins, and inflammatory cytokines stimulate its production [51]. Mice unable to produce G-CSF, G-CSF knockout mice, and mice lacking the normal gene for its receptor have severe neutropenia, but are otherwise hematologically normal [52, 53]. Recombinant human G-CSF stimulates neutrophil production in normal subjects [19] and in cancer patients after chemotherapy [54, 55]. G-CSF both increases the number of cells produced by the marrow and accelerates new cell formation [19]. It also extends the life span of neutrophils through its anti-apoptotic effects [56]. The acceleration of new cell formation is particularly important because it decreases the duration of neutropenia and prevents fever and infections. GM-CSF-deficient mice have normal hematopoiesis [57]. GM-CSF can stimulate neutrophil production, but it is much less potent, has more adverse effects and is used clinically far less than G-CSF.

In response to infections or the administration of endotoxin, endogenous G-CSF levels rise promptly to very high levels, but both the natural and the recombinant cytokine are rapidly cleared from the blood. Clearance occurs by both renal mechanisms and by binding to the G-CSF receptor on myeloid cells. Based on these principles, it is usually necessary to administer recombinant human G-CSF at frequent, usually daily, intervals for its maximal effects. Daily drug administration was used in the original clinical trials that led to its approval and in almost all subsequent randomized clinical studies. Both glycosylated and non-glycosylated G-CSF have similar biological properties and clinical uses.

After the utility of daily therapy with recombinant G-CSF was established, this glycoprotein was modified to increase its size and thereby reduce its renal clearance [58]. Clearance of the new molecule, pegylated G-CSF (peg G-CSF), depended primarily on it binding to receptors on myeloid cells. Because patients with myelosuppression after cancer chemotherapy have a reduced myeloid mass, the drug's blood half-life was greatly extended. Clinical trials established that a single injection was sufficient to permit once per cycle supportive therapy [59, 60]. For patients with chemotherapy sensitive malignancies, particularly breast cancer and non-Hodgkin's lymphoma, supportive care with peg G-CSF is currently quite common. Other modifications such as pegylation at different sites and increased glycosylation at various sites are alternative ways to extend the biological half-life of G-CSF [61].

Clinical use of G-CSF to accelerate marrow recovery after cancer chemotherapy was originally based on two randomized controlled trials in patients receiving combination chemotherapy for lung cancer [54, 55]. The findings are supported by a large body of additional evidence [55]. The original trials demonstrated that patients receiving chemotherapy without G-CSF had approximately 6 days of severe neutropenia and that G-CSF added the day after completion of chemotherapy infusion decreased the duration of severe neutropenia to approximately 3 days. This acceleration of recovery closely parallels studies in normal subjects showing that G-CSF could reduce the transit time of neutrophils through the post-mitotic pool from 6 to 3 days [19]. From a clinical perspective the most important effect was a reduction of the risk of febrile neutropenia and proven infections by approximately 50%. The importance of timing was subsequently demonstrated in trials in which G-CSF treatment was delayed. If therapy with G-CSF was delayed until neutropenia, the benefit was diminished [62], and if treatment was delayed until the time of fever and neutropenia the benefit was lost [63]. Randomized clinical trials in breast cancer patients administered once per cycle peg G-CSF compared to daily non-peg G-CSF demonstrated equal efficacy [59, 60]. Filgrastim (recombinant G-CSF) and pegfilgrastim (pegylated recombinant G-CSF) are the principal products used currently in the USA. There are now several forms of G-CSF approved and licensed in other countries. GM-CSF is approved in the USA to reduce the risk of death from infections in patients over age 55 with acute myeloid leukemia and to accelerate neutrophil recovery after autologous or allogeneic bone marrow transplantation. Both G-CSF and GM-CSF are approved to facilitate collection of hematopoietic stem cells for bone marrow transplantation.

The American Society for Clinical Oncology (ASCO; http://www.asco.org), the NCCN (http://www.nccn.org), the ESMO (http://www.esmo.org), and other groups have practice guidelines for the myeloid growth factors. These guidelines bridge from evidence in randomized controlled trials and meta-analysis of these trials to practical advice for patient care. Because the myeloid growth factors, like other aspects of cancer care, are expensive, the goal is to guide clinicians toward evidence-based and cost-effective cancer management. These organizations regularly update their guidelines based upon new evidence and changing practices.

ASCO guidelines recommend the use of myeloid growth factors as primary prophylaxis after cancer chemotherapy, that is with the first cycle and all subsequent cycles, for patients with an expected risk of febrile neutropenia of 20% or greater. ASCO also recommends primary prophylaxis for patients given "dose dense" chemotherapy, short interval treatment requiring the stimulus of the myeloid growth factor or marrow to recover, e.g., 14-day regimens. The NCCN guidelines also recommend primary prophylaxis with the expected risk of febrile neutropenia greater than 20% when treatment is being given with curative intent, to prolong survival, to maintain quality of life, or to avoid symptoms and complications associated with febrile neutropenia.

Erythropoietin

Erythropoietin is the 34 kD glycoprotein that regulates erythropoiesis [37]. Mice not expressing the erythropoietin gene die in utero with severe anemia. Erythropoietin is produced primarily by the juxtaglomerular cells in the kidney, and levels of erythropoietin are inversely related to hemoglobin levels and tissue delivery of oxygen.

Anemia is a common consequence of cancer and is often multifactorial—attributable to inflammatory effects of cancer, anorexia and poor nutrition, iron and other nutrient deficiencies. It is also a frequent complication of the direct effects of cancer chemotherapy on hematopoiesis. In relatively healthy cancer patients, the anemia is often mild and the suppressive effects of chemotherapy on erythropoiesis only moderately intensify the anemia. Most patients are asymptomatic and recover from the anemia without complications. Because anemia can be severe and clinically significant, there have been numerous studies on the treatment of cancer-related anemia with recombinant human erythropoietin, both recombinant human erythropoietin and a modified erythropoietin called darbepoietin [64–66]. These clinical trials have focused on increasing hemoglobin and hematocrit, prevention of transfusions, improvement of quality of life, reduced fatigue, and other functional assessments. Administration of either epoetin alpha or darbepoietin by various schedules increases circulating hemoglobin and provides functional improvement during chemotherapy. In the USA the use of erythropoietin decreased substantially in 2008 because clinical studies suggested that the erythroid-stimulating agents (ESAs) may shorten overall survival or increase the risk of cancer progression or recurrence for some types of cancer [67]. For this reason the original indications for the ESAs were revised to state that these agents are no longer indicated for patients receiving myelosuppressive chemotherapy given with curative intent. The threshold for administration was also revised; the guidelines state that ESAs should not be administered to patients with hemoglobin levels above 10 g/dL and should be targeted to the lowest hemoglobin that will avoid transfusion and should not be given to exceed hemoglobin above 12 g/dL. The benefits, risks, and appropriate use of the ESAs to overcome the myelotoxicity of cancer chemotherapy, however, remain controversial. Current guidelines for appropriate use of the ESAs are available at the websites for ASCO, NCCN, EORTC (http://www.eortc.be/), and other organizations.

Thrombopoietin

Normal production of platelets is regulated by TPO, a glycoprotein primarily produced by the liver. Megakaryocytes produce platelets in response to TPO, and platelet counts vary inversely with circulating TPO levels. Several TPO receptor agonists are now available or in clinical development [68–71]. A key feature of the physiology of platelet formation is the time delay between the TPO stimulus and the response to increased platelets in the blood. The physiology of platelet production has made it difficult to conduct clinical trials in patients receiving cyclic chemotherapy.

Thrombocytopenia frequently occurs in cancer patients but it is often mild and requires no specific therapy. Platelet transfusions are standard supportive care in those patients who develop severe thrombocytopenia and bleeding, with the threshold of transfusion usually of 10,000–20,000/dL and usually with evidence of bleeding or risk of serious bleeding.

Early clinical trials in cancer patients used a recombinant TPO called megakaryocytic growth and development factor (MGDF). These studies did not demonstrate a benefit of MGDF to shorten the duration of severe thrombocytopenia or reduce the patients' requirements for platelet transfusion. Other thrombopoietic agents have now been successfully developed for the treatment of idiopathic thrombocytopenia and new clinical trials in cancer patients are in progress.

Interleukin 11

Recombinant IL-11 (oprelvekin) is an alternative agent to increase platelet counts [72]. It has been shown to be effective for primary and secondary prevention of the need for platelet transfusions in women with breast cancer treated with doxorubicin and cyclophosphamide. Adverse effects associated with the oprelvekin included anemia, myalgias, edema, and shortness of breath, and this adverse events profile has limited the use of this agent [73].

Other Hematopoietic Stimulatory Agents

A number of other hematopoietic growth factors and combinations of growth factors have been investigated. For several years there was a high level of interest in ligands for the Kit receptor because this receptor is expressed on primitive hematopoietic cells. A recombinant full-length cytokine called "stem cell factor" was extensively tested. It did not prove useful for management of myelosuppression but it was approved for use in Australia and Canada as an adjunct to G-CSF or collection of hematopoietic progenitor cells.

References

1. Zubrod CG, Schepartz SA, Carter SK. Historical background of the National Cancer Institute's drug development thrust. Natl Cancer Inst Monogr. 1977;45:7–11.
2. DeVita Jr VT, Chu E. A history of cancer chemotherapy. Cancer Res. 2008;68:8643–53.
3. Goodman LS, Wintrobe MM, Dameshek W, et al. Nitrogen mustard therapy. Use of methyl-bis(beta-chloroethyl)amine hydrochloride and tris(beta-chloroethyl)amine hydrochloride for Hodgkin's disease, lymphosarcoma, leukemia and certain allied and miscellaneous disorders. JAMA. 1984;251:2255–61.
4. Farber S. Some observations on the effect of folic acid antagonists on acute leukemia and other forms of incurable cancer. Blood. 1949;4:160–7.
5. Farber S, Diamond LK. Temporary remissions in acute leukemia in children produced by folic acid antagonist, 4-aminopteroyl-glutamic acid. N Engl J Med. 1948;238:787–93.
6. DeVita Jr VT, Lewis BJ, Rozencweig M, et al. The chemotherapy of Hodgkin's disease: past experiences and future directions. Cancer. 1978;42(2S):979–90.
7. Lyman GH, Lyman CH, Agboola O. Risk models for predicting chemotherapy-induced neutropenia. Oncologist. 2005;10:427–37.
8. National Comprehensive Cancer Network. http://www.nccn.com/ (2012). Accessed 18 Feb 2012.
9. MedlinePlus® (2012) Cancer Chemotherapy. US National Library of Medicine/NIH National Institutes of Health. http://www.nlm.nih.gov/medlineplus/cancerchemotherapy.html. Accessed 18 Feb 2012.
10. Catlin SN, Busque L, Gale RE, et al. The replication rate of human hematopoietic stem cells in vivo. Blood. 2011;117:4460–6.
11. Mercier FE, Ragu C, Scadden DT. The bone marrow at the crossroads of blood and immunity. Nat Rev Immunol. 2011;12:49–60.
12. Hoggatt J, Pelus LM. Many mechanisms mediating mobilization: an alliterative review. Curr Opin Hematol. 2011;18:231–8.
13. Keating GM. Plerixafor: a review of its use in stem-cell mobilization in patients with lymphoma or multiple myeloma. Drugs. 2011;71:1623–47.
14. Demirer T, Buckner CD, Gooley T, et al. Factors influencing collection of peripheral blood stem cells in patients with multiple myeloma. Bone Marrow Transplant. 1996;17:937–41.
15. To LB, Levesque JP, Herbert KE. How I treat patients who mobilize hematopoietic stem cells poorly. Blood. 2011;118:4530–40.
16. Cronkite EP, Fliedner TM. Granulocytopoiesis. N Engl J Med. 1964;270:1403–8.
17. Perry S, Godwin HA, Zimmerman TS. Physiology of the granulocyte. 1. JAMA. 1968;203:937–44.
18. Dancey JT, Deubelbeiss KA, Harker LA, et al. Neutrophil kinetics in man. J Clin Invest. 1976;58:705–15.
19. Price TH, Chatta GS, Dale DC. Effect of recombinant granulocyte colony-stimulating factor on neutrophil kinetics in normal young and elderly humans. Blood. 1996;88:335–40.
20. Dale DC, Liles WC, Llewellyn C, et al. Effects of granulocyte-macrophage colony-stimulating factor (GM-CSF) on neutrophil kinetics and function in normal human volunteers. Am J Hematol. 1998;57:7–15.
21. Cartwright GE, Athens JW, Wintrobe MM. The kinetics of granulopoiesis in normal man. Blood. 1964;24:780–803.

22. Dale DC, Fauci AS, Guerry IV D, et al. Comparison of agents producing a neutrophilic leukocytosis in man. Hydrocortisone, prednisone, endotoxin, and etiocholanolone. J Clin Invest. 1975;56:808–13.
23. Hansen PB, Knudsen LM, Johnsen HE, et al. Stimulation tests for the bone marrow neutrophil pool in malignancies. Leuk Lymphoma. 1995;16:237–46.
24. Hansen PB, Johnsen HE, Ralfkiaer E, et al. Blood neutrophil increment after a single injection of rhG-CSF or rhGM-CSF correlates with marrow cellularity and may predict the grade of neutropenia after chemotherapy. Br J Haematol. 1993;84:581–5.
25. Friberg LE, Henningsson A, Maas H, et al. Model of chemotherapy-induced myelosuppression with parameter consistency across drugs. J Clin Oncol. 2002;20:4713–21.
26. Hansson EK, Friberg LE. The shape of the myelosuppression time profile is related to the probability of developing neutropenic fever in patients with docetaxel-induced grade IV neutropenia. Cancer Chemother Pharmacol. 2012;69(4):881–90.
27. Athlin L, Domellöf L, Björkstén B. The effect of combination chemotherapy on neutrophil function in cancer patients. Cancer Lett. 1979;7:21–5.
28. Lejeune M, Ferster A, Cantinieaux B, et al. Prolonged but reversible neutrophil dysfunctions differentially sensitive to granulocyte colony-stimulating factor in children with acute lymphoblastic leukaemia. Br J Haematol. 1998;102:1284–91.
29. Lejeune M, Sariban E, Cantinieaux B, et al. Granulocyte functions in children with cancer are differentially sensitive to the toxic effect of chemotherapy. Pediatr Res. 1996;39:835–42.
30. Wijayahadi N, Haron MR, Stanslas J, et al. Changes in cellular immunity during chemotherapy for primary breast cancer with anthracycline regimens. J Chemother. 2007;19:716–23.
31. Mendonça MA, Cunha FQ, Murta EF, et al. Failure of neutrophil chemotactic function in breast cancer patients treated with chemotherapy. Cancer. 2006;57:663–70.
32. Lower EE, Baughman RP. The effect of cancer and chemotherapy on monocyte function. J Clin Lab Immunol. 1990;31:121–5.
33. Markowicz S, Walewski J, Zajda K, et al. Recovery of dendritic cell counts and function in peripheral blood of cancer patients after chemotherapy. Cytokines Cell Mol Ther. 2002;7:15–24.
34. Kondo M, Oshita F, Kato Y, et al. Early monocytopenia after chemotherapy as a risk factor for neutropenia. Am J Clin Oncol. 1999;22:103–5.
35. Thirumala R, Ramaswamy M, Chawla S. Diagnosis and management of infectious complications in critically ill patients with cancer. Crit Care Clin. 2010;26:59–91.
36. Hattangadi SM, Wong P, Zhang L, et al. From stem cell to red cell: regulation of erythropoiesis at multiple levels by multiple proteins, RNAs, and chromatin modifications. Blood. 2011;118:6258–68.
37. Erslev AJ. Erythropoietin and anemia of cancer. Eur J Haematol. 2000;64:353–8.
38. Kuter DJ. The physiology of platelet production. Stem Cells. 1996;14 Suppl 1:88–101.
39. Sneed TB, Kantarjian HM, Talpaz M, et al. The significance of myelosuppression during therapy with imatinib mesylate in patients with chronic myelogenous leukemia in chronic phase. Cancer. 2004;100:116–21.
40. Ruchatz H, Puttini M, Cleris L, et al. Effect of imatinib on haematopoietic recovery following idarubicin exposure. Leukemia. 2003;17:298–304.
41. Zaucha JM, Wyrowinska E, Prejzner W, et al. Imatinib-associated neutropenia may not be overcome by filgrastim treatment in patients with blastic phase of chronic myeloid leukaemia. Clin Lab Haematol. 2006;28:208–10.
42. O'Shaughnessy JA. Management of febrile neutropenia and cardiac toxicity in the adjuvant treatment of breast cancer. Clin Breast Cancer. 2007;8 Suppl 1:S11–21.
43. Gianni L, Pienkowski T, Im YH, et al. Efficacy and safety of neoadjuvant pertuzumab and trastuzumab in women with locally advanced, inflammatory, or early HER2-positive breast cancer (NeoSphere): a randomised multicentre, open-label, phase 2 trial. Lancet Oncol. 2012;13:25–32.
44. Servitja S, Ramos M, Gil M, et al. Multicenter, phase II, nonrandomized study of docetaxel plus trastuzumab every 21 days as the primary therapy in metastatic breast cancer overexpressing HER2. Anticancer Drugs. 2012;23:239–46.

45. Baselga J, Cortés J, Kim SB, et al. CLEOPATRA Study Group. Pertuzumab plus trastuzumab plus docetaxel for metastatic breast cancer. N Engl J Med. 2012;366:109–19.
46. Wolach O, Bairey O, Lahav M. Late-onset neutropenia after rituximab treatment: case series and comprehensive review of the literature. Medicine. 2010;89:308–18.
47. Aapro MS, Bohlius J, Cameron DA, et al. 2010 update of EORTC guidelines for the use of granulocyte-colony stimulating factor to reduce the incidence of chemotherapy-induced febrile neutropenia in adult patients with lymphoproliferative disorders and solid tumours. Eur J Cancer. 2011;47:8–32.
48. Gethins M. International cancer conference: targeted therapies boom expected in 3–5 years. J Natl Cancer Inst. 2011;103:1491–3.
49. Langer CJ. Individualized therapy for patients with non-small cell lung cancer: emerging trends and challenges. Crit Rev Oncol Hematol. 2012;83(1):130–44.
50. Kuderer NM, Lyman GH. Personalized medicine and cancer supportive care: appropriate use of colony-stimulating factor support of chemotherapy. J Natl Cancer Inst. 2011;103(12):910–3.
51. Page AV, Liles WC. Colony-stimulating factors in the prevention and management of infectious diseases. Infect Dis Clin North Am. 2011;25:803–17.
52. Lieschke GJ, Grail D, Hodgson G, et al. Mice lacking granulocyte colony-stimulating factor have chronic neutropenia, granulocyte and macrophage progenitor cell deficiency, and impaired neutrophil mobilization. Blood. 1994;84:1737–46.
53. Liu F, Wu HY, Wesselschmidt R, et al. Impaired production and increased apoptosis of neutrophils in granulocyte colony-stimulating factor receptor-deficient mice. Immunity. 1996;5:491–501.
54. Crawford J, Ozer H, Stoller R, et al. Reduction by granulocyte colony-stimulating factor of fever and neutropenia induced by chemotherapy in patients with small-cell lung cancer. N Engl J Med. 1991;325:164–70.
55. Cooper KL, Madan J, Whyte S, et al. Granulocyte colony-stimulating factors for febrile neutropenia prophylaxis following chemotherapy: systematic review and meta-analysis. BMC Cancer. 2011;11:404.
56. van Raam BJ, Drewniak A, Groenewold V, et al. Granulocyte colony-stimulating factor delays neutrophil apoptosis by inhibition of calpains upstream of caspase-3. Blood. 2008;112:2046–54.
57. Stanley E, Lieschke GJ, Grail D, et al. Granulocyte/macrophage colony-stimulating factor-deficient mice show no major perturbation of hematopoiesis but develop a characteristic pulmonary pathology. Proc Natl Acad Sci U S A. 1994;91:5592–6.
58. Molineux G. The design and development of pegfilgrastim (PEG-rmetHuG-CSF, Neulasta). Curr Pharm Des. 2004;10:1235–44.
59. Holmes FA, Jones SE, O'Shaughnessy J, et al. Comparable efficacy and safety profiles of once-per-cycle pegfilgrastim and daily injection filgrastim in chemotherapy-induced neutropenia: a multicenter dose-finding study in women with breast cancer. Ann Oncol. 2002;13:903–9.
60. Green MD, Koelbl H, Baselga J, et al., International Pegfilgrastim 749 Study Group. A randomized double-blind multicenter phase III study of fixed-dose single-administration pegfilgrastim versus daily filgrastim in patients receiving myelosuppressive chemotherapy. Ann Oncol. 2003;14:29–35.
61. Gravel P, Naik A, Le Cotonnec JY. Biosimilar rhG-CSFs: how similar are they? Target Oncol Suppl. 2012;1:3–16.
62. Hartmann LC, Tschetter LK, Habermann TM, et al. Granulocyte colony-stimulating factor in severe chemotherapy-induced afebrile neutropenia. N Engl J Med. 1997;336:1776–80.
63. Pérez Velasco R. Review of granulocyte colony-stimulating factors in the treatment of established febrile neutropenia. J Oncol Pharm Pract. 2011;17:225–32.
64. Glaspy J, Crawford J, Vansteenkiste J, et al. Erythropoiesis-stimulating agents in oncology: a study-level meta-analysis of survival and other safety outcomes. Br J Cancer. 2010;102:301–15.
65. Ludwig H, Aapro M, Bokemeyer C, et al. Treatment patterns and outcomes in the management of anaemia in cancer patients in Europe: findings from the Anaemia Cancer Treatment (ACT) study. Eur J Cancer. 2009;45:1603–15.

66. Vansteenkiste J, Glaspy J, Henry D, et al. Benefits and risks of using erythropoiesis-stimulating agents (ESAs) in lung cancer patients: study-level and patient-level meta-analyses. Lung Cancer. 2012;76(3):478–85.
67. Rizzo JD, Somerfield MR, Hagerty KL, et al. Use of epoetin and darbepoetin in patients with cancer: 2007 American Society of Hematology/American Society of Clinical Oncology clinical practice guideline update. Blood. 2008;111:25–41.
68. Kuter DJ. New thrombopoietic growth factors. Clin Lymphoma Myeloma. 2009;9 Suppl 3:S347–56.
69. Cheng G, Saleh MN, Marcher C, et al. Eltrombopag for management of chronic immune thrombocytopenia (RAISE): a 6-month, randomised, phase 3 study. Lancet. 2011;377:393–402.
70. Cohn CS, Bussel JB. Romiplostim: a second-generation thrombopoietin agonist. Drugs Today. 2009;45:175–88.
71. Psaila B, Bussel JB, Linden MD, et al. In vivo effects of eltrombopag on platelet function in immune thrombocytopenia: no evidence of platelet activation. Blood. 2012;119(17):4066–72.
72. Isaacs C, Robert NJ, Bailey FA, et al. Randomized placebo-controlled study of recombinant human interleukin-11 to prevent chemotherapy-induced thrombocytopenia in patients with breast cancer receiving dose-intensive cyclophosphamide and doxorubicin. J Clin Oncol. 1997;15:3368–77.
73. Dorner AJ, Goldman SJ, Keith Jr JC. Interleukin-11. BioDrugs. 1997;8:18–29.

Neuropathy

Elizabeth K. Joseph

Abstract Advancement in *cancer* research and relevant application of it *in the clinic* has increased the life span and quality of life *for* many *individuals diagnosed with the disease*. But the existing chemotherapeutic agents are double edged swords. Cancer chemotherapy-induced peripheral neuropathy is a clinical challenge for *all oncologists*. As research develops, numerous drugs with varying mechanisms of action are developed, *but none are devoid* of neurotoxicities. Assessment of neurotoxicity is complicated, and opioids remain the main stay of analgesic treatment *for cancer pain*, even though they do not alleviate pain in most patients. Presently, basic and clinical researchers are faced with the responsibility of developing novel cancer therapies selective for the treatment of different types of cancers with little or no neurotoxicity. In parallel, analgesics that appropriately *relieve pain associated with the particular* neurotoxic effects also need to be developed. This article is an attempt to highlight the different classes of chemotherapeutic agents, their neurotoxicities, various classes of *analgesics* used to alleviate *cancer* pain and the emerging trends in research.

Introduction

The incidence and prevalence of cancer is on the increase and cancer has become one of the leading causes of death world-wide [1]. Surgery, radiation, and chemotherapy alone or in combination remain the common treatment choices for cancer. As of today, none of these interventions is fully satisfactory; rather they leave the

E.K. Joseph, M.Pharm, Ph.D. (✉)
University of California San Francisco,
521 Parnassus Avenue, Box # 0440, San Francisco, CA 94143, USA
e-mail: Elizabeth.Joseph@ucsf.edu

S.T. Sonis and D.M. Keefe (eds.), *Pathobiology of Cancer Regimen-Related Toxicities*,
DOI 10.1007/978-1-4614-5438-0_11,

patients with high toxic effects. Painful peripheral neuropathy is an adverse effect accompanying the administration of most effective chemotherapeutic agents, including taxanes (paclitaxel, docetaxel), vinca alkaloids (vincristine, vinblastine), platinum compounds (oxaliplatin, cisplatin, carboplatin), and the molecularly targeted proteosome inhibitors [2–7]. There are over 100 different types of cancers, but the most prevalent ones in the United States are: cancers of the lung, breast, bladder, colon and rectum, endometrium, prostate, kidney, thyroid as well as lymphoma [8]. Several neuroprotective agents (neurotrophic factors, thiols, anticonvulsants, antioxidants) have been tested in preclinical and clinical studies and none has proven to give satisfactory protection [9–11]. The available anticancer drugs have distinct mechanisms of action which may lead to different effects on various normal and cancer cells. Most, but not all of these chemotherapy drugs interfere with RNA and DNA synthesis or cell division and affect cell growth [12, 13]. Though newer molecularly targeted, biological agents were expected to have fewer side effects, they are not devoid of neurotoxicity [14]. *Since most cancer drugs induce chronic neurotoxicities*, it is highly important to develop chemotherapeutic agents with little or no neurotoxicity

Neurotoxicities

Pathophysiological mechanisms of neuropathy differ depending on the chemotherapeutic agent, and prevalence of neurotoxicity varies with the dose and schedules of drug treatment [15]. Very often the side effects are overlooked and the patients are either over-sedated or given any or all kinds of alternatives, which may or may not be beneficial. Regardless of the advancement made in the diagnosis, treatment, and cure of certain cancers, there is a general attitude among the population that cancer is incurable *and educating the public to remove unnecessary fear is essential.*

Assessment of chemotherapy-induced neurotoxicity is complicated *as symptoms and severity may vary with age, sex, demography, genetics, comorbidities, and use of other drugs/alcohol. Regardless of the use of the scoring scales and criteria recommended by WHO (World Health Organization), ECOG (Eastern Cooperative Oncology Group), and NCIC (National Cancer Institute of Canada) to assess the chemotherapy-induced neurotoxicity in patients, inter-scale and interobserver variability still prevails. Further, each individual is different and neurotoxicity or symptoms of the same drug treatment reported by patients may vary.* Therapeutic doses of *all cancer* drugs cause peripheral neuropathy in nearly all *patients* [2]. The mechanisms by which different drugs induce neurotoxicity are not clearly understood, and satisfactory elucidation will require the concerted effort of different disciplines. Advances in neurobiology (structure and function of neurons and cells) have improved our understanding of the different expressions of neurotoxicity and susceptibilities of the nervous system. The numerous anticancer drugs available on the market today attack the cancer cells differently. The most cytotoxic chemotherapeutic agents are vinca alkaloids, taxanes, and platins.

Vinca Alkaloids: Vincristine, Vinblastin, and Vinorelbine

Vinca alkaloids prevent microtubule assembly and cell division [16] and have a broad spectrum of activity against tumors. Vincristine is used in combination with prednisone to induce remission in childhood leukemia [17]. Vincristine neurotoxicity is cumulative; the higher the drug concentration per dose, the shorter the intervals between the doses and the longer the therapy is continued, the greater is the degree of neurotoxicity [18]. Secondly, the patient's age is believed to be related to the degree of neurotoxicity—children are less susceptible than infants, adolescents, and adults [19]. In animal (rats) studies, age (young) and repeated dosing of vincristine were found to have more toxicity (personal observation). Other possible predisposing factors include poor nutritional condition, impaired performance status, liver dysfunction [19], and prior disorders of the peripheral nervous system [20]. The combined use of vincristine and other chemotherapeutic agents may lead to synergistic neurotoxicity [21].

Vinblastine is used in combination with Bleomycin and Cisplatin for metastatic testicular tumors [22]. All the drugs in this family produce dose-dependent sensorimotor neuropathy. Vinorelbine is used to treat non-small cell lung cancer and breast cancer [23]. Vincristine neurotoxicity affects motor and sensory neurons [24, 25]. Vinorelbine is a "third generation" vinca alkaloid used in non-small cell lung cancer either as a monotherapy or in combination with other agents [26]. Vincristine induces hyperalgesia/allodynia in animal models and the neurotoxicity is dose-limiting [27, 28]. Vincristine treatment in rats demonstrates abnormal microtubule assemblies, axonal swelling, and hyper-responsiveness of C-fibers to mechanical stimulation [29]. Tactile allodynia induced by vincristine treatment is associated with an increase in macrophages in the sciatic nerve and lumbar dorsal root ganglia (DRG) and interleukin-6 (IL6) expression [30].

Taxanes: Paclitaxel, Docetaxel, and Epothilones

Taxanes enhance all aspects of tubulin polymerization, an action that is the opposite to that of vinca alkaloids, which are also cytotoxic, thereby emphasizing the dynamic importance of tubulin polymerization as a target for cytotoxic drugs. Paclitaxel is used in ovarian, breast, head, and neck cancers as well as in non-small cell lung cancers [31]. It produces sensory *neuropathy* [32, 33]. Docetaxel is a second-generation taxane used in the treatment of breast, non-small cell lung, stomach, and androgen-independent prostate cancers [34]. Though its neurotoxicity is rarely dose-limiting, it is suggested to have neurotoxic effects on DRG [15]. The taxanes cause damage to large myelinated fibers and axonal atrophy with reduction in nerve conduction velocity [35–37]. Epothilones A and B, the naturally occurring microtubule stabilizers, are used in different types of multidrug-resistant cell lines *in vitro* and multidrug-resistant tumors *in vivo*.

Platinum Compounds: Cisplatin, Carboplatin, and Oxaliplatin

All the platinum compounds induce dose-limiting neurotoxic effects [10, 38, 39]. Cisplatin binds to guanine in DNA and RNA, and the interaction is stabilized by hydrogen bonding. The molecular mechanism of action is unwinding and shortening of the DNA helix. Cisplatin has efficacy against a wide range of neoplasms. It is given intravenously as a first-line drug for testicular, ovarian, and bladder cancer and it is also useful in the treatment of melanoma [40]. Carboplatin has a similar spectrum of activity, but it is approved only as a second-line drug for ovarian cancer [41]. Oxaliplatin is commonly used in colorectal cancer [42, 43] and patients experience acute transient cold hyperalgesia and chronic sensory neuropathy with cumulative dose [44, 45]. A number of studies have indicated that acute oxaliplatin-induced neurotoxicity is related to dysfunction of axonal Na^+ channels [46]. There are incidences of patients not having neuropathy while receiving oxaliplatin but developing it 2–3 weeks or years after the treatment [47]. In an acute model, oxaliplatin produced heat/cold alodynia and hyperalgesia [10]. Oxaliplatin hyperalgesia was not attenuated by the inhibitors of protein kinase A, protein kinase C, NO, Ca^{2+}, and caspase, but the antioxidants (acetyl L-carnitine, α-lipoic acid, vitamin C) and inhibitors of mitochondrial electron transport complex IV (sodium cyanide) and complex V (oligomycin) did attenuate it [38].

Proteasome Inhibitors

Introduction of proteasome inhibitors has improved treatment for multiple myeloma patients. Bortezomib, a novel proteasome inhibitor, used in multiple myeloma is not devoid of neurotoxicity [48, 49]. Repeated treatment with bortezomib increases the incidence of peripheral neuropathy [50]. The discontinuation rate of the treatment is reduced when bortezomib is combined with thalidomide [51]. However, unlike other anticancer drugs, proteasome inhibitors are more toxic to proliferating malignant cells than to normal cells [52]. Bortezomib down regulates nuclear factor-kappa B [53], which is a transcription factor that plays a pivotal role in carcinogenesis [54]. Clinically bortezomib is used in the treatment of solid tumor cancers, including metastatic head and neck cancer, prostate cancer, kidney and colorectal cancer, non-small cell lung cancer, gastric, breast, and pancreatic cancers [51, 55]. NPI-0052, a second-generation proteasome inhibitor, is demonstrated to have synergistic activity with bortezomib in the treatment of multiple myeloma [56]. The use of this drug also produces severe neuropathy [57]. Carfilzomib is another second-generation proteasome inhibitor with increased activity in comparison to bortezomib in preclinical models of multiple myeloma [58] and with no incidence of neuropathic pain [59].

Neuroprotective Measures

In general, symptoms or etiology-based approaches are taken to treat peripheral neuropathy. Opioids (μ-opioid receptor-targeting analgesics) continue to be the primary option to treat the pain associated with peripheral neuropathy [60]. Anticonvulsants [61], antidepressants [62], local anesthetics [63], and corticosteroids are adjuvant agents employed. Erythropoietin [64], supplements such as vitamin E [65], acetyl-L-carnitine, and alpha-lipoic acid are additional interventions used to treat chemotherapy-induced neurotoxicity [66, 67]. Glutamic acid and neurotrophic factors are also tested for their ability to prevent neuropathy [28, 68].

Mitochondrial Dynamics and Neurotoxicity

Peripheral neuropathy involves degeneration of the longest axons in the body and emerging evidence implicates the involvement of mitochondrial dynamics (fusion and fission) in peripheral disease [69]. Mitochondria travel in axons provides ATP necessary for the maintenance of ionic gradients for firing action potentials [70] and regulation of calcium homeostasis [71]. Defects in mitochondrial morphology affect mitochondrial movement [72]. Mitochondria are not randomly distributed in axons, there are high concentrations at specific regions based on energy demand and in pathological conditions they accumulate in demyelinated axonal segments [73]. The concept that mitochondria play an important role in neuropathy is an emerging concept and is based on their selective localization at sites of high ATP demand and high numbers of mitochondria in sensory organs of the skin.

Conclusions

Detailed studies have to be pursued to avoid or minimize neurotoxicity caused by chemotherapeutic drugs. There have been many studies done examining DNA repair and drug disposition genes in various patient populations treated with a number of different drugs. Despite these efforts, the precise association between chemotherapeutic agents and sensory neurotoxicity is not fully elucidated. Studies on drug disposition and determination of specific phenotype may be of help to find the correlation between the incidence of neurotoxicity and tumor response rate. Eliminating or decreasing neurotoxicity should not reduce the effectiveness of the drug. It will take a more concerted effort to discover meaningful predictors of this serious adverse drug effect.

Invariably there are many distinct mechanisms in the peripheral and central nervous systems that function simultaneously causing pain. Experimental results from different laboratories demonstrate the involvement of one or the other. A cohort attempt to elucidate all the mechanisms that simultaneously or sequentially function to generate pain is lacking. Both preclinical and clinical assessment of the neurotoxicities is limited in investigation and details are very often omitted.

Different drugs as well as different models are employed in independent research under varying conditions on which drug development is based. Treating the person as a whole regardless of the diseased condition is essential for the success of the discipline and for a better tomorrow.

Prevention is Better than Cure

Like any other disease, certain cancers can be prevented. Various factors, such as one's lifestyle, heredity, and environment, are responsible for causing cancer in a person. In high-income countries, excessive use of tobacco (lung cancer), alcohol (liver cancer), and being overweight or obese are major risk factors, while in low-income countries, cervical cancer, which is caused by HPV is a leading cause of cancer death among women [74]. Indoor smoke, air-pollution, exposure to ultraviolet radiation from the sun (skin cancer), exposure to other carcinogens from work place and environment (chemicals, asbestos, nickel, cadmium, uranium, radon, vinyl chloride, benzidene, and benzene), and secondhand smoking are contributing factors that increase a person's risk. Most people who get the disease may not have been exposed to any of the above mentioned risk factors. However, the public has to be aware and take responsible measures for prevention. Dietary factors are also reported to play a role in the development of different types (breast, colon, prostate, pancreas, uterus, ovary) of cancer, while foods containing fiber and certain nutrients may help protect against some types of cancer [75]. Low fruit and vegetable intake, and lack of physical activity are contributing factors.

References

1. Pisani P, Bray F, Parkin DM. Estimates of the world-wide prevalence of cancer for 25 sites in the adult population. Int J Cancer. 2002;97(1):72–81.
2. Baker J, et al. Docetaxel-related side effects and their management. Eur J Oncol Nurs. 2008;12(3):253–68.
3. Kannarkat G, Lasher EE, Schiff D. Neurologic complications of chemotherapy agents. Curr Opin Neurol. 2007;20(6):719–25.
4. Ocean AJ, Vahdat LT. Chemotherapy-induced peripheral neuropathy: pathogenesis and emerging therapies. Support Care Cancer. 2004;12(9):619–25.
5. Park SB, et al. Mechanisms underlying chemotherapy-induced neurotoxicity and the potential for neuroprotective strategies. Curr Med Chem. 2008;15(29):3081–94.
6. Walker M, Ni O. Neuroprotection during chemotherapy: a systematic review. Am J Clin Oncol. 2007;30(1):82–92.

7. Wolf S, et al. Chemotherapy-induced peripheral neuropathy: prevention and treatment strategies. Eur J Cancer. 2008;44(11):1507–15.
8. American Cancer Society. Cancer facts and figures. Atlanta: American Cancer Society; 2010.
9. Cavaletti G, Marmiroli P. The role of growth factors in the prevention and treatment of chemotherapy-induced peripheral neurotoxicity. Curr Drug Saf. 2006;1(1):35–42.
10. Joseph EK, et al. Oxaliplatin acts on IB4-positive nociceptors to induce an oxidative stress-dependent acute painful peripheral neuropathy. J Pain. 2008;9(5):463–72.
11. Rodriguez-Menendez V, et al. Valproate protective effects on cisplatin-induced peripheral neuropathy: an in vitro and in vivo study. Anticancer Res. 2008;28(1A):335–42.
12. Allan DJ, Harmon BV. The morphologic categorization of cell death induced by mild hyperthermia and comparison with death induced by ionizing radiation and cytotoxic drugs. Scan Electron Microsc. 1986;(Pt 3):1121–33.
13. Jaiswal AS, et al. DNA polymerase B as a novel target for chemotherapeutic intervention of colorectal cancer. PLoS One. 2011;26(2):16691.
14. de Bono JS, Ashworth A. Translating cancer research into targeted therapeutics. Nature. 2010;467(7315):543–9.
15. Persohn E, et al. Morphological and morphometric analysis of paclitaxel and docetaxel-induced peripheral neuropathy in rats. Eur J Cancer. 2005;41(10):1460–6.
16. Jordan MA, Thrower D, Wilson L. Mechanism of inhibition of cell proliferation by vinca alkaloids. Cancer Res. 1991;51(8):2212–22.
17. Rana ZA, et al. Outcome of childhood acute lymphoblastic leukaemia after induction therapy—3 years experience at a single paediatric oncology centre. J Ayub Med Coll Abbottabad. 2009;21(4):150–3.
18. Gidding CE, et al. Vincristine pharmacokinetics after repetitive dosing in children. Cancer Chemother Pharmacol. 1999;44(3):203–9.
19. Allen AM. Epidemiologic methods in dermatology, part 1: describing the occurrence of disease in human populations. Int J Dermatol. 1978;17(3):186–93.
20. Naumann R, et al. Early recognition of hereditary motor and sensory neuropathy type 1 can avoid life-threatening vincristine neurotoxicity. Br J Haematol. 2001;115(2):323–5.
21. Aisner J, et al. Combination chemotherapy for small cell carcinoma of the lung: continuous versus alternating non-cross-resistant combinations. Cancer Treat Rep. 1982;66(2):221–30.
22. Nakamura T, Miki T. Recent strategy for the management of testicular cancer. Int J Urol. 2010;17(2):148–57.
23. Jahnke K, et al. Randomized phase II study of paclitaxel and carboplatin or vinorelbine in advanced non-small cell lung cancer. Anticancer Res. 2011;31(1):317–23.
24. Quasthoff S, Hartung HP. Chemotherapy-induced peripheral neuropathy. J Neurol. 2002;249(1):9–17.
25. Wilkes G. Antiangiogenic agents. Oncology (Williston Park). 2007;21(2 Suppl Nurse Ed):48.
26. Rossi D, et al. A phase II study of single-agent oral vinorelbine in patients with pretreated advanced non-small-cell lung cancer. Clin Lung Cancer. 2007;8(6):382–5.
27. Aley KO, Reichling DB, Levine JD. Vincristine hyperalgesia in the rat: a model of painful vincristine neuropathy in humans. Neuroscience. 1996;73(1):259–65.
28. Verstappen CC, et al. Dose-related vincristine-induced peripheral neuropathy with unexpected off-therapy worsening. Neurology. 2005;64(6):1076–7.
29. Tanner KD, et al. Altered temporal pattern of evoked afferent activity in a rat model of vincristine-induced painful peripheral neuropathy. Neuroscience. 2003;118(3):809–17.
30. Kiguchi N, et al. The critical role of invading peripheral macrophage-derived interleukin-6 in vincristine-induced mechanical allodynia in mice. Eur J Pharmacol. 2008;592(1–3):87–92.
31. Mekhail TM, Markman M. Paclitaxel in cancer therapy. Expert Opin Pharmacother. 2002;3(6):755–66.
32. Rowinsky EK, et al. Neurotoxicity of Taxol. J Natl Cancer Inst Monogr. 1993;15:107–15.
33. Smith RE, et al. Randomized trial of 3-hour versus 24-hour infusion of high-dose paclitaxel in patients with metastatic or locally advanced breast cancer: National Surgical Adjuvant Breast and Bowel Project Protocol B-26. J Clin Oncol. 1999;17(11):3403–11.

34. Engels FK, Verweij J. Docetaxel administration schedule: from fever to tears? a review of randomised studies. Eur J Cancer. 2005;41(8):1117–26.
35. Cavaletti G, et al. Peripheral neurotoxicity of taxol in patients previously treated with cisplatin. Cancer. 1995;75(5):1141–50.
36. Cavaletti G, et al. Experimental peripheral neuropathy induced in adult rats by repeated intraperitoneal administration of taxol. Exp Neurol. 1995;133(1):64–72.
37. Zhuang SH, et al. A Phase I clinical trial of ixabepilone (BMS-247550), an epothilone B analog, administered intravenously on a daily schedule for 3 days. Cancer. 2005;103(9):1932–8.
38. Joseph EK, Levine JD. Comparison of oxaliplatin- and cisplatin-induced painful peripheral neuropathy in the rat. J Pain. 2009;10(5):534–41.
39. McKeage MJ. Comparative adverse effect profiles of platinum drugs. Drug Saf. 1995;13(4):228–44.
40. Wheate NJ, et al. The status of platinum anticancer drugs in the clinic and in clinical trials. Dalton Trans. 2010;39(35):8113–27.
41. Boulikas T, Vougiouka M. Recent clinical trials using cisplatin, carboplatin and their combination chemotherapy drugs (review). Oncol Rep. 2004;11(3):559–95.
42. Andre T, et al. Oxaliplatin, fluorouracil, and leucovorin as adjuvant treatment for colon cancer. N Engl J Med. 2004;350(23):2343–51.
43. Andre T, de Gramont A. An overview of adjuvant systemic chemotherapy for colon cancer. Clin Colorectal Cancer. 2004;4 Suppl 1:S22–8.
44. Gamelin E, et al. Clinical aspects and molecular basis of oxaliplatin neurotoxicity: current management and development of preventive measures. Semin Oncol. 2002;29(5 Suppl 15):21–33.
45. Pietrangeli A, et al. Persistence of high-dose oxaliplatin-induced neuropathy at long-term follow-up. Eur Neurol. 2006;56(1):13–6.
46. Adelsberger H, et al. The chemotherapeutic oxaliplatin alters voltage-gated Na(+) channel kinetics on rat sensory neurons. Eur J Pharmacol. 2000;406(1):25–32.
47. Tofthagen C. Surviving chemotherapy for colon cancer and living with the consequences. J Palliat Med. 2010;13(11):1389–91.
48. Hideshima T, Richardson P, Anderson KC. Novel therapeutic approaches for multiple myeloma. Immunol Rev. 2003;194:164–76.
49. Richardson PG, Hideshima T, Anderson KC. Bortezomib (PS-341): a novel, first-in-class proteasome inhibitor for the treatment of multiple myeloma and other cancers. Cancer Control. 2003;10(5):361–9.
50. Corso A, et al. Bortezomib plus dexamethasone is highly effective in relapsed and refractory myeloma patients but responses are short-lived. Eur J Haematol. 2009;83(5):449–54.
51. Badros A, et al. Neurotoxicity of bortezomib therapy in multiple myeloma: a single-center experience and review of the literature. Cancer. 2007;110(5):1042–9.
52. Masdehors P, et al. Increased sensitivity of CLL-derived lymphocytes to apoptotic death activation by the proteasome-specific inhibitor lactacystin. Br J Haematol. 1999;105(3):752–7.
53. Mitsiades N, et al. Biologic sequelae of nuclear factor-kappaB blockade in multiple myeloma: therapeutic applications. Blood. 2002;99(11):4079–86.
54. Van Waes C. Nuclear factor-kappaB in development, prevention, and therapy of cancer. Clin Cancer Res. 2007;13(4):1076–82.
55. Utecht KN, Kolesar J. Bortezomib: a novel chemotherapeutic agent for hematologic malignancies. Am J Health Syst Pharm. 2008;65(13):1221–31.
56. Chauhan D, et al. Combination of proteasome inhibitors bortezomib and NPI-0052 trigger in vivo synergistic cytotoxicity in multiple myeloma. Blood. 2008;111(3):1654–64.
57. Corso A, et al. Bortezomib-induced peripheral neuropathy in multiple myeloma: a comparison between previously treated and untreated patients. Leuk Res. 2010;34(4):471–4.
58. Kuhn DJ, et al. Potent activity of carfilzomib, a novel, irreversible inhibitor of the ubiquitin-proteasome pathway, against preclinical models of multiple myeloma. Blood. 2007;110(9):3281–90.
59. Stewart AK, et al. A practical guide to defining high-risk myeloma for clinical trials, patient counseling and choice of therapy. Leukemia. 2007;21(3):529–34.

60. de Leon-Casasola OA. Current developments in opioid therapy for management of cancer pain. Clin J Pain. 2008;24 Suppl 10:S3–7.
61. Saif MW, Hashmi S. Successful amelioration of oxaliplatin-induced hyperexcitability syndrome with the antiepileptic pregabalin in a patient with pancreatic cancer. Cancer Chemother Pharmacol. 2008;61(3):349–54.
62. Jann MW, Slade JH. Antidepressant agents for the treatment of chronic pain and depression. Pharmacotherapy. 2007;27(11):1571–87.
63. Mao J, Chen LL. Systemic lidocaine for neuropathic pain relief. Pain. 2000;87(1):7–17.
64. Keswani SC, et al. A novel endogenous erythropoietin mediated pathway prevents axonal degeneration. Ann Neurol. 2004;56(6):815–26.
65. Pace A, et al. Neuroprotective effect of vitamin E supplementation in patients treated with cisplatin chemotherapy. J Clin Oncol. 2003;21(5):927–31.
66. Gedlicka C, et al. Amelioration of docetaxel/cisplatin induced polyneuropathy by alpha-lipoic acid. Ann Oncol. 2003;14(2):339–40.
67. Maestri A, et al. A pilot study on the effect of acetyl-L-carnitine in paclitaxel- and cisplatin-induced peripheral neuropathy. Tumori. 2005;91(2):135–8.
68. Apfel SC. Managing the neurotoxicity of paclitaxel (Taxol) and docetaxel (Taxotere) with neurotrophic factors. Cancer Invest. 2000;18(6):564–73.
69. Chan DC. Mitochondria: dynamic organelles in disease, aging, and development. Cell. 2006;125(7):1241–52.
70. Hollenbeck PJ, Saxton WM. The axonal transport of mitochondria. J Cell Sci. 2005;118(Pt 23):5411–9.
71. Ly CV, Verstreken P. Mitochondria at the synapse. Neuroscientist. 2006;12(4):291–9.
72. Verstreken P, et al. Synaptic mitochondria are critical for mobilization of reserve pool vesicles at Drosophila neuromuscular junctions. Neuron. 2005;47(3):365–78.
73. Mutsaers SE, Carroll WM. Focal accumulation of intra-axonal mitochondria in demyelination of the cat optic nerve. Acta Neuropathol. 1998;96(2):139–43.
74. Juarez-Albarran AC, Juarez-Gamez CA. [Vaccine against human papilloma virus]. Rev Med Inst Mex Seguro Soc. 2008;46(6):631–7.
75. Hedelin M, et al. Dietary phytoestrogens and the risk of ovarian cancer in the women's lifestyle and Health Cohort Study. Cancer Epidemiol Biomarkers Prev. 2011;20(2):308–17.

Mechanisms of Cancer Related Fatigue

Jonathan A. Zombeck

Abstract One of the most prevalent and distressing symptoms associated with cancer and cancer therapy is fatigue. Fatigue significantly impairs patients' quality of life and in some cases may last years following treatment. Despite a clear need, there is currently no pharmaceutical therapy approved to treat fatigue. Development of treatments for fatigue would benefit from an understanding of the mechanisms driving fatigue. While the mechanism of fatigue is unknown, recent research has led to several plausible hypotheses. This chapter reviews the hypothesized mechanisms of fatigue and the strength of evidence for each. There are multiple factors that are likely to influence fatigue including tumor burden, treatment, and psychological reactions associated with cancer diagnosis. Each of these influences has several physiological effects which may act independently or may converge on related pathways to produce fatigue. Anemia, cytokine dysregulation, dysregulation of the hypothalamic–pituitary–adrenal (HPA) axis, serotonin dysregulation, adenosine triphosphate depletion, and impaired neurogenesis are all possible outcomes of cancer and cancer treatment which may impact fatigue. Of these, the interplay between cytokines and the HPA axis are particularly intriguing. The influence of cytokines on behavior has been extensively studied in animal models which have demonstrated that an increase in proinflammatory cytokines leads to a suite of symptoms called "sickness behaviors" which are similar to human symptoms of fatigue (i.e., listlessness, lethargy, and decreased consumatory behavior). Cytokines are partially regulated by the HPA axis. The HPA axis is part of the neuroendocrine system which controls reactions to stressors and regulates homeostasis of body processes including the immune system, mood, and energy storage. Given the large number of stressors related to a diagnosis of cancer, patients may experience dysregulation of the HPA axis due to chronic stress. Therefore, increases in cytokines,

J.A. Zombeck, Ph.D. (✉)
Biomodels, LLC, Watertown, MA, USA
e-mail: jzombeck@biomodels.com

S.T. Sonis and D.M. Keefe (eds.), *Pathobiology of Cancer Regimen-Related Toxicities*,
DOI 10.1007/978-1-4614-5438-0_12,

combined with deregulation of cytokines from the HPA axis, and chronic activation of the HPA axis may cause fatigue. Clinically, this hypothesis is supported by evidence of elevated cytokines and abnormal levels of the stress hormone cortisol in cancer patients. However, further evidence is needed to conclusively determine the mechanisms of fatigue.

Overview of the Clinical Condition

Prevalence and Impact

Fatigue associated with cancer and cancer therapy is among the most prevalent and distressing symptoms reported by cancer patients [1–3]. The incidence rate of fatigue varies by report and methodology for screening, but it has been estimated at around 70% of cancer patients [2, 4, 5] and has been cited as the most commonly reported side effect [6–8]. In addition to being prevalent, fatigue is often long lasting. The temporal pattern of fatigue varies with treatment. Patients receiving chemotherapy report the most severe fatigue in the days following treatment and diminished symptoms the weeks after [9]. Peak fatigue is commonly delayed in radiotherapy patients as compared to chemotherapy patients with the most severe symptoms occurring midway through treatment [10, 11]. Fatigue may resolve within a few weeks for some patients, however, for others fatigue may last months or even years [12].

The impact of fatigue on patients' life can be profound. Cancer patients often report fatigue as the most distressing symptom they experience during and following treatment [13]. The influence of fatigue is broad and can negatively affect cognitive performance, mood, and physical function [2]. Accordingly, it is not surprising that fatigue has been shown to negatively impact quality of life [2, 8]. In some cases, the impact of fatigue, combined with other side effects of cancer treatment, is so great that it influences a patients' decision to continue treatment [13]. Given the clear ramifications fatigue has on patients, it is important to study fatigue and innovate therapies to combat the negative effects of fatigue.

Definition and Diagnosis of Fatigue

Generally, cancer-related fatigue is defined as a state of extreme tiredness that impedes physical and mental functioning regardless of recent activity or rest [1, 14]. While useful as a definition of fatigue, this description lacks the specificity required for clinical diagnosis or as an objective measurement for research studies. Cella et al. [15] developed a set of diagnostic criteria in order to better define fatigue as a clinical condition. The format of diagnosis is akin to that of the American Psychiatric

Association in the Diagnostic and Statistical Manual for Mental Disorders and has successfully indentified cancer patients experiencing severe fatigue symptoms [16, 17]. The selection criteria cover a broad range of symptoms ranging from physical fatigue, mental fatigue, and emotional or affective fatigue. This highlights that fatigue can be a complex symptom with many correlated effects.

Clinically, fatigue is primarily measured by patient self-report. The length and type of scales used are highly variable among studies and at least 22 different fatigue scales have been implemented [18]. In general, patients report their perception of fatigue using a Likert or numerical scale. Scales vary in the number of questions asked (generally between 1 and 30), the dimensions of fatigue experienced (i.e., physical, affective, and cognitive), and in which cancer populations the scale was validated. No one scale is considered the "standard" or is ideal for all purposes [14]. Clinicians and researchers should choose a scale based upon their specific diagnostic or study requirements.

Progenitors of Fatigue

Overview

Cancer-related fatigue originates from at least one of the three main factors. The first factor is the influence of a tumor itself. Following diagnosis, the course and type of treatment the patient receives is also likely to impact fatigue. Both these factors lead to stressors in a patient's life which may exacerbate or cause fatigue. This section will review the strength of the relationship between each of these factors and fatigue.

Tumor

Few studies have examined the impact of tumor burden directly. Clinically, patients often undergo treatment shortly after diagnosis, making clinical studies separating tumor burden from treatment challenging. Anecdotally, patients often report fatigue as a reason they sought medical care. Furthermore, fatigue is often most severe in advanced-stage cancer patients [19]; however, the mechanisms driving fatigue may be different from those associated with early stage cancer as progressive cancer can affect several organs and muscle control.

Preclinical studies using mice have examined the impact of tumor burden on fatigue. Wood et al. [20] measured voluntary wheel running activity as an index of fatigue in mice with and without an injection of Lewis Lung Cancer cells. Authors reported significantly less wheel running behavior in mice with tumor burden as compared to those without, suggesting a tumor may induce fatigue in the absence of other treatments or conditions.

Further research is needed to establish the strength of the relationship between tumor burden and fatigue. Nevertheless, the current research suggests tumor burden likely contributes to at least the initial experience of fatigue prior to and during treatment. It is important to note that fatigue may persist even years following the removal of the tumor [12, 21]. Therefore a tumor is unlikely the sole cause of fatigue.

Treatment

Surgery

Tumor resection is often the first method of treatment for solid tumors. The location and size of the tumor dictates the type of surgery, but most procedures are considered major surgeries. Chemotherapy and/or radiotherapy are often administered to patients following surgery making it difficult to assess the impact of surgery vs. other therapies in cancer patients. Therefore, it is useful to examine postoperative fatigue in non-cancer patient populations to assess the impact of surgery in the absence of a tumor or other treatments that may contribute to fatigue. Postoperative fatigue is common in the first week following the procedure, but may last as long as 1–3 months [22, 23]. The extent and duration of fatigue are likely influenced by the type of operation. Fatigue has been shown to be greatest following major abdominal and cardiac surgeries [24]. In a study of fatigue following middle-ear surgery or abdominal surgery, abdominal surgery was observed to result in greater subjective feelings of fatigue [25]. This difference did not correspond to procedure duration and fatigue severity was not correlated to age or sex. Overall, the evidence gathered from studies in non-cancer populations suggests that surgery is a likely contributor to cancer-related fatigue.

Chemotherapy

The relationship between chemotherapy and fatigue has been widely examined, particularly in breast cancer patients. In order to determine the contribution of chemotherapy vs. other possible contributing factors (e.g., tumor burden, surgery, radiotherapy, and psychosocial), it is useful to compare differences in fatigue severity following different chemotherapy protocols. In postsurgical breast cancer patients, women receiving doxorubicin-based chemotherapy protocols demonstrated significantly less activity than women receiving non-doxorubicin-based protocols as measured by wrist actigraphy [26]. Similarly, different chemotherapy regimens have resulted in varying levels of fatigue in small cell lung carcinoma patients [27].

Additionally, the pattern of fatigue may be informative in determining the influence of chemotherapy. Fatigue has been shown to be concomitant with chemotherapy administration. Reports of fatigue were higher following treatment than at the midpoint between treatments in breast cancer patients [26]. Genre et al. [28]

found that shortening the time between doxorubicin plus cyclophosphamide cycles resulted in diminished quality of life, including increased fatigue. Overall, chemotherapy is a likely contributor to fatigue as evidenced by variation in fatigue among chemotherapeutic agents and the correlation between fatigue and timing of chemotherapy administration.

Radiotherapy

Radiotherapy is administered either alone or in combination with surgery and/or chemotherapy. Evidence suggests the temporal pattern of fatigue associated with radiotherapy is unique from chemotherapy. While fatigue scores tend to peak around the days following chemotherapy treatment, fatigue associated with radiation therapy tends to accumulate over the course of treatment with peak reports midway through treatment [9–11]. It is unclear whether radiotherapy or chemotherapy induces more severe fatigue. Chan and Molassiotis [29] reported that chemotherapy patients described greater fatigue severity than radiotherapy patients 2 weeks following commencement of their treatment. Conversely, Berglund et al. [30] found radiotherapy patients reported lower quality of life and decreased stamina than chemotherapy patients 2–10 years following surgery for breast cancer. Given the temporal differences in fatigue between chemotherapy and radiotherapy, direct comparisons are difficult and likely contribute to the variability in the literature. However, both are likely to contribute to fatigue and combined treatment is known to increase the risk of fatigue [31].

Psychological Factors

Diagnosis of cancer itself is likely to influence fatigue. Patients must cope with the diagnosis, alter their schedule to accommodate treatment schedules, and endure other related disruptions to their everyday life. These potential stressors may result in emotional distress, which has been shown to predict fatigue severity [32]. Poor coping strategies have also been shown to correlate with greater fatigue severity [33]. Following initiation of treatment, common side effects of cancer therapy (e.g., nausea, hair loss, pain) are likely to exacerbate anxiety and fatigue. The severity of nausea and pain have both been shown to correlate with fatigue severity [34, 35]. This has led researchers to suggest that fatigue may diminish with aggressive management of these common side effects of cancer treatment [36].

It is important to separate fatigue related to cancer diagnosis and treatment from mental illness, such as depression. The phenomenology among symptom clusters is often similar. For example, anxiety, weight loss, insomnia, and malaise are all symptoms of both cancer-related fatigue and major depressive disorder [37]. The distinction is further complicated by the finding that these symptoms may share common mechanisms [38–40]. However, fatigue and depression are two unique disorders.

Fatigue has been demonstrated in patients without mood disorders [41]. Furthermore, to meet the criteria for diagnosis of cancer-related fatigue, the symptoms must not be a result of a comorbid psychiatric condition [15].

Putative Mechanisms of Fatigue

Conceptual Model

The evidence outlined in section "Progenitors of Fatigue" suggests there are several possible progenitors of fatigue. One hypothesis is that each of these factors acts independently with separate mechanisms to reach a common phenotype. An alternative hypothesis is that several of these factors converge on related pathways. This section will examine the physiological effects of each of the factors and how they may induce the symptoms of fatigue. The framework here is a conceptual model based on review of the literature (see Fig. 1). Additional evidence is needed to validate the model and determine the strengths of each of the relationships.

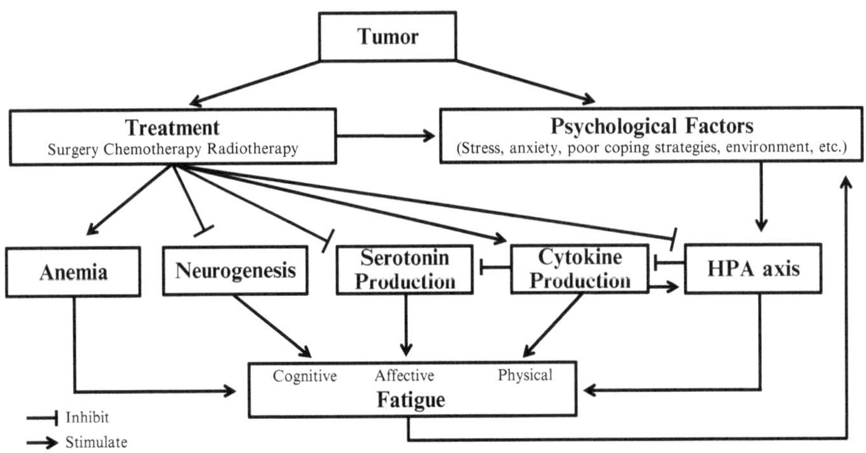

Fig. 1 Diagram outlines the hypothesized contributors to fatigue. Diagnosis of a tumor is a stressful event for many patients. Treatment for cancer is likely to add to the psychological stress of the diagnosis. Stress can manifest as both emotional and physical components of fatigue. Furthermore, chronic stress can alter the body's stress response and prevent proper maintenance of other bodily systems such as cytokine production. Cancer treatments can have a broad impact, including anemia, impairment of neurogenesis, serotonin dysregulation, increased cytokine production, and impairment of the HPA axis. All these factors are likely to contribute to the cognitive, affective, and physical components of fatigue. Furthermore, fatigue itself may be a self-perpetuating loop if the patient becomes anxious about being fatigued. Further research is needed to establish the strength of the relationship between each of these systems and fatigue, but this framework provides an overview of the current hypotheses

Anemia

Chemotherapy and radiotherapy both can lead to anemia. Anemia occurs in approximately 10–40% of cancer patients [42–44]. However, the occurrence is dependent on treatment and cancer type. The relationship between anemia and fatigue is well established and uncontroversial [45–48]. Anemia has been demonstrated to result in fatigue in both patients without cancer and cancer patient populations [46, 48]. The mechanism of anemia-related fatigue is unknown; however, it is hypothesized that fatigue results from lack of oxygen delivery to brain, muscle, and other organs [49]. Fatigue in anemic patients is reduced following increased hemoglobin by administration of erythropoietin therapy [50]. However, fatigue has been observed in patients independently of anemia, suggesting alternative mechanisms also contribute to fatigue [51].

Cytokine Dysregulation

Recent studies suggest that elevation of inflammatory cytokines may influence fatigue. A review of clinical studies examining inflammatory markers and fatigue in cancer patients revealed significant positive correlations between fatigue and interleukin (IL)-6, IL-1, and neopterin [52]. The role of cytokines in behavioral changes has been extensively studied in mice using a component of the cell wall of gram-negative bacteria, lipopolysaccharide (LPS), to induce sickness behavior (i.e., listlessness, lethargy, and decreased consumatory behavior) [53]. In this model, LPS binds to toll-like receptors on innate immune cells which results in the production of the proinflammatory cytokines, IL-1α and IL-1β. IL-1 then promotes the synthesis of tumor necrosis factor (TNF)-α and IL-6 which potentiate its action. The central nervous system (CNS) responds to peripheral cytokines via two modes of transmission. The first mode is via IL-1 receptors expressed on afferent nerves, particularly the vagus nerve [54]. The main recipient of vagal afferents is the nucleus tractus solitarii (NTS) of the dorsal vagal complex. The inflammatory signal is then propagated to multiple forebrain nuclei to precipitate sickness behavior. The second mode in which peripheral cytokines signal the CNS is via activation of macrophage-like cells in circumventricular organs and endothelial cells of brain vessels [53]. These cells in turn produce cytokines and molecular intermediates, such as prostaglandins, which diffuse into the local brain parenchyma. Activation of the hypothalamic–pituitary–adrenal (HPA) axis, which is responsible for the control of stress hormones, is mediated by prostaglandin. Both the neural and humoral modes of transmission influence cytokine-induced behavioral changes.

HPA Axis

Dysregulation of the HPA axis is an intriguing target for fatigue because of its involvement in regulating systems associated with fatigue. The HPA axis is part of the neuroendocrine system which controls reactions to stressors and regulates homeostasis of body processes including the immune system, mood, and energy storage. The HPA axis regulates the release of the stress hormone cortisol in response to physical or psychosocial stressors [49, 55, 56]. Cortisol influences a number of biological processes including blood pressure, cardiovascular function, carbohydrate metabolism, and immune function. Chronic stress may result in down-regulation of central corticotrophin-releasing hormone (CRH) and CRH receptors leading to fatigue [49]. This hypothesis is supported by research with breast cancer patients which demonstrated fatigued patients had an abnormal circadian rhythm of cortisol levels compared to non-fatigued patients [57].

The interactions between the HPA axis, immune system, and psychological outcomes are complex. Activation of the HPA axis is thought to inhibit the immune system and expression of inflammatory cytokines [58]. Reciprocally, cytokines stimulate the HPA axis resulting in self-regulation of the immune system [59]. Chronic stress or some cancer treatments may lead to a suppression of the HPA axis [60–62]. Suppression of the HPA axis may inhibit cytokine regulation resulting in an overall increase in cytokine signaling and fatigue. Further evidence is needed to elucidate the role and interactions of the HPA axis and cytokines, but their involvement is likely given the large number of stressors related to a diagnosis of cancer.

Serotonin Dysregulation

Dysregulation of serotonin signaling has been proposed as a possible explanation for cancer-related fatigue [49, 63, 64]. Serotonin is thought to modulate feelings of fatigue and depression. One hypothesis is that there is an overexpression of serotonin signaling. This idea is supported in rodents where sustained exercise resulted in increased serotonin which in turn correlated with fatigue [65, 66]. An alternative hypothesis is that attenuated serotonin signaling is responsible for the symptoms. Proinflammatory cytokines may influence serotonin metabolism. Interleukin, interferon, and TNF-α may alter serotonin metabolism by stimulating indoleamine 2,3-dioxygenase which results in a depletion of the serotonin precursor, tryptophan [67]. The serotonin hypothesis was most directly tested in two clinical trials of a selective serotonin reuptake inhibitor, paroxetine hydrochloride (Paxil) [68, 69]. Both studies demonstrated a positive effect of the drug in reducing depressed mood, however no effect on fatigue was observed. Therefore, it is unlikely serotonin dysregulation is the primary mechanism of cancer-related fatigue.

Energy Metabolism

Reports of peripheral fatigue, such as limb weakness or heaviness, have been supported by objective measurements of physical performance [70–73]. Adenosine triphosphate (ATP) is the major source of energy for muscle contraction; therefore a reduction in its availability may explain fatigue symptoms. Increased mRNA of uncoupling proteins, which increase energy expenditure, has been observed in skeletal muscle of cancer patients as compared to healthy controls, suggesting metabolism may be increased in cancer populations [74]. Additionally, patients often have reduced caloric intake due to common side effects of therapy (e.g., nausea, anorexia, cachexia). The ATP hypothesis has not been tested clinically in cancer populations; however, administration of the precursor to nicotinamide adenine dinucleotide (NAD), a coenzyme which facilitates the formation of ATP, was found to decrease fatigue in chronic fatigue syndrome patients [75]. Further evidence is needed to determine the role of ATP availability in cancer-related fatigue.

Neurogenesis

Neurogenesis refers to the formation of new neurons. It is thought to occur in only two areas of the CNS during adulthood, the olfactory bulb and the dentate gyrus of the hippocampus [76]. Chemotherapeutic agents that pass the blood–brain barrier or radiation targeted at the brain are likely to reduce neurogenesis. While neurogenesis cannot be measured in vivo in humans, animal studies have found attenuated neurogenesis in the hippocampus of rodents treated with radiation or chemotherapy [77–79].

Depressed mood and cognitive impairments are features of cancer-related fatigue which may be related to diminished neurogenesis. Reductions in neurogenesis have been associated with cognitive dysfunction and depression. Smaller hippocampal volumes have been observed in patients with depression compared to those without [80, 81]. In rodents, ablation of progenitor cells in the hippocampus by targeted irradiation resulted in impaired performance in learning and memory tasks [77]. Similar reductions in learning and memory were observed in rodents treated with chemotherapy [78, 79, 82–85].

Exercise may ameliorate fatigue and is often recommended to patients experiencing fatigue [86–88]. Clinically, exercise has been shown to increase hippocampal volume in elderly patients [89]. In rodents, voluntary wheel running exercise has been shown to increase neurogenesis and improve performance on learning and memory tasks [77]. Further research is needed to conclude changes in hippocampal volume are a result of neurogenesis as the preclinical research suggests. Overall the evidence suggests reductions in neurogenesis may contribute to cancer treatment-related fatigue, cognitive dysfunction, and depression.

Treatment Options

Current Standard of Care

There is currently a large need for therapies to combat cancer-related fatigue. No pharmaceutical compound is presently approved by the United States Federal Drug Administration (FDA) for the treatment of cancer-related fatigue. However, open label clinical trials have been performed for multiple psychostimulant medications as a possible intervention. These trials have resulted in limited success. Of the psychostimulant medications, methylphenidate is one of the most studied treatments for cancer-related fatigue. Methylphenidate has been shown to be effective in treating fatigue in some studies [90–92], but not others [93–96]. When examined in a meta-analysis, methylphenidate was favored over placebo for the treatment of fatigue [97]. Despite the potential efficacy, the clinical utility of methylphenidate in treating fatigue may be limited by adverse reactions to the drug. Several studies cited patient complaints or even withdrawal from the study following side effects including headache, nausea, dry mouth, and blood pressure elevation [90–92].

Recently interest in modafinil for the treatment of fatigue has risen. Modafinil is an attractive candidate because it is thought to have a different mechanism of action than other psychostimulants such as methylphenidate [98], and in two pilot studies it was shown to be well tolerated and potentially efficacious [99, 100]. A phase 3 randomized, placebo-controlled, double-blind, clinical trial in 631 patients showed that modafinil was effective in treating patients with severe, but not mild or moderate, fatigue [101]. Further research is needed before modafinil can gain FDA approval; however, the results so far are promising.

Anemia is a possible side effect of chemotherapy and radiation therapy which is thought to influence fatigue. Erythropoietin-stimulating agents (ESAs) are effective in increasing hemoglobin and therefore combating anemia. Interestingly, they are also effective in decreasing reports of fatigue and improving quality of life [50]. However, studies have shown that ESAs increase the risk of cardiovascular problems due to blood clots. This has prompted the FDA to advise against the use of ESAs for the treatment of fatigue [97].

The most common treatments prescribed to patients suffering from fatigue are non-pharmaceutical. For some patients cognitive behavioral therapy may be appropriate. Cognitive behavioral therapy was shown effective in reducing fatigue in cancer patients in a randomized controlled trial [102]. In this study, mixed-type cancer survivors were divided into an intervention or wait-list control group. Cognitive behavioral therapy was focused on treating one or more of the following factors thought to perpetuate fatigue: insufficient coping with the experience of cancer, fear of disease recurrence, dysfunctional cognitions regarding fatigue, dysregulation of sleep, dysregulation of activity, low social support, and/or negative social interactions. Clinically significant reductions in fatigue severity and functional impairment were observed in the cognitive behavioral therapy group as compared to wait-list controls.

Recent evidence suggests exercise may be beneficial in treating fatigue. Meta-analyses of studies examining the effect of exercise treatments have shown exercise to be more effective than control in treating fatigue during and following cancer therapy [86–88]. The beneficial effects of exercise are not limited to specific patient populations. While the majority of studies have examined breast cancer patients, exercise has been found to be effective in germ cell cancer, lung carcinoma, and leukemia patients as well [103–105]. Additionally, multiple forms of exercise have been shown effective. Most studies have used aerobic exercise programs focused on endurance (e.g., treadmill running); however, incorporation of anabolic resistance training (e.g., weight lifting) has also been shown to reduce fatigue and improve quality of life [87, 103, 106]. Currently, exercise treatment is limited by patients' ability to exercise and adherence to exercise programs. Compliance is a serious issue as evidenced by Swenson et al. [107] who found approximately 1 in 4 breast cancer patients failed to adhere to a 10,000 steps per day protocol. Further research is needed to increase patient compliance with exercise programs and to develop methods to ensure that all patients are successfully treated.

Conclusion

Cancer-related fatigue is now a well-known symptom associated with cancer and cancer care. Tumor burden, cancer treatment, and stress associated with dealing with the diagnosis of cancer are all likely to contribute to fatigue. The mechanism of fatigue is currently unknown; however, several putative mechanisms are hypothesized. Anemia is known to induce fatigue, however fatigue has been observed in non-anemic patients, suggesting alternative mechanisms contribute. Inflammatory cytokines have been suggested as a potential mechanism. Animal research has provided support for this hypothesis by detailing the pathways involved in "sickness behavior" following increases in cytokines. The stress response is also likely to contribute. Cytokine production is partially regulated by the HPA axis which is sensitive to chronic stress, chemotherapy, and circulating cytokines. Additional mechanisms, including serotonin dysregulation, ATP depletion, and impaired neurogenesis, may also contribute to fatigue.

Depending on the primary driving factors of fatigue, any one therapy may not act as a panacea. Given the number of putative mechanisms and the complexity of fatigue, it is likely that several factors contribute to fatigue symptoms. Improved screening tools may help inform clinicians what is driving fatigue in patients is so they can more accurately treat fatigue. Assays to measure cytokines and cortisol are commercially available, while measurements of ATP and neurogenesis are more difficult.

Current treatment options for fatigue are limited. ESAs are useful for treating fatigue related to anemia; however, the cardiovascular risk involved with the drugs prevents them from widespread use. Currently, exercise and cognitive behavioral therapy are suggested for fatigue. Psychostimulant medications are being investigated for potential efficacy for fatigue. Results with modafinil look promising; however, further testing is required before FDA approval is achieved. There is still a large need for a pharmaceutical therapy for fatigue and as the understanding of the mechanisms of fatigue grows so will the opportunity to develop targeted therapies.

References

1. Campos MP, et al. Cancer-related fatigue: a practical review. Ann Oncol. 2011;22(6): 1273–9.
2. Curt GA. The impact of fatigue on patients with cancer: overview of FATIGUE 1 and 2. Oncologist. 2000;5 Suppl 2:9–12.
3. Patarca-Montero R. Handbook of cancer-related fatigue. Haworth research series on malaise, fatigue, and debilitation. New York: Haworth Medical Press; 2004, xi. p. 483.
4. Meyerowitz BE, Sparks FC, Spears IK. Adjuvant chemotherapy for breast carcinoma: psychosocial implications. Cancer. 1979;43(5):1613–8.
5. Richardson A, Ream EK. Self-care behaviours initiated by chemotherapy patients in response to fatigue. Int J Nurs Stud. 1997;34(1):35–43.
6. Ashbury FD, et al. A Canadian survey of cancer patients' experiences: are their needs being met? J Pain Symptom Manage. 1998;16(5):298–306.
7. Bernhard J, Ganz PA. Psychosocial issues in lung cancer patients (Part 2). Chest. 1991;99(2):480–5.
8. Hurny C, et al. "Fatigue and malaise" as a quality-of-life indicator in small-cell lung cancer patients. The Swiss Group for Clinical Cancer Research (SAKK). Support Care Cancer. 1993;1(6):316–20.
9. Schwartz AL, et al. Fatigue patterns observed in patients receiving chemotherapy and radiotherapy. Cancer Invest. 2000;18(1):11–9.
10. Dhruva A, et al. Trajectories of fatigue in patients with breast cancer before, during, and after radiation therapy. Cancer Nurs. 2010;33(3):201–12.
11. Miaskowski C, et al. Trajectories of fatigue in men with prostate cancer before, during, and after radiation therapy. J Pain Symptom Manage. 2008;35(6):632–43.
12. Bower JE, et al. Fatigue in long-term breast carcinoma survivors: a longitudinal investigation. Cancer. 2006;106(4):751–8.
13. Redmond K. Advances in supportive care. Eur J Cancer Care (Engl). 1996;5(2 Suppl): 1–7.
14. Jacobsen PB. Assessment of fatigue in cancer patients. J Natl Cancer Inst Monogr. 2004;32:93–7.
15. Cella D, et al. Progress toward guidelines for the management of fatigue. Oncology (Williston Park). 1998;12(11A):369–77.
16. Cella D, et al. Cancer-related fatigue: prevalence of proposed diagnostic criteria in a United States sample of cancer survivors. J Clin Oncol. 2001;19(14):3385–91.
17. Sadler IJ, et al. Preliminary evaluation of a clinical syndrome approach to assessing cancer-related fatigue. J Pain Symptom Manage. 2002;23(5):406–16.
18. Minton O, Stone P. A systematic review of the scales used for the measurement of cancer-related fatigue (CRF). Ann Oncol. 2009;20(1):17–25.
19. Glaus A. Fatigue in patients with cancer. Analysis and assessment. Recent Results Cancer Res. 1998;145:I–XI, 1–172.
20. Wood LJ, et al. The cancer chemotherapy drug etoposide (VP-16) induces proinflammatory cytokine production and sickness behavior-like symptoms in a mouse model of cancer chemotherapy-related symptoms. Biol Res Nurs. 2006;8(2):157–69.
21. Servaes P, Verhagen S, Bleijenberg G. Determinants of chronic fatigue in disease-free breast cancer patients: a cross-sectional study. Ann Oncol. 2002;13(4):589–98.
22. Rubin GJ, Hardy R, Hotopf M. A systematic review and meta-analysis of the incidence and severity of postoperative fatigue. J Psychosom Res. 2004;57(3):317–26.
23. Schroeder D, Hill GL. Postoperative fatigue: a prospective physiological study of patients undergoing major abdominal surgery. Aust N Z J Surg. 1991;61(10):774–9.
24. Rubin GJ, Cleare A, Hotopf M. Psychological factors in postoperative fatigue. Psychosom Med. 2004;66(6):959–64.

25. Christensen T, Hougard F, Kehlet H. Influence of pre- and intra-operative factors on the occurrence of postoperative fatigue. Br J Surg. 1985;72(1):63–5.
26. Berger AM. Patterns of fatigue and activity and rest during adjuvant breast cancer chemotherapy. Oncol Nurs Forum. 1998;25(1):51–62.
27. Silberfarb PM, et al. Psychological response of patients receiving two drug regimens for lung carcinoma. Am J Psychiatry. 1983;140(1):110–1.
28. Genre D, et al. Quality of life of breast cancer patients receiving high-dose-intensity chemotherapy: impact of length of cycles. Support Care Cancer. 2002;10(3):222–30.
29. Chan CW, Molassiotis A. The impact of fatigue on Chinese cancer patients in Hong Kong. Support Care Cancer. 2001;9(1):18–24.
30. Berglund G, et al. Late effects of adjuvant chemotherapy and postoperative radiotherapy on quality of life among breast cancer patients. Eur J Cancer. 1991;27(9):1075–81.
31. Jereczek-Fossa BA, Marsiglia HR, Orecchia R. Radiotherapy-related fatigue. Crit Rev Oncol Hematol. 2002;41(3):317–25.
32. Higgins SC, et al. Effect of pretreatment distress on daily fatigue after chemotherapy for breast cancer. J Oncol Pract. 2008;4(2):59–63.
33. Broeckel JA, et al. Characteristics and correlates of fatigue after adjuvant chemotherapy for breast cancer. J Clin Oncol. 1998;16(5):1689–96.
34. Burrows M, Dibble SL, Miaskowski C. Differences in outcomes among patients experiencing different types of cancer-related pain. Oncol Nurs Forum. 1998;25(4):735–41.
35. Wang XS, et al. Clinical factors associated with cancer-related fatigue in patients being treated for leukemia and non-Hodgkin's lymphoma. J Clin Oncol. 2002;20(5):1319–28.
36. Jacobsen PB, Stein K. Is fatigue a long-term side effect of breast cancer treatment? Cancer Control. 1999;6(3):256–63.
37. Jacobsen PB, Donovan KA, Weitzner MA. Distinguishing fatigue and depression in patients with cancer. Semin Clin Neuropsychiatry. 2003;8(4):229–40.
38. Bower JE, et al. Inflammation and behavioral symptoms after breast cancer treatment: do fatigue, depression, and sleep disturbance share a common underlying mechanism? J Clin Oncol. 2011;29(26):3517–22.
39. Lee BN, et al. A cytokine-based neuroimmunologic mechanism of cancer-related symptoms. Neuroimmunomodulation. 2004;11(5):279–92.
40. Miller AH, et al. Neuroendocrine-immune mechanisms of behavioral comorbidities in patients with cancer. J Clin Oncol. 2008;26(6):971–82.
41. Goldstein D, et al. Fatigue states after cancer treatment occur both in association with, and independent of, mood disorder: a longitudinal study. BMC Cancer. 2006;6:240.
42. Coiffier B. The impact and management of anaemia in haematological malignancies. Med Oncol. 2000;17 Suppl 1:S2–10.
43. Groopman JE, Itri LM. Chemotherapy-induced anemia in adults: incidence and treatment. J Natl Cancer Inst. 1999;91(19):1616–34.
44. Monti M, et al. Use of red blood cell transfusions in terminally ill cancer patients admitted to a palliative care unit. J Pain Symptom Manage. 1996;12(1):18–22.
45. Cella D. The Functional Assessment of Cancer Therapy-Anemia (FACT-An) Scale: a new tool for the assessment of outcomes in cancer anemia and fatigue. Semin Hematol. 1997;34 (3 Suppl 2):13–9.
46. Demetri GD, et al. Quality-of-life benefit in chemotherapy patients treated with epoetin alfa is independent of disease response or tumor type: results from a prospective community oncology study. Procrit Study Group. J Clin Oncol. 1998;16(10):3412–25.
47. Yellen SB, et al. Measuring fatigue and other anemia-related symptoms with the Functional Assessment of Cancer Therapy (FACT) measurement system. J Pain Symptom Manage. 1997;13(2):63–74.
48. Agnihotri P, et al. Chronic anemia and fatigue in elderly patients: results of a randomized, double-blind, placebo-controlled, crossover exploratory study with epoetin alfa. J Am Geriatr Soc. 2007;55(10):1557–65.

49. Ryan JL, et al. Mechanisms of cancer-related fatigue. Oncologist. 2007;12 Suppl 1:22–34.
50. Charu V, et al. Efficacy and safety of every-2-week darbepoetin alfa in patients with anemia of cancer: a controlled, randomized, open-label phase II trial. Oncologist. 2007;12(6):727–37.
51. Holzner B, et al. The impact of hemoglobin levels on fatigue and quality of life in cancer patients. Ann Oncol. 2002;13(6):965–73.
52. Schubert C, et al. The association between fatigue and inflammatory marker levels in cancer patients: a quantitative review. Brain Behav Immun. 2007;21(4):413–27.
53. Dantzer R. Cytokine, sickness behavior, and depression. Neurol Clin. 2006;24(3):441–60.
54. Ek M, et al. Activation of vagal afferents after intravenous injection of interleukin-1beta: role of endogenous prostaglandins. J Neurosci. 1998;18(22):9471–9.
55. Bonfiglio JJ, et al. The corticotropin-releasing hormone network and the hypothalamic-pituitary-adrenal axis: molecular and cellular mechanisms involved. Neuroendocrinology. 2011;94(1):12–20.
56. Smith SM, Vale WW. The role of the hypothalamic-pituitary-adrenal axis in neuroendocrine responses to stress. Dialogues Clin Neurosci. 2006;8(4):383–95.
57. Bower JE, Ganz PA, Aziz N. Altered cortisol response to psychologic stress in breast cancer survivors with persistent fatigue. Psychosom Med. 2005;67(2):277–80.
58. Petrovsky N, McNair P, Harrison LC. Diurnal rhythms of pro-inflammatory cytokines: regulation by plasma cortisol and therapeutic implications. Cytokine. 1998;10(4):307–12.
59. Vgontzas AN, Chrousos GP. Sleep, the hypothalamic-pituitary-adrenal axis, and cytokines: multiple interactions and disturbances in sleep disorders. Endocrinol Metab Clin North Am. 2002;31(1):15–36.
60. Del Priore G, et al. Adrenal function following high-dose steroids in ovarian cancer patients. Gynecol Oncol. 1995;59(1):102–4.
61. Morrow GR, et al. Reduction in serum cortisol after platinum based chemotherapy for cancer: a role for the HPA axis in treatment-related nausea? Psychophysiology. 2002;39(4):491–5.
62. Schmiegelow M, et al. Assessment of the hypothalamo-pituitary-adrenal axis in patients treated with radiotherapy and chemotherapy for childhood brain tumor. J Clin Endocrinol Metab. 2003;88(7):3149–54.
63. Barsevick A, et al. I'm so tired: biological and genetic mechanisms of cancer-related fatigue. Qual Life Res. 2010;19(10):1419–27.
64. Wang XS. Pathophysiology of cancer-related fatigue. Clin J Oncol Nurs. 2008;12(5 Suppl): 11–20.
65. Bailey SP, Davis JM, Ahlborn EN. Neuroendocrine and substrate responses to altered brain 5-HT activity during prolonged exercise to fatigue. J Appl Physiol. 1993;74(6):3006–12.
66. Blomstrand E, et al. Effect of sustained exercise on plasma amino acid concentrations and on 5-hydroxytryptamine metabolism in six different brain regions in the rat. Acta Physiol Scand. 1989;136(3):473–81.
67. Wichers M, Maes M. The psychoneuroimmuno-pathophysiology of cytokine-induced depression in humans. Int J Neuropsychopharmacol. 2002;5(4):375–88.
68. Roscoe JA, et al. Effect of paroxetine hydrochloride (Paxil) on fatigue and depression in breast cancer patients receiving chemotherapy. Breast Cancer Res Treat. 2005;89(3):243–9.
69. Morrow GR, et al. Differential effects of paroxetine on fatigue and depression: a randomized, double-blind trial from the University of Rochester Cancer Center Community Clinical Oncology Program. J Clin Oncol. 2003;21(24):4635–41.
70. Dimeo F, et al. Correlation between physical performance and fatigue in cancer patients. Ann Oncol. 1997;8(12):1251–5.
71. Akechi T, et al. Fatigue and its associated factors in ambulatory cancer patients: a preliminary study. J Pain Symptom Manage. 1999;17(1):42–8.
72. Brown DJ, McMillan DC, Milroy R. The correlation between fatigue, physical function, the systemic inflammatory response, and psychological distress in patients with advanced lung cancer. Cancer. 2005;103(2):377–82.
73. Lee JQ, et al. Differences in physical performance between men and women with and without lymphoma. Arch Phys Med Rehabil. 2003;84(12):1747–52.

74. Collins P, et al. Muscle UCP-3 mRNA levels are elevated in weight loss associated with gastrointestinal adenocarcinoma in humans. Br J Cancer. 2002;86(3):372–5.
75. Forsyth LM, et al. Therapeutic effects of oral NADH on the symptoms of patients with chronic fatigue syndrome. Ann Allergy Asthma Immunol. 1999;82(2):185–91.
76. Abrous DN, Koehl M, Le Moal M. Adult neurogenesis: from precursors to network and physiology. Physiol Rev. 2005;85(2):523–69.
77. Clark PJ, et al. Intact neurogenesis is required for benefits of exercise on spatial memory but not motor performance or contextual fear conditioning in C57BL/6J mice. Neuroscience. 2008;155(4):1048–58.
78. Mustafa S, et al. 5-Fluorouracil chemotherapy affects spatial working memory and newborn neurons in the adult rat hippocampus. Eur J Neurosci. 2008;28(2):323–30.
79. Seigers R, et al. Long-lasting suppression of hippocampal cell proliferation and impaired cognitive performance by methotrexate in the rat. Behav Brain Res. 2008;186(2):168–75.
80. Bremner JD, et al. Hippocampal volume reduction in major depression. Am J Psychiatry. 2000;157(1):115–8.
81. Colla M, et al. Hippocampal volume reduction and HPA-system activity in major depression. J Psychiatr Res. 2007;41(7):553–60.
82. ElBeltagy M, et al. Fluoxetine improves the memory deficits caused by the chemotherapy agent 5-fluorouracil. Behav Brain Res. 2010;208(1):112–7.
83. Konat GW, et al. Cognitive dysfunction induced by chronic administration of common cancer chemotherapeutics in rats. Metab Brain Dis. 2008;23(3):325–33.
84. Macleod JE, et al. Cancer chemotherapy impairs contextual but not cue-specific fear memory. Behav Brain Res. 2007;181(1):168–72.
85. Winocur G, et al. The effects of the anti-cancer drugs, methotrexate and 5-fluorouracil, on cognitive function in mice. Pharmacol Biochem Behav. 2006;85(1):66–75.
86. Cramp F, Daniel J. Exercise for the management of cancer-related fatigue in adults. Cochrane Database Syst Rev. 2008(2):CD006145.
87. Galvao DA, Newton RU. Review of exercise intervention studies in cancer patients. J Clin Oncol. 2005;23(4):899–909.
88. Kangas M, Bovbjerg DH, Montgomery GH. Cancer-related fatigue: a systematic and meta-analytic review of non-pharmacological therapies for cancer patients. Psychol Bull. 2008;134(5):700–41.
89. Kramer AF, Erickson KI, Colcombe SJ. Exercise, cognition, and the aging brain. J Appl Physiol. 2006;101(4):1237–42.
90. Hanna A, et al. A phase II study of methylphenidate for the treatment of fatigue. Support Care Cancer. 2006;14(3):210–5.
91. Lower EE, et al. Efficacy of dexmethylphenidate for the treatment of fatigue after cancer chemotherapy: a randomized clinical trial. J Pain Symptom Manage. 2009;38(5):650–62.
92. Roth AJ, et al. Methylphenidate for fatigue in ambulatory men with prostate cancer. Cancer. 2010;116(21):5102–10.
93. Butler Jr JM, et al. A phase III, double-blind, placebo-controlled prospective randomized clinical trial of d-threo-methylphenidate HCl in brain tumor patients receiving radiation therapy. Int J Radiat Oncol Biol Phys. 2007;69(5):1496–501.
94. Mar Fan HG, et al. A randomised, placebo-controlled, double-blind trial of the effects of d-methylphenidate on fatigue and cognitive dysfunction in women undergoing adjuvant chemotherapy for breast cancer. Support Care Cancer. 2008;16(6):577–83.
95. Bruera E, et al. Patient-controlled methylphenidate for cancer fatigue: a double-blind, randomized, placebo-controlled trial. J Clin Oncol. 2006;24(13):2073–8.
96. Moraska AR, et al. Phase III, randomized, double-blind, placebo-controlled study of long-acting methylphenidate for cancer-related fatigue: North Central Cancer Treatment Group NCCTG-N05C7 trial. J Clin Oncol. 2010;28(23):3673–9.
97. Minton O, et al. Drug therapy for the management of cancer-related fatigue. Cochrane Database Syst Rev. 2010(7):CD006704.

98. Engber TM, et al. Differential patterns of regional c-Fos induction in the rat brain by amphetamine and the novel wakefulness-promoting agent modafinil. Neurosci Lett. 1998; 241(2–3): 95–8.
99. Blackhall L, et al. A pilot study evaluating the safety and efficacy of modafinal for cancer-related fatigue. J Palliat Med. 2009;12(5):433–9.
100. Spathis A, et al. Modafinil for the treatment of fatigue in lung cancer: a pilot study. Palliat Med. 2009;23(4):325–31.
101. Jean-Pierre P, et al. A phase 3 randomized, placebo-controlled, double-blind, clinical trial of the effect of modafinil on cancer-related fatigue among 631 patients receiving chemotherapy: a University of Rochester Cancer Center Community Clinical Oncology Program Research base study. Cancer. 2010;116(14):3513–20.
102. Gielissen MF, et al. Effects of cognitive behavior therapy in severely fatigued disease-free cancer patients compared with patients waiting for cognitive behavior therapy: a randomized controlled trial. J Clin Oncol. 2006;24(30):4882–7.
103. Dimeo F, et al. Effects of an endurance and resistance exercise program on persistent cancer-related fatigue after treatment. Ann Oncol. 2008;19(8):1495–9.
104. Dimeo F, et al. Effects of aerobic exercise on the physical performance and incidence of treatment-related complications after high-dose chemotherapy. Blood. 1997;90(9):3390–4.
105. Dimeo FC, et al. Aerobic exercise in the rehabilitation of cancer patients after high dose chemotherapy and autologous peripheral stem cell transplantation. Cancer. 1997;79(9): 1717–22.
106. Schneider CM, et al. Effects of supervised exercise training on cardiopulmonary function and fatigue in breast cancer survivors during and after treatment. Cancer. 2007;110(4):918–25.
107. Swenson KK, Nissen MJ, Henly SJ. Physical activity in women receiving chemotherapy for breast cancer: adherence to a walking intervention. Oncol Nurs Forum. 2010;37(3):321–30.

Xerostomia

Ana P. Cotrim, Changyu Zheng, and Bruce J. Baum

Abstract Xerostomia, the sensation of a dry mouth, most often results from a decreased secretion of the salivary glands (salivary hypofunction). Long-term, moderate to severe xerostomia has significant, negative effects on the quality of life of surviving cancer patients. In this chapter we focus on salivary hypofunction resulting from therapeutic radiation received by head and neck cancer patients. Specifically, we address the frequency of its occurrence, the mechanisms by which it likely occurs, as well as available and future approaches for prevention and treatment.

Introduction

Xerostomia is a symptom indicating the presence of a dry mouth, i.e., it is based on a patient's subjective impression. While most often associated with salivary gland hypofunction, i.e., the reduced production of saliva, it can result from other causes, e.g., oral sensory or cognitive deficits. Theoretically, it could also result from a qualitative change in the secreted saliva, e.g., the absence or modification of a key lubricatory protein, although there are few, if any, concrete examples of this possibility. It is also important to recognize that the complaint of a dry mouth may or may not be associated with an obvious reduction in salivary flow. In a general sense, there are three main causes of xerostomia: common prescription medications, autoimmune disease, and cancer treatment-associated toxicity. The focus of this chapter will be limited to xerostomia due to salivary hypofunction resulting from cancer therapy.

A.P. Cotrim, D.D.S., Ph.D. • C. Zheng, M.D., Ph.D. • B.J. Baum, D.M.D., Ph.D. (✉)
Molecular Physiology and Therapeutics Branch, National Institute of Dental and Craniofacial Research, NIH, Building 10, Room 1A01, MSC-1190, Betehsda, MD 20892-1190, USA
e-mail: bbaum@mail.nih.gov

S.T. Sonis and D.M. Keefe (eds.), *Pathobiology of Cancer Regimen-Related Toxicities*,
DOI 10.1007/978-1-4614-5438-0_13, © Springer Science+Business Media New York 2013

Table 1 Late radiation toxicities in salivary glands [3]

Grade	Characteristics
1	Slight mouth dryness; good response to stimulation
2	Moderate mouth dryness; poor response to stimulation
3	Complete mouth dryness; no response to stimulation
4	Gland fibrosis

The prevalence of xerostomia resulting from cancer therapy varies widely. It is near universal for patients during treatment for head and neck cancers, but occurs less frequently in those being treated for thyroid cancer, patients with hematopoietic malignancies treated with stem cell transplantation, or in other cancer patients receiving chemotherapy [1]. Interested readers should note that the Jensen et al. [1] reference and a companion article [2] are cited frequently in this chapter. These are two excellent, systematic reviews that have extensively examined published research reports related to the prevalence, severity, impact on quality of life, management strategies, and economic impact of xerostomia and salivary hypofunction. Both include many valuable primary references, which cannot be cited herein for economy of space. All interested readers of this chapter should read both Jensen et al. [1, 2] reviews.

There are many studies evaluating xerostomia in patients being treated for head and neck cancers. Xerostomia is extremely common in such patients while receiving conventional high dose radiotherapy (>90%, ~60 Gy), declining somewhat in frequency once treatment is completed (~75–80% [1]). The Radiation Therapy Oncology Group (RTOG) describes four grades of late radiation toxicity to salivary glands (Table 1 [3]). Grades 1 and 2 xerostomia (slight, moderate) occur most frequently, ~60–75% [1]. When whole salivary flow rates are examined in such patients, both unstimulated and stimulated flow rates dramatically fall (by >80% on average) from pretreatment levels once radiotherapy begins [1]. While some return of salivary output will occur 3–12 months after radiation is completed, on average whole salivary flows remain markedly depressed, and the subjective complaint of xerostomia improves modestly [1]. At >1 year post-radiotherapy, the prevalence of grades 2 and 3 xerostomia is ~45% [1].

There are few studies of xerostomia in patients treated for thyroid cancer (with radioactive iodine; [4]) or hematopoietic malignancies (with stem cell transplants). Jensen et al. [1] estimate the prevalence of posttreatment xerostomia to be ~35% in the former, while there are too few studies to assess the latter reasonably. Additionally, there are too few studies to make any definitive general conclusions about salivary flow rates for both patient groups. Interestingly, in other cancer patients receiving chemotherapy, the prevalence of xerostomia appears to be fairly high during treatment (~50%), but this does not correspond to any change in unstimulated or stimulated whole saliva output from pretreatment levels [1]. There are not enough studies examining long-term effects, if any, of chemotherapy on xerostomia in cancer patients; however, salivary flow rates appear to be unchanged or slightly increased more than 1 year after treatment in such patients [1].

Table 2 Main physiological functions of saliva[a]

Function	Some key salivary components involved
Mucosal protection, lubrication	Mucins, basic proline-rich glycoprotein, epidermal growth factor, water
pH control	Bicarbonate, histatins
Re-mineralization of teeth	Statherin, anionic proline-rich proteins, Ca^{2+}, PO_4^{3-}
Microbial control	Histatins, immunoglobulins, lysozyme, lactoferrin, lactoperoxidase, mucins
Alimentation, bolus formation, mastication	Mucins, water
Digestion	Amylase, proteases, lipase, DNase, RNase
Taste mediation (solvent and delivery system)	Water

[a]Modified from Kaplan and Baum [84] and Nieuw Amerongen and Veerman [5]

Table 3 Highest and lowest tenth percentile of salivary flow rates[a]

	Parotid		Submandibular	
	Male	Female	Male	Female
Unstimulated				
Lowest	0.000	0.000	0.020	0.025
Highest	0.160	0.156	0.223	0.216
Stimulated				
Lowest	0.201	0.180	0.133	0.107
Highest	0.866	0.838	0.644	0.716

[a]These data are from Ship et al. [8] and are expressed as mL/min gland. Individuals in the lowest tenth percentile for salivary flow were at or below values shown, and those in the highest tenth percentile were at or above values shown

Why is the symptom of a dry mouth significant for a cancer patient? Cancer patients experience many dramatic changes in their life following diagnosis, most seemingly much more important than oral dryness. Certainly during treatment, patients are willing to endure much, including reduced saliva levels and oral dryness, in order to be free of disease. However, once the latter is achieved the significance of xerostomia is realized. In order to appreciate this significance it is necessary for the reader to understand (1) the function of saliva in normal physiology (Table 2), (2) how much saliva is required for such normal functioning (Table 3), and (3) how much saliva needs to be lost before xerostomia occurs.

Saliva has been called the "defender of the oral cavity" [5] and, indeed, one glance at the functions listed in Table 2 shows that it is. It literally and figuratively touches every exposed oral tissue and plays a seminal role in protecting, maintaining, and repairing all hard and soft tissues in the mouth. It is no surprise that patients who lack saliva, for whatever reason, experience dysphagia, excessive tooth decay, candidiasis, slow healing of oral mucosal ulcers, and considerable discomfort. They also may experience difficulty in speaking, as well as dysgeusia. When considered together, it is apparent that many common, but important, socializing activities require the presence of saliva. For example, without adequate salivation, one cannot enjoy a good meal and conversation with family and friends. Thus, it is not surprising that cancer treatment survivors with xerostomia have a significant reduction in their quality of life [1, 6, 7]. The title of the 1960s soul song by William Bell describes most people's every day attention to saliva: "you don't miss your water 'til the well runs dry."

How much saliva is required for normal function and how much saliva needs to be lost before xerostomia occurs? Surprisingly, these two questions have not been often or easily addressed. Almost all oral exams, whether performed by a dentist or a physician, do not include a quantified assessment of salivary output. At best, the examiner recognizes the presence or absence of a sublingual salivary pool and the tone of the mucosa (moist or dry). This means that at a time when someone's salivary glands are functioning normally and the person is paying little attention to his or her saliva or level of oral comfort, there is no record of how much saliva is being produced. It is only when a patient has a xerostomia-related complaint that a clinician may assess salivary flow and, unfortunately, there is no basis of comparison with that patient's normal value. Additionally, there are no longitudinal studies examining salivary function that extend long enough, or are adequately powered, to evaluate the transitions occurring in salivary secretion and perception between normal and xerostomic conditions.

As a consequence, these two questions have been addressed only indirectly and with relatively small numbers of subjects. For example, Ship et al. [8] examined the first question by determining major gland salivary flow rates, objective measurements of oral health status, and subjective assessments of oral health in different-aged men ($n=50$) and women ($n=52$) who were healthy, i.e., using no prescription medications and having no systemic conditions for which they were receiving medical treatment. Within this population, individuals with salivary flow rates in the highest and lowest tenth percentile were identified. There were dramatic differences between these two groups (Table 3). However, when objective and subjective assessments of oral health in these two groups were compared, there were few differences, leading to the conclusion that a wide range of salivary flow rates supports oral health. Thus, individuals can be free from oral complaints and problems despite huge differences in their salivary flow rate, i.e., each one of us has our own normal level of saliva production, and it is important to monitor saliva in individuals over time to interpret the effects of salivary changes on oral health [8].

To address the question of how much saliva needs to be lost before xerostomia occurs, two types of approaches have been used. The first employed the administration of differing doses of anti-cholinergic drugs to healthy individuals and monitoring salivary flow rates and the occurrence of xerostomic symptoms. Both Dawes [9] and Wolff and Kleinberg [10] found that oral dryness occurred when salivary flow rates were reduced to ~50% of an individual's baseline. In these pharmacological titrations of salivary output, symptoms were recognized within 30 min of drug administration [10]. The other approach assessed the residual saliva volume present in the mouth after an individual swallowed ([11]; $n=25$), and found that this was reduced by ~30% in patients complaining of a very dry mouth. Both approaches agree that patients with significant salivary dysfunction do not have a complete lack of oral fluid. Thus, normal sensations of oral moisture represent an accommodation by individuals to the level of saliva they routinely produce.

Mechanisms of Salivary Gland Damage

This section will only review possible mechanisms involved in salivary gland damage induced by ionizing radiation, since there has been little study of xerostomia related to chemotherapy. It is important to recognize that the exact mechanisms leading to radiation-induced xerostomia are not well understood and, indeed, there likely is no "single" causative mechanism. Rather, it is most probable that multiple gland sites are damaged by radiation (Table 4) and, in some unknown combination, lead to a permanent xerostomia in many patients.

It is widely recognized that ionizing radiation can cause many types of cell damage, including that to DNA, nuclear and cell membranes, and various cellular organelles. These can occur at different frequencies, depending upon the cell type targeted and the dose of radiation to which cells were exposed. Radiation produces fast electrons, which can directly (e.g., breakage of chemical bonds in DNA) or indirectly (via radiolysis of water, generating diffusible free radicals) lead to biological damage in critical cellular targets [12]. Free radicals, e.g., hydrogen (H^-), hydroxyl (OH^-), superoxide (O_2^-), and hydroperoxyl (HO_2^-), although short-lived ($t½=\sim10^{-9}$ s), are highly reactive [12].

DNA is considered to be the primary target of radiation-induced free radical damage. The nature of this damage varies considerably and can include base pair deletions, cross-linking, single or double strand breaks, as well as lesions that result in chromosomal aberrations, including deletions, duplications, inversions, insertions, translocations, chromatid exchanges, etc. In response, cells will activate DNA repair pathways to determine the extent of radiation damage suffered and if, and how, the damage can be repaired. If it cannot be repaired, multiple cell death pathways will be activated [13], resulting in accelerated senescence, autophagy, apoptosis, or necrosis [14]. Radiation can also cause sublethal damage that can be addressed through the degradative or repair processes in normal metabolic pathways, e.g., β-oxidation of lipids, aerobic respiration of carbohydrates, or acid hydrolysis of proteins [12].

Given that DNA is such a critical target, generally the most radiosensitive cells are those that are actively dividing, e.g., normal mucosal epithelial cells or undifferentiated malignant cells. Typically, radio-resistant cells are not in active cell division, are well differentiated and have low metabolic rates. Salivary gland cells are considered to be highly differentiated and have low rates of cell division. Thus, it is surprising that salivary glands are relatively radiosensitive.

Salivary glands contain two principal cell types, acinar and duct. Salivary acinar cells, the only cells in a gland able to secrete fluid, are much more sensitive to radiation than duct cells [15, 16]. Additionally, irradiation damage to serous acini occurs more quickly than to the mucous acini [15, 16]. In human glands, an acute

Table 4 Possible key radiation target sites in salivary glands

Acinar cell DNA
Acinar cell exocrine granules
Acinar cell plasma membranes
Microvascular endothelial cells
Autonomic innervation
Salivary-specific stem or progenitor cells

inflammatory response occurs within hours after radiation [17]. Salivary amylase rapidly increases (10–80-fold) in serum, reaching a peak within 12–36 h. In all likelihood, this results from a disruption of parotid acinar cell membranes and immediate cell death [17–21]. Unstimulated and stimulated salivary secretion decreases dramatically soon after conventional radiotherapy, within a week [11, 22–25], and an early loss of parotid acini from apoptosis has been observed both in monkeys [26] and rats [27].

Given this background, the high radiation sensitivity of salivary glands remains an enigma, and no single mechanism adequately explains it. We now consider that multiple radiation damage steps and/or target sites can lead to an accumulation of insults (Table 4). The aggregate result of these insults is acinar cell loss and a dramatic reduction in salivary fluid secretion. Since acinar cells divide slowly, it is unlikely that many cells are lost in the short term due directly to DNA damage. The rapid detection of salivary amylase in serum suggests that initially some cells undergo necrosis or that their membranes are at least rendered highly leaky. There is some evidence in rodents that the high heavy metal (Zn, Mn, Fe) content of exocrine secretory granules may promote free radical formation and enhance radiosensitivity of serous cells [28]. Indeed, in rats, salivary fluid secretion can be protected via inducing acinar cell degranulation prior to irradiation [28].

The occurrence of an acute inflammatory response in human glands within hours of irradiation suggests that another key site for radiation damage is gland endothelium. Vascular endothelium divides slowly and is considered to be of intermediate radiosensitivity [12]. Nonetheless, many have suggested that radiation can directly cause tissue dysfunction by damaging endothelium. For example, radiation damage to the microvascular endothelium has been suggested to cause central nervous system injury [29], radiation pneumonitis [30], and intestinal crypt cell damage, subsequent organ failure, and death from GI syndrome [31]. Recently, we showed in a mouse model that salivary gland radiation decreased microvascular density in submandibular glands by ~45% within 4 h. However, the loss in microvascular density was significantly reduced, and salivary secretion preserved, if mice were pretreated with adenoviral vectors encoding an angiogenic factor [32]. A similar phenomenon occurs in miniature pig parotid glands after radiation, i.e., rapid loss of microvascular endothelial cells and reduction in local parotid gland blood flow [33]. This was associated with enhanced activity of glandular acid and neutral Mg^{2+}-dependent sphingomyelinases and the presence of CD31 positive apoptotic cells in the glands [33].

Another possible non-glandular target possibly affected by radiation therapy is the autonomic innervation of the gland. While there is not an abundance of experimental evidence supporting this notion, there is some (e.g., [34, 35]). Furthermore, recent studies by Knox et al. [36] suggest an important role of parasympathetic innervation in facilitating normal salivary gland development, and by inference in gland repair and maintenance. Indeed, the studies of Knox et al. [36] support a fine orchestration of signals between gland epithelium, adjacent mesenchyme and the associated parasympathetic nerves, and microvessels in normal gland development.

Finally, another cell population, salivary gland-specific stem or progenitor cells, also may be critical targets for radiation damage. In rats, it previously was recognized that loss of acinar cells occurred 30 days postirradiation, which was probably due to death of cells that normally proliferate [37]. In mice, when salivary gland stem or progenitor cells were cultured as salispheres, the numbers of such cells obtained after irradiation were significantly decreased from those in nonirradiated glands. Furthermore, if mice received an injection of keratinocyte growth factor prior to irradiation, the baseline number of salivary stem or progenitor cells was increased, enabling the subsequently irradiated mice to become more resistant to salivary gland damage [38] and preserve saliva production. The latter experiments strongly indicate that salivary stem or progenitor cell damage reduces the ability of irradiated glands to generate new gland cells and repair, on a tissue level, the radiation-induced xerostomia [38]. Recently, we also demonstrated that ionizing radiation reduced the numbers of BrdU positive (replicating) and c-kit (stem or progenitor) positive cells in mouse submandibular glands. However, if mice were treated prior to radiation with an adenoviral vector encoding keratinocyte growth factor, salivary flow rate was preserved [39], consistent with Lombaert et al. [38].

There is, however, an important caveat to this discussion of mechanisms: the majority of head and neck cancers today are treated with a combination of radiation and chemotherapy, with or without surgery. A main goal of chemotherapy in this context is to sensitize tumor cells to radiation; however, it may also affect normal tissue. It remains controversial as to whether patients undergoing chemoradiation therapy develop xerostomia more frequently or severely than patients treated with radiotherapy alone.

Prevention of Salivary Hypofunction

A key factor in establishing the future quality of life for patients receiving radiotherapy for a head and neck cancer is if they experience long-term xerostomia. Recently, several technical advances have been made that help to limit the amount of radiation exposure these patients receive to their salivary glands and, therefore, minimize the loss of salivary function.

Conformal radiotherapy, intensity-modulated radiotherapy (IMRT), image-guided radiotherapy, and proton radiotherapy rely on imaging the tumor and normal tissue, usually via tomography, and precisely delineating its area. This allows a higher amount of radiation to target the tumor while a lower amount is delivered to the adjacent normal tissue. As a consequence, all of these techniques diminish the normal tissue toxicity associated with radiation treatments. IMRT has an advantage over the other techniques as it uses combinations of intensity-modulated fields to produce a custom dose distribution around the tumor, with higher dose gradients at the transition to adjacent normal tissues. IMRT can improve local control of tumor and reduce salivary hypofunction, resulting in patients treated with this modality having a higher health-related quality of life than other patients treated with conventional therapy [40]. However, tumor location, e.g., tonsils and the proximity of involved lymph nodes, can impede sparing, even with highly conformal approaches.

Although IMRT is a very significant improvement for the treatment of cancer in general, when treating cancer there is a persistent need to irradiate some normal tissue as a safety margin. This is due to uncertainties as to the extent of microscopic disease present, the reliability of pre-radiotherapy examinations and imaging studies, and concerns about possible positioning setup errors. For head and neck tumors, typically ~2 cm of normal tissue surrounding the tumor will be included in the higher dose radiation field [41]. Thus, ~40% of patients treated with parotid sparing IMRT still present with xerostomic complaints [42], and presumably, for the foreseeable future, normal tissue toxicity will remain a possible dose limiting factor and a side effect.

Salivary gland displacement surgery before radiotherapy can limit radiation exposure. This procedure transfers the submandibular gland, usually to the submental space, where it would be shielded or receive a lower dose of radiation. Patients who had their submandibular glands displaced before radiotherapy maintain a higher salivary flow, and in general experience a better quality of life with less xerostomic complaints, than other patients. Until recently only patients undergoing tumor resection surgery before radiation treatment were eligible for the salivary gland displacement. However, new minimally invasive and intraoral surgical approaches are being developed [43] allowing a faster healing time and lowering the risk of delaying the start of radiotherapy due to this procedure. Nonetheless, the transfer of the submandibular gland is not feasible for patients with malignant tumors in the oral cavity, or patients who have involvement of the submental or submandibular lymph nodes with metastatic disease, as in these patients the submental space typically is included in the radiation field.

Several pharmaceutical agents also have been used to prevent the loss of salivary function during radiotherapy. Sialogogues, e.g., pilocarpine and cevimeline, are drugs that increase the production of saliva, and are usually used after radiation treatment in patients with enough responsive salivary gland tissue, i.e., RTOG grade 1 (Table 1; see below). However pilocarpine has also been used during the radiotherapy in an attempt to prevent salivary hypofunction. Pilocarpine acts, non-selectively, at muscarinic receptors present in several tissues, including salivary gland cells, stimulating secretion. However, the benefits of administering pilocarpine to prevent xerostomia are limited, and several clinical trials show conflicting results, e.g., beneficial in preserving some salivary flow or showing no benefit in placebo-controlled trials [44].

Currently the only FDA radioprotector approved for salivary glands is Amifostine. Amifostine is an organic thiophosphate that accumulates in higher concentrations in normal tissue than in tumor. Also, it is a pro-drug, whose activity depends on cellular alkaline phosphatase for conversion to its active form. This enzyme is more efficient in the alkaline environment of normal tissue than in the acidic tumor environment. Amifostine acts by scavenging free radicals at the time of radiation treatment [45]. The guidelines of the American Society of Clinical Oncology, however, do not support the use of Amifostine for chemotherapy in combination with radiation. In addition, concerns about tumor protection and toxicity have led to controversy about its use [46]. Indeed, Amifostine has a high toxicity potential. The most common adverse effects are emesis and hypotension and, even when administered subcutaneously, the rate of adverse effects remains higher than 10% [47]. Of note, Amifostine is not radioprotective for salivary glands in differentiated thyroid cancer patients treated with high-dose radioactive iodine [48].

Nitroxides are a class of compounds that offer selective radioprotection to normal tissue. The most promising drug in this category is Tempol. Tempol is a stable free radical nitroxide that reacts with free radicals generated at the time of radiation, preventing them from causing normal tissue damage [49]. Tempol has been shown in mouse studies to preserve salivary flow when administered systemically [50] and also when a topical gel was applied to the oral cavity [51]. Interestingly, studies in guinea pigs showed that Tempol was able to prevent radiation-induced alopecia [52]. Those results were replicated in a clinical trial in which topical application of Tempol prevented alopecia in patients receiving whole brain radiation [53]; Tempol has not shown side effects that would limit its topical oral administration.

Among other strategies being developed to prevent radiation-induced salivary hypofunction, one with seemingly great potential is to expand the pool of stem cells in salivary glands pharmacologically prior to, or to decrease the rate of apoptosis and increase survival rate of acinar cells during, radiotherapy. As noted above, Lombaert et al. [38, 54] described the ability to grow salivary gland stem cells in vitro. Administration of keratinocyte growth factor to mice before or after radiation treatment decreased salivary hypofunction either by increasing the starting number, or surviving number, of stem cells.

Treatment of Existing Salivary Hypofunction

RTOG Grade-1: Minimal Treatment Using Sialogogues or Electrostimulation

Patients in RTOG grade 1 have had minimal radiation-damage to the salivary glands, i.e., they have enough functional gland tissue remaining to respond to a stimulus from sialogogues, i.e., pilocarpine (Salagen) or cevimeline (Evoxac). Pilocarpine has been used clinically for the alleviation of dry mouth since 1964, and several clinical trials established its efficacy in doses ranging from 5 to 10 mg, 3–4 times daily. Cevemiline is a more specific sialogogue that acts as a high affinity agonist of M1 and M3 muscarinic receptors. Several clinical trials have shown cevemiline to be effective alleviating xerostomia at a typical dose of 30 mg every 8 h. The most common side effects of both drugs, though occurring more commonly with pilocarpine, are excessive sweating, increased urinary frequency and flushing. Both drugs are contraindicated for patients with uncontrolled asthma, hypertension, acute iritis, narrow angle glaucoma or being treated with beta-blockers [55].

A quite different type of stimulus that may benefit these patients, as well as possibly some grade 2 patients, was recently shown to be safe and effective in a multicenter randomized clinical trial [56]. This novel approach involves the use of an intraoral electrostimulation device, which delivers an electrical signal through the oral mucosa to the lingual nerve, thus enhancing the salivary reflex. Although this trial treated mainly patients with xerostomia resulting from Sjögren's syndrome, 11% of the studied patients had a dry mouth due to radiotherapy. Overall, the results showed a cumulative positive effect from daily use of the electrostimulation device over the study period [56].

RTOG Grade-2/3: Gene Therapy or Progenitor Cells

Patients, whose level of xerostomia corresponds to RTOG grades 2 or 3, experience little to no beneficial response from the administration of sialogogues. Unfortunately, for these patients, there are no other conventional therapies available. As described above, such patients have lost most of their acinar cells and, thus, have few epithelial cells remaining that are able to secrete fluid. To address the needs of these patients, there has been an effort to develop novel biological therapies capable of restoring function to the damaged glands. In particular, two approaches have shown promise and will be briefly considered here: gene therapy and progenitor cells.

Twenty years ago, our laboratory began to examine the feasibility of direct gene transfer to salivary glands [57]. Once feasibility was established, we evaluated a strategy to repair irradiation-damaged glands involving transfer of the human aquaporin-1 gene (hAQP1 [58]). hAQP1 is the archetypal water channel discovered by Preston and Agre [59]. We hypothesized that the relatively water-impermeable duct cells surviving after radiation could generate an osmotic gradient (lumen > interstitium) in the absence of significant acinar cell primary fluid formation, i.e., the situation in irradiation-damaged glands. However, while this osmotic gradient was capable of driving transepithelial fluid secretion, fluid secretion could not occur because duct cells lacked functional water channels in their plasma membranes. Initially, we demonstrated that such a strategy could restore fluid secretory ability to damaged salivary glands in irradiated rats using an adenoviral vector encoding hAQP1 (AdhAQP1 [58]). Thereafter, we showed that the strategy was also effective in a large animal model, the irradiated miniature pig parotid gland [60].

Based on those promising preclinical efficacy studies, as well as a large safety and biodistribution study administering AdhAQP1 to rats using FDA Good Laboratory Practice standards [61], we submitted a clinical protocol that received all required approvals by mid-2007 (NIH protocol 06-D-0206; [62]; http://www.clinicaltrials.gov/ct/show/NCT00372320?order=). Eleven patients were treated with AdhAQP1 in this Phase I clinical trial (3/dose cohort at either 4.8×10^7, 2.9×10^8, 1.3×10^9, and two patients at 5.8×10^9, vector particles/gland). Initial safety and efficacy results have recently been reported, and this strategy proved beneficial to five of the eleven treated subjects [62a].

A few research groups have taken a different therapeutic approach to repairing the damaged salivary glands; the use of progenitor cells. These studies have been conducted in rodents (mice, rats) and employed progenitor cells obtained either from bone marrow [63, 64] or the salivary glands [54, 65].

While there is considerable precedent for the use of bone marrow progenitor cells in treating hematologic malignancies, it is becoming common to find that these cells also can be used to treat non-hematopoietic disorders (e.g., see Burt et al. [66]). Recently, Tran et al. [67] made an important observation that is consistent with the notion that bone marrow-derived progenitor cells might be useful in repairing damaged salivary glands. In that study, they examined minor salivary gland biopsies from five female patients who were treated for hematologic malignancies with bone marrow-derived stem cells from their brothers 1–16 years previously. In all patient samples there was evidence for microchimerism (Y-chromosome presence) in salivary epithelial cells [67]. Thus, in humans there already exist data to suggest bone marrow progenitor cells can find their way into salivary glands and trans-differentiate into epithelial cells. In that context, the studies reported by Lombaert et al. [63] and Sumita et al. [64] are particularly encouraging. The two studies were methodologically quite different (Lombaert et al. [63] mobilized bone marrow progenitor cells by treatment with granulocyte colony-stimulating factor, while Sumita et al. [64] delivered gender mismatched cells by direct tail vein injection), but both led to significant improvements in salivary flow rates in previously irradiated mice.

Lombaert et al. [54] and Yaniv et al. [65] have isolated progenitor cells directly from adult rodent salivary glands (mouse and rat, respectively), and Lombaert et al. [54] have used those cells to treat irradiation-damaged salivary glands. Following direct injection of relatively low numbers of cells into murine submandibular glands, they observed increased salivary flow rates and increased acinar area in glands. Additional studies from Coppes and colleagues [68], and from Tatsuishi et al. [69] suggest that progenitor cells similar to those found in rodent salivary glands can be isolated from human salivary glands, supporting their potential for therapeutic use.

RTOG Grade-4: Artificial Saliva

If the patient has a contraindication for one of the above sialogogues, or is in RTOG grade 4, i.e., with fibrosis and no salivary gland parenchymal tissue remaining, the only alternative treatment available for temporary symptomatic relief is the use of salivary substitutes (artificial salivas). Several formulations of artificial salivas are available. A gel formulation that will enhance lubrication of the oral cavity for a longer time is indicated for patients with a more accentuated dry mouth or for use during the night. A more fluid oral moisturizer can be helpful for patients with moderate xerostomic complaints [70]. However, most patients discontinue the use of salivary substitutes, opting instead for the constant company of a water bottle.

RTOG Grade-4: Artificial Salivary Gland

Efforts to develop an artificial salivary gland were stimulated in the mid-1990s by two things: (1) the clear recognition that artificial saliva preparations provided inadequate benefit for most RTOG-4 patients, as noted above, and (2) considerable progress in the general field of tissue engineering [71]. As viewed from a biomaterials approach [72], these efforts have evolved in two directions: (1) use of inductive materials with transplanted cells and (2) use of programmable materials to control the organization of transplanted or endogenous cells. Although neither strategy has yet been successful, it seems reasonable to be optimistic that such a device can be produced and function within a decade or so (e.g., see Atala et al. [73]).

The first direction involved the use of various biodegradable substrata, coated or not with extracellular matrix protein, upon which were seeded a monolayer of epithelial cells capable of unidirectional fluid movement (e.g., see refs. [74–78]). Our own efforts were directed at creation of a simple device, shaped as a blind-end tube [79], and identified components and cellular manipulations to underlie secretory function in vitro [80]. However, using this strategy no group has yet demonstrated in vivo function of a prototype.

The second direction, the use of programmable materials, exists only theoretically, based on the work done with different tissues, e.g., the control of vascular development [81]. Such efforts to direct the growth of either transplantable cells or endogenous cells leading ultimately to the formation of a functioning salivary gland requires more fundamental understanding of the molecules and regulatory systems involved in normal gland development. While there has been significant progress towards such an understanding (e.g., [36, 82, 83]), it is not yet sufficient to allow testing of this second strategy in vivo.

Concluding Remarks

Xerostomia due to salivary hypofunction occurs quite frequently following radiotherapy of patients being treated for a head and neck cancer. Roughly half of all surviving patients develop long-term oral dryness that results in dysphagia, frequent oral infections, considerable pain and discomfort, and a diminished quality of life. For some of these patients (RTOG grade 1) a suitable treatment, e.g., use of a sialogogue or an electrostimulation device, is available. However, for many others

(RTOG categories 2–4) there are no suitable conventional treatments. For these latter patients, recent advances in translational science have raised the possibility that novel therapies may become available. Equally important, progress in the technical means by which radiation is administered, as well as modest advances in understanding the mechanistic events leading to salivary gland damage, has made it possible to develop novel approaches to prevent radiation-induced changes. Some of these are available clinically and others should reach the clinic within a decade or so. The radiation oncologist who today is preparing to treat patients with a head and neck cancer or the various clinicians who today are managing surviving patients with irradiation-induced xerostomic complaints have more tools available than they did two decades ago. While progress has been modest, it is reasonable to expect exponential advances on the horizon.

Acknowledgement The Division of Intramural Research of the National Institute of Dental and Craniofacial Research has supported all of the authors' research. We also thank Drs. John A. Chiorini and Deborah E. Citrin for their helpful comments on an earlier version of this chapter.

References

1. Jensen SB, Pedersen AML, Vissink A, et al. A systematic review of salivary gland hypofunction and xerostomia induced by cancer therapies: prevalence, severity and impact on quality of life. Support Care Cancer. 2010;18:1039–60.
2. Jensen SB, Pedersen AML, Vissink A, et al. A systematic review of salivary gland hypofunction and xerostomia induced by cancer therapies: management strategies and economic impact. Support Care Cancer. 2010;18:1061–79.
3. Cox JD, Steitz J, Pajak TF. Toxicity criteria of the Radiation Therapy Oncology Group (RTOG) and the European Organization for research and treatment of cancer. Int J Radiat Oncol Biol Phys. 1995;31:1341–6.
4. Van Nostrand D. Sialoadenitis secondary to ^{131}I therapy for well-differentiated thyroid cancer. Oral Dis. 2011;17:154–61.
5. Nieuw Amerongen A, Veerman EC. Saliva—the defender of the oral cavity. Oral Dis. 2002;8:12–22.
6. Langendijk JA, Doornaert P, Verdonck-de Leeuw IM, et al. Impact of late treatment-related toxicity on quality of life among patients with head and neck cancer treated with radiotherapy. J Clin Oncol. 2008;26:3770–6.
7. Ho KF, Farnell DJJ, Routledge JA, et al. Developing a CTCAEs patient questionnaire for late toxicity after head and neck radiotherapy. Eur J Cancer. 2009;45:1992–8.
8. Ship JA, Fox PC, Baum BJ. How much saliva is enough? J Am Dent Assoc. 1991;122:63–9.
9. Dawes C. Physiological factors affecting salivary flow rate, oral sugar clearance, and the sensation of dry mouth in man. J Dent Res. 1987;66:648–53.
10. Wolff MS, Kleinberg I. The effect of ammonium glycopyrrolate (Robinul®)-induced xerostomia on oral mucosal wetness and flow of gingival crevicular fluid in humans. Arch Oral Biol. 1999;44:97–102.
11. Dawes C, Odlum O. Salivary status in a treated head and neck cancer patient group. J Can Dent Assoc. 2004;70:397–400.
12. Hall EJ. Radiobiology for the radiologist. 5th ed. Philadelphia: Lippincott Williams & Wilkins; 2000.
13. Zhou L, Yuan R, Lanata S. Molecular mechanisms of irradiation-induced apoptosis. Front Biosci. 2003;8:d9–19.

14. Verheij M. Clinical biomarkers and imaging for radiotherapy-induced cell death. Cancer Metastasis Rev. 2008;27:471–80.
15. Vissink A, 's-Gravenmade EJ, Ligeon EE, et al. A functional and chemical study of radiation effects on rat parotid and submandibular/sublingual glands. Radiat Res. 1990;124:259–65.
16. O'Connell AC. Natural history and prevention of radiation injury. Adv Dent Res. 2000;14:57–61.
17. Kashima H, Kirkham W, Andrews J. Postirradiation sialadenitis. Am J Roentgenol. 1965;94:271–91.
18. Wolf RO, Taylor LL, Brace K. Effects of irradiation of the parotid gland and pancreas on human isoamylases. Am J Clin Pathol. 1970;54:214–8.
19. Anderson MW, Izutsu KT, Rice JC. Parotid gland pathophysiology after mixed gamma and neutron irradiation of cancer patients. Oral Surg Oral Med Oral Pathol. 1981;52:495–500.
20. Baum BJ, Bodner L, Fox PC, et al. Therapy-induced dysfunction of salivary glands: implications for oral health. Spec Care Dentist. 1985;5:274–2747.
21. Maier H, Bihl H. Effect of radioactive iodine therapy on parotid gland function. Acta Otolaryngol. 1987;103:318–24.
22. Epstein JB, Chin EA, Jacobson JJ, et al. The relationships among fluoride, cariogenic oral flora, and salivary flow rate during radiation therapy. Oral Surg Oral Med Oral Pathol Oral Radiol Endod. 1998;86:286–92.
23. Moller P, Perrier M, Ozsahin M, et al. A prospective study of salivary gland function in patients undergoing radiotherapy for squamous cell carcinoma of the oropharynx. Oral Surg Oral Med Oral Pathol Oral Radiol Endod. 2004;97:173–89.
24. Braam PM, Roesink JM, Raaijmakers CP, et al. Quality of life and salivary output in patients with head-and-neck cancer five years after radiotherapy. Radiat Oncol. 2007;2(3):1–8.
25. Chambers MS, Tomsett KL, Artopoulou II, et al. Salivary flow rates measured during radiation therapy in head and neck cancer patients: a pilot study assessing salivary sediment formation. J Prosthet Dent. 2008;100:142–6.
26. Stephens LC, Ang KK, Schultheiss TE, et al. Target cell and mode of radiation injury in rhesus salivary glands. Radiother Oncol. 1986;7:165–74.
27. Vissink A, Kalicharan D, S-Gravenmade EJ, et al. Acute irradiation effects on morphology and function of rat submandibular glands. J Oral Pathol Med. 1991;20:449–56.
28. Nagler RM. Effects of head and neck radiotherapy on major salivary glands—animal studies and human implications. In Vivo. 2003;17:369–75.
29. Pena LA, Fuks Z, Kolesnick RN. Radiation- induced apoptosis of endothelial cells in the murine central nervous system: protection by fibroblast growth factor and sphingomyelinase deficiency. Cancer Res. 2000;60:321–7.
30. Fuks Z, Alfieri A, Haimovitz-Friedman A, et al. Intravenous basic fibroblast growth factor protects the lung but not mediastinal organs against radiation-induced apoptosis in vivo. Cancer J Sci Am. 1995;1:62–72.
31. Paris F, Fuks Z, Kang A, Capodieci P, Juan G, Ehleiter D, et al. Endothelial apoptosis as the primary lesion initiating intestinal radiation damage in mice. Science. 2001;293:293–7.
32. Cotrim AP, Sowers A, Mitchell JB, Baum BJ. Prevention of irradiation-induced salivary hypofunction by microvessel protection in mouse salivary glands. Mol Ther. 2007;15:2101–6.
33. Xu J, Yan X, Gao R, Mao L, Cotrim AP, Zheng C, et al. Effect of irradiation on microvascular endothelial cells of parotid glands in the miniature pig. Int J Radiat Oncol Biol Phys. 2010;78:897–903.
34. Franzen L, Forsgren S, Gustafsson H, et al. Irradiation-induced effects on the innervation of rat salivary glands: changes in enkephalin- and bombesin-like immunoreactivity in ganglionic cells and intraglandular nerves. Cell Tissue Res. 1993;271:529–36.
35. Aalto Y, Forsgren S, Kjorell U, et al. Time- and dose-related changes in the expression of substance P in salivary glands in response to fractionated irradiation. Int J Radiat Oncol Biol Phys. 1995;30:297–305.
36. Knox SM, Lombaert IMA, Reed X, et al. Parasympathetic innervation maintains epithelial progenitor cells during salivary organogenesis. Science. 2010;329:1645–7.
37. Zeilstra LJW, Vissink A, Konings AWT, Coppes RP. Radiation induced cell loss in rat submandibular gland and its relation to gland function. Int J Radiat Biol. 2000;76:419–29.

38. Lombaert IM, Brunsting JF, Wierenga PK, et al. Keratinocyte growth factor prevents radiation damage to salivary glands by expansion of the stem/progenitor pool. Stem Cells. 2008;26: 2595–601.
39. Zheng C, Cotrim AP, Rowzee A, Swaim W, Sowers A, Mitchell JB, et al. Prevention of radiation-induced salivary hypofunction following hKGF gene delivery to murine submandibular glands. Clin Cancer Res. 2011;17:2842–51.
40. Leung SW, Lee TF, Chien CY, et al. Health-related Quality of life in 640 head and neck cancer survivors after radiotherapy using EORTC QLQ-C30 and QLQ-H&N35 questionnaires. BMC Cancer. 2011;11:128.
41. Genden EM, Varvaris MA. Head and neck cancer: an evidence-based team approach. New York: Thieme Medical Publishers; 2008.
42. Nutting CM, Morden JP, Harrington KJ, et al. Parotid-sparing intensity modulated versus conventional radiotherapy in head and neck cancer (PARSPORT): a phase 3 multicentre randomised controlled trial. Lancet Oncol. 2011;12:127–36.
43. Stachiw N, Hornig J, Gillespie MB. Minimally-invasive submandibular transfer (MIST) for prevention of radiation-induced xerostomia. Laryngoscope. 2010;20 Suppl 4:S184.
44. Atkinson JC, Grisius M, Massey W. Salivary hypofunction and xerostomia: diagnosis and treatment. Dent Clin N Am. 2005;49:309–26.
45. Citrin D, Cotrim AP, Hyodo F, et al. Radioprotectors and mitigators of radiation-induced normal tissue injury. Oncologist. 2010;15:360–71.
46. Brizel DM, Overgaard J. Does amifostine have a role in chemoradiation treatment? Lancet Oncol. 2003;4:378–81.
47. Anné PR, Machtay M, Rosenthal DI, et al. A Phase II of subcutaneous amifostine and radiation therapy in patients with head-and-neck cancer. Int J Radiat Oncol Biol Phys. 2007;67:445–52.
48. Ma C, Xie J, Chen Q, et al. Amifostine for salivary glands in high-dose radioactive iodine treated differentiated thyroid cancer. Cochrane Database Syst Rev. 2009;7:4.
49. Mitchell JB, DeGraff W, Kaufman D, et al. Inhibition of oxygen-dependent radiation-induced damage by the nitroxide superoxide dismutase mimic, tempol. Arch Biochem Biophys. 1991;289:62–71.
50. Vitolo JM, Cotrim AP, Sowers AL, et al. The stable nitroxide tempol facilitates salivary gland protection during head and neck irradiation in a mouse model. Clin Cancer Res. 2004;10:1807–12.
51. Cotrim AP, Hyodo F, Matsumoto K, et al. Differential radiation protection of salivary glands versus tumor by tempol with accompanying tissue assessment of tempol by magnetic resonance imaging. Clin Cancer Res. 2007;13:4928–33.
52. Goffman T, Cuscela D, Glass J, et al. Topical application of nitroxide protects radiation-induced alopecia in guinea pigs. Int J Radiat Oncol Biol Phys. 1992;22(4):803–6.
53. Metz JM, Smith D, Mick R, et al. A phase I study of topical Tempol for the prevention of alopecia induced by whole brain radiotherapy. Clin Cancer Res. 2004;10:6411–7.
54. Lombaert IMA, Brunsting JF, Wierenga PK, et al. Rescue of salivary gland function after stem cell transplantation in irradiated glands. PLoS One. 2008;3(e2063):1–8.
55. Sreebny LM, Vissink A. Dry mouth—the malevolent symptom: a clinical guide. Ames: Wiley-Blackwell; 2010.
56. Strietzel FP, Lafaurie GI, Bautista GR, et al. Efficacy and safety of an intraoral electrostimulation device for xerostomia relief. Arthritis Rheum. 2011;63:180–90.
57. Mastrangeli A, O'Connell B, Aladib W, et al. Direct in vivo adenovirus-mediated gene transfer to salivary glands. Am J Physiol. 1994;266:G1146–55.
58. Delporte C, O'Connell BC, He X, et al. Increased fluid secretion after adenoviral-mediated transfer of the aquaporin-1 cDNA to irradiated rat salivary glands. Proc Natl Acad Sci USA. 1997;3268–73.
59. Preston GM, Agre P. Isolation of the cDNA for erythrocyte integral membrane protein of 28 kilodaltons: member of an ancient channel family. Proc Natl Acad Sci USA. 1991;88:11110–4.
60. Shan Z, Li J, Zheng C, et al. Increased fluid secretion after adenoviral-mediated transfer of the human aquaporin-1 cDNA to irradiated miniature pig parotid glands. Mol Ther. 2005;11:444–51.
61. Zheng C, Goldsmith CM, Mineshiba F, et al. Toxicity and biodistribution of a first-generation recombinant adenoviral vector, encoding aquaporin-1, after retroductal delivery to a single rat submandibular gland. Hum Gene Ther. 2006;17:1122–33.

62. Baum BJ, Zheng C, Alevizos I, et al. Development of a gene transfer-based treatment for radiation-induced salivary hypofunction. Oral Oncol. 2010;46:4–8.
62a. Baum BJ, Alevizos I, Zheng C, et al. Early responses to adenoviral-mediated transfer of the aquaporin-1 cDNA for radiation-induced salivary hypofunction. Proc Natl Acad Sci USA 2012; [Epub ahead of print].
63. Lombaert IMA, Wierenga PK, Kok T, et al. Mobilization of bone marrow stem cells by granulocyte colony-stimulating factor ameliorates radiation-induced damage to salivary glands. Clin Cancer Res. 2006;12:1804–12.
64. Sumita Y, Liu Y, Khalili S, et al. Bone marrow-derived cells rescue salivary gland function in mice with head and neck irradiation. Int J Biochem Cell Biol. 2011;43:80–7.
65. Yaniv A, Neumann Y, David R, et al. Establishment of immortal multipotent rat salivary progenitor cell line toward salivary gland regeneration. Tissue Eng. 2011;17C:69–78.
66. Burt RK, Loh Y, Pearce W, et al. Clinical applications of blood-derived and marrow-derived stem cells for nonmalignant diseases. JAMA. 2008;299:925–36.
67. Tran SD, Redman RS, Barrett AJ, et al. Microchimerism in salivary glands after blood- and marrow-derived stem cell transplantation. Biol Blood Marrow Transplant. 2011;17:429–33.
68. Feng J, van der Zwaag M, Stokman MA, et al. Isolation and characterization of human salivary gland cells for stem cell transplantation to reduce radiation-induced hyposalivation. Radiother Oncol. 2009;92:466–71.
69. Tatsuishi Y, Hirota M, Kishi T, et al. Human salivary gland stem/progenitor cells remain dormant even after irradiation. Int J Mol Med. 2009;24:361–6.
70. Vissink A, Mitchell KB, Baum BJ, et al. Clinical management of salivary gland hypofunction and xerostomia in head and neck cancer patients: successes and barriers. Int J Radiat Oncol Biol Phys. 2010;78:983–91.
71. Langer R, Vacanti JP. Tissue engineering. Science. 1993;260:920–6.
72. Arany PR, Mooney DJ. At the edge of translation—materials to program cells for directed differentiation. Oral Dis. 2011;17:241–51.
73. Atala A, Bauer SB, Soker S, et al. Tissue-engineered autologous bladders for patients needing cystoplasty. Lancet. 2006;367:1241–6.
74. Wang S, Cukierman E, Swaim WD, et al. Extracellular matrix protein-induced changes in human salivary epithelial cell organization and proliferation on a model biological substratum. Biomaterials. 1999;20:1043–9.
75. Aframian DJ, Cukierman E, Nikolovski MS, et al. The growth and morphological behavior of salivary epithelial cells on matrix protein-coated biodegradable substrata. Tissue Eng. 2000;9:209–15.
76. Chen MH, Hsu YH, Lin CP, et al. Interactions of acinar cells on biomaterials with various surface properties. J Biomed Mater Res A. 2005;74:254–62.
77. Joraku A, Sullivan CA, Yoo JJ, et al. Tissue engineering of functional salivary gland tissue. Laryngoscope. 2005;115:244–8.
78. Aframian DJ, Palmon A. Current status of the development of an artificial salivary gland. Tissue Eng. 2008;14:187–98.
79. Baum BJ, Wang S, Cukierman E, et al. Re-engineering the functions of a terminally differentiated epithelial cell in vivo. Ann N Y Acad Sci. 1999;875:294–300.
80. Tran SD, Sugito T, Dipasquale G, et al. Re-engineering primary epithelial cells from rhesus monkey parotid glands for use in developing an artificial salivary gland. Tissue Eng. 2006;12:2939–48.
81. Yuen WW, Du NR, Chan CH, et al. Mimicking nature by codelivery of stimulant and inhibitor to create temporally stable and spatially restricted angiogenic zones. Proc Natl Acad Sci USA. 2010;107:17933–8.
82. Lu P, Werb Z. Patterning mechanisms of branched organs. Science. 2008;322:1506–8.
83. Onodera T, Sakai T, Hsu JC, et al. Btbd7 regulates epithelial cell dynamics and branching morphogenesis. Science. 2010;329:562–5.
84. Kaplan MD, Baum BJ. The functions of saliva. Dysphagia. 1993;8:225–9.

Osteonecrosis

Caterina Musolino and Alessandro Allegra

Abstract Bisphosphonates (BPs) are a widely used class of drugs with known efficacy in the prevention and treatment of postmenopausal and steroid-induced osteoporosis; Paget's disease of bone; hypercalcemia of malignancy; osteolytic lesions of multiple myeloma; and bone metastases associated with breast, prostate, lung, and other soft tissue tumors. When compared to many other therapies, especially in the cancer setting, the severity of adverse events related to BPs is generally mild, and side effects are infrequent. Thus, the benefits of treatment with BPs almost always outweigh the risks. However, several side effects such as upper aerodigestive tract issues, effects concerning renal function, and acute-phase reactions have been described. In 2003, a fourth adverse effect, bisphosphonate-associated osteonecrosis of the jaw, was described for the first time and has been diagnosed with increasing frequency. However several other heretofore unrecognized side effects of BPs were recently reported such as ocular side effects, atrial fibrillation, esophageal cancer, cutaneous reactions, atypical fractures of the femoral diaphysis, hypocalcemia, and hepatitis.

The present work provides a literature review and update on bisphosphonate treatment to explore the risk of adverse drug reactions in patients treated with BPs.

General Considerations

Bisphosphonates (BPs) are drugs made up of two phosphonic acids joined to a carbon plus two side chains designated R^1 and R^2 (Fig. 1). They have wide use as antiscaling agents because of their physical-chemical property of complexing with divalent cations (e.g., calcium and magnesium). The P–C–P structure acts as a bone

C. Musolino (✉) • A. Allegra
Division of Hematology, University of Messina,
Via Consolare Valeria, 98125 Messina, Italy
e-mail: cmusolino@unime.it; alessandro.allegra@tin.it

S.T. Sonis and D.M. Keefe (eds.), *Pathobiology of Cancer Regimen-Related Toxicities*,
DOI 10.1007/978-1-4614-5438-0_14,

Fig. 1 Bisphosphonate structure

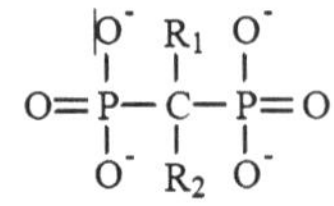

Table 1 Medical indications for BPs use

Postmenopausal women with vertebral compression fractures
Postmenopausal women with total hip bone density *T*-score below −2.5
Elderly men with non-traumatic fractures
Some patients with secondary osteoporosis due to corticosteroids
Paget's disease
Cancer metastatic to bone
Other bone diseases with high bone resorption

hook that causes these compounds to bind avidly to hydroxyapatite crystals on bone surfaces, particularly at sites of active bone remodeling.

Due to their great affinity to calcium phosphate they were used to perform osseous gammagraphies with Tc99m. They were also used as toothpaste ingredients, given their anti-tartar and anti-plaque effects.

However, in 1968 Fleisch et al. described the *in vitro* and *in vivo* influence of bisphosphonates—erroneously called diphosphonates at that time—on the precipitation and dissolution of calcium phosphate [1, 2]. This research paved the way to the study of the effects of these agents on hydroxyapatite, and their first therapeutic application in a patient with myositis ossificans just 1 year later [3, 4].

Today, BPs are a widely used class of drugs with known efficacy in the prevention and treatment of postmenopausal and steroid-induced osteoporosis; Paget's disease of bone; hypercalcemia of malignancy; osteolytic lesions of multiple myeloma; and bone metastases associated with breast, prostate, lung, and other soft tissue tumors [5–8] (Table 1).

Bisphosphonates have the ability to bind divalent ions such as Ca^{2+} and hence are rapidly cleared from the circulation and adsorbed into active bone mineral surfaces undergoing osteoclastic bone resorption. This results in an effective uptake of bisphosphonates into osteoclasts by endocytosis and explains their selective action on these cells. Other cell types which internalize bisphosphonates by endocytosis are osteoblasts, macrophages, epithelial and endothelial cells, circulating monocytes, and also neoplastic cells such as myeloma and prostate tumor cells.

Depending on the ligands, bisphosphonates are grouped into nitrogen-containing and non-nitrogen-containing bisphosphonates (Table 2). The former inhibit the mevalonate pathway, and the latter are built into the ATP-molecule; both result in cytotoxic effects to the osteoclast [9, 10].

The first-generation of non-nitrogen-containing bisphosphonates, are metabolized intracellularly to analogs of ATP. These metabolites inhibit bone resorption by inducing osteoclast apoptosis, very likely by inhibiting ATP-dependent enzymes. In contrast to the first-generation bisphosphonates, the nitrogen-containing bisphosphonates are not metabolized to ATP analogs. They act by inhibiting farnesyl

Table 2 Nitrogen-containing and non-nitrogen-containing bisphosphonates

Non-nitrogenous
Non-N-containing bisphosphonates
Etidronate
Clodronate
Tiludronate
Nitrogenous
N-containing bisphosphonates
Pamidronate
Neridronate
Olpadronate
Alendronate
Ibandronate
Risedronate
Zoledronate

diphosphate (FPP) synthase, a key enzyme of the mevalonate pathway. This, among other complex biochemical changes, leads to intracellular accumulation of isopentenyl diphosphate (IPP). The inhibition of FPP synthase prevents the prenylation of small GTPases. Prenylated small GTPases are important signaling proteins and this inhibition of protein prenylation affects a variety of cell functions. The inhibition of osteoclasts by the nitrogen-containing bisphosphonates is very likely mediated by their action on the FPP synthase, which leads to protracted apoptosis of these cells [11, 12].

Bisphosphonate Pharmacology

Oral bisphosphonate absorption is poor (about 1% of the formulated dose). Due to their avid affinity for the bone, between 27 and 62% of the drug in the blood rapidly binds to bone mineral. Any remaining bisphosphonate is excreted via the kidneys, predominantly within the first few hours after administration [13–16]. Renal excretion occurs by both passive glomerular filtration and active transportation in renal proximal tubular cells [17, 18]. Following intravenous administration of bisphosphonates, approximately 50% of the dose is incorporated into the bone. Apart from a negligible amount of drug transiently exposed to other tissues, most of the remainder is also excreted unchanged in urine via the same filtration and proximal tubular secretion pathways as oral bisphosphonates. Bisphosphonates persist in bone for long durations, are slowly released during cycles of bone remodeling, and can reenter the systemic circulation, and also the kidney, with no change observed in their molecular structure or metabolic activity. In general bisphosphonates do not undergo any detectable metabolism, do not induce or inhibit cytochrome P450 activity, and do not use the anionic or cationic renal transport systems involved in the excretion of many other drugs [19–24].

Significant differences in binding affinity to bone mineral have been found among the bisphosphonates in vitro, with a rank order of highest to lowest: zoledronate > alendronate > ibandronate > risedronate [25, 26]. Similarly, the degree to which bisphosphonates reduce osteoclastic activity by inhibition of farnesyl pyrophosphate synthase also differs among them, with a rank of order from highest to lowest as follows: zoledronate > risedronate > ibandronate > alendronate [27–29].

The intravenous bisphosphonates commonly used for the treatment of malignancies are pamidronate (90 mg infused over 2 h), ibandronate (6 mg infused over 1 h), and zoledronic acid (4 mg infused over 15 min). They are usually administered every 3–4 weeks. Intravenous ibandronate (3 mg every 3 months) and zoledronic acid (5 mg once-yearly) have also been approved for the treatment of osteoporosis.

When taken orally, BPs must be taken after a prolonged fast, with water only, followed by 30–60 min with nothing else by mouth to allow for adequate absorption. Taking a bisphosphonate with food or anything containing divalent cations will completely block its absorption.

The half-life in plasma is short. Fifty percent of the absorbed dose binds to bone surfaces, mostly avidly at sites of active remodeling. The skeletal capacity is large and the binding sites are virtually unsaturable. The 50% or so that does not bind to bone is excreted rapidly by the kidneys.

Bisphosphonates Tolerability and Safety

Bisphosphonates are one of the most thoroughly studied groups of drugs used in medicine. Since their first use in 1969, their efficacy and safety have been established through a series of comprehensive trials for their various clinical uses. This includes the trials with nitrogen-containing BPs in which over 60,000 participants were followed for 3 years at least. All approved nitrogen-containing BPs have a favorable tolerability and safety profile [30].

After treatment with alendronate (ALN) (3–10 years), risedronate (3–5 years), or ibandronate (3 years), the incidence of overall adverse events as well as withdrawal rates due to adverse events were similar between each treatment and its respective placebo arm [31–37].

However there may be confusion about the risks associated with BPs use for the treatment of osteoporosis vs. treatment of other conditions, such as cancer, which typically involves a very different patient population and different doses or frequency of drug administration [38].

These adverse effects can be separated into three main groups: upper aerodigestive tract issues, effects concerning renal function, and acute-phase reactions [39]. In 2003, a fourth adverse effect, bisphosphonate-associated osteonecrosis of the jaw (BP-ONJ), was described for the first time [40] and has been diagnosed with increasing frequency. However several other heretofore unrecognized side effects of BPs were recently reported such as ocular side effects, atrial fibrillation, esophagus cancer, cutaneous reactions, atypical fractures of the femoral diaphysis, hypocalcemia, and hepatitis (Table 3).

Table 3 Bisphosphonates adverse effects

Gastrointestinal adverse effects	Nausea Vomiting Epigastric pain Dyspepsia Esophagitis
Renal toxicity	Minimal change disease Focal segmental glomerulosclerosis Collapsing variant
Acute phase response	Fever Influenza-like symptoms Fatigue Malaise Arthralgia
Bone pain	
Ocular side effects	Nonspecific conjunctivitis Ptosis Optic neuritis Periorbital edema Eyelid edema Uveitis Scleritis
Esophageal cancer	
Atrial fibrillation	
Atypical fractures of the femoral diaphysis	
Hypocalcemia	
Cutaneous reactions	
Hepatitis	
ONJ	

Gastrointestinal Adverse Effects

Gastrointestinal (GI) problems are often considered to be an inevitable consequence associated with the oral use of BPs which are prescribed for the prevention and treatment of osteoporosis. Adverse events such as nausea, vomiting, epigastric pain, and dyspepsia were reported. These adverse events are due to mucosal irritation of the upper GI tract caused by the BPs. Several cases of esophagitis with esophageal erosions or ulcerations associated with the use of ALN were also reported [41–43].

However, the results from the major prospective RCTs assessing the reduction of fractures are notable in not showing an excess of GI problems. The multi-center study by Eisman et al. and the study of Greenspan et al. concluded that ALN 70 mg administered once-weekly to women and men with osteoporosis has an upper GI and overall tolerability profile similar to that of placebo [44, 45].

Risedronate and ibandronate seem to have a GI tolerability similar to that of ALN [46, 47].

However, oral BPs are in general poorly absorbed (<1%) from the intestinal tract, which means that potentially large amounts of bisphosphonates can come into contact with the GI mucosa at all levels of the GI tract. Indeed diarrhea was recognized as a side effect of giving large amounts of the early bisphosphonates, etidronate, and clodronate, and seems not to be an issue with the current clinically used BPs, which are given in smaller amounts.

Studies with preclinical models have suggested that esophageal irritation was greater in animals exposed to generic ALN than in those exposed to branded ALN [48]. Differences in the disintegration profiles between the branded and the many different generic products available may result in GI problems, and many studies have demonstrated differences in the in vitro disintegration properties between generic and branded ALN products [49–51].

Recommendations to reduce the risk of esophagitis include swallowing ALN with 180–240 mL of water on arising in the morning, remaining upright for at least 30 min after swallowing the tablet and until the first food of the day has been ingested.

Renal Toxicity

Bisphosphonates are excreted unaltered through the kidneys via filtration and possibly by proximal tubular secretion.

Several publications indicate that the use of oral bisphosphonates may be safe and effective in patients with glomerular filtration rates less than 30 mL/min [52, 53], or down to estimated glomerular filtration rates of 15 mL/min [54–56].

However, IV administration produces exposure to high initial concentrations of BPs in the kidney and may be associated with acute renal injury. Permanent renal damage has been reported in case reports using the first-generation IV bisphosphonates, but these data were not validated by randomized, controlled clinical trials [57–60].

Assessment of changes in serum creatinine concentrations at 9–11 days after the infusion in a group of the total randomized population showed that a small percentage of patients receiving zoledronate had an increase in serum creatinine concentration of 0.5 mg/dL or more compared with patients given placebo (1.2% vs. 0.4%, $P<0.05$) only after the second infusion [61]. In all of these patients increases in serum creatinine returned to baseline levels before the next annual infusion.

However, in clinical practice, the administration of bisphosphonates is contraindicated for use in patients with severe renal impairment (creatinine clearance <30 mL/min).

Although BPs generally induce clinically reversible forms of acute renal failure, they can trigger functional alterations that are sometimes persistent and lead to chronic kidney disease (CKD), necessitating dialysis [60].

Patterns of nephrotoxicity with these drugs include toxic acute tubular necrosis and glomerulosclerosis [62].

Renal biopsy may reveal any of several patterns, including minimal change disease, focal segmental glomerulosclerosis, and the collapsing variant of focal segmental glomerulosclerosis.

In a study conducted on a cohort of pamidronate-treated patients, Markowitz found that a high percentage of patients developed the nephritic syndrome, with a picture of membranoproliferative glomerulopathy at histologic examination; the typical features are focal glomerulosclerosis and a glomerular basement membrane reaction [63, 64].

The renal toxicity of the intravenous bisphosphonates used in patients with malignant diseases involving bone is probably enhanced by risk factors for kidney function which may be present in these patients: preexisting CKD, acidosis, furosemide, multiple myeloma, hypercalcemia, hypertension, diabetes mellitus, advanced age, chemotherapy, or previous treatment with a bisphosphonate [65, 66].

However, other factors which may increase renal toxicity of the intravenous bisphosphonates are higher dose, shorter infusion times, or dose interval lower than recommended [67]. The total dose of the drug which has been administered during a long-term treatment may also play a role because of its probable cumulative effect.

Data regarding rapid infusions of IV pamidronate and zoledronate provided additional evidence of direct renal effects and suggest that any renal adverse effect may be related to the C_{max}, rather than the area under the curve (AUC). Rapid infusion of zoledronate in neoplastic patients was shown to induce acute increases in serum creatinine concentrations in some patients [68, 69].

Although many cases of renal toxicity were reported with the newer intravenous bisphosphonates, pamidronate and zoledronic acid, they were not with ibandronate [70, 71].

The renal safety profile of ibandronate in fact, appears to be quite different to that of zoledronic acid. In patients with advanced breast cancer, the renal safety profile of intravenous ibandronate was comparable with placebo, both in the short- and long-term.

The favorable renal safety profile of ibandronate is reflected in its product labeling; there is no requirement for dose reductions due to mild-to-moderate renal impairment or mandatory kidney function tests [72]. Preliminary studies indicate that ibandronate has a similarly good renal safety profile in multiple myeloma. In an open-label study of 40 patients with normal renal function or preexisting renal insufficiency intravenously administered ibandronate did not result in significant changes in serum creatinine [73, 74].

On the contrary, renal function testing is mandatory before and during treatment with zoledronic acid and doses higher than the standard 4 mg every 3–4 weeks are prohibited for renal safety reasons. For patients with creatinine clearance between 30 and 60 mL/min, dose reductions are required [75]. Creatinine levels should be assessed 7–10 days before the start of therapy and before every subsequent drug administration session.

Renal damage is mostly frequently found when BP treatment is given for a period of more than 2 years. In randomized studies the effects of the prolonged administration of pamidronate and zoledronic acid have been studied in more than 3,000 oncologic patients. In one study no significant difference was found between creatinine levels in the patients receiving zoledronic acid and those in the controls receiving a placebo. In another trial the percentage for renal function decrease was 15.2 and 11.5 in the zoledronic acid- and placebo-treated groups [76, 77].

However we have to point out that the American Society of Clinical Oncology (ASCO) guidelines suggest than in presence of bone metastases BP therapy should be continued until a worsening in the patient's general status occurs and regardless of any renal adverse effects [78].

Acute Phase Response and Bone Pain

Intravenous nitrogen-containing bisphosphonates are known to cause an adverse event resembling the acute phase response (AFR). Clinically, this systemic reaction is present in about 30–35% of patients and is characterized by fever, and influenza-like symptoms such as fatigue, malaise, myalgia, arthralgia, and bone pain. This effect is dose-dependent, occurs mainly after the first infusion of a nitrogen-containing bisphosphonate and is rare in subsequent infusions of the drug. The AFR is maximally expressed 28–36 h after IV administration of nitrogen-containing bisphosphonates and subsides 2–3 days later despite continuous treatment [79, 80].

The underlying mechanisms have been studied and appear to involve a rapid and profuse production of proinflammatory cytokines by peripheral blood gamma delta ($\gamma\delta$) T cells. This is mediated by the release of isopentenyl pyrophosphate (IPP) as a result of the blockade of the mevalonate pathway. IPP would subsequently activate/expand $\gamma\delta$ T cells by acting on a receptor that recognizes phosphate compounds [81–84].

The bisphosphonate-induced AFR in fact is mediated by interleukin 6 [85], tumor necrosis factor α, and other proinflammatory cytokines released by receptor-activated $\gamma\delta$T cells and macrophages. The activation of $\gamma\delta$T cells can be inhibited by statins in vitro, although atorvastatin was ineffective on AFR following bisphosphonate infusion in children [86]. The bisphosphonate-induced AFR is usually benign and self-limited and can be treated with antipyretics.

Oral acetaminophen/paracetamol or ibuprofen administered 4 h post-infusion reduced the proportion of patients with increased oral temperature [87].

There is a possibility of severe bone, joint, and/or muscle pain after BPs therapy. The severe musculoskeletal pain may occur within days, months, or years after starting a bisphosphonate. Some patients have reported complete relief of symptoms after discontinuing the bisphosphonate, whereas others have reported slow or incomplete resolution [88].

Bock et al. studied the occurrence of musculoskeletal events in patients who were treated with ALN or risedronate daily or once-weekly. No patient treated with ALN 10 mg/day or risedronate 5 mg/day experienced significant musculoskeletal events. In contrast, 20% of patients treated initially with ALN 70 mg/week and 25%

of patients treated initially with risedronate 35 mg/week experienced events such as arthralgia (12.6%), back pain (9.1%), myalgia (4.2%), bone pain (4.2%), chest pain (1.8%), and fever (1.2%) [89].

The pathological basis of the delayed painful bisphosphonate-associated adverse events remains undetermined in most cases [88, 90], but a hypothetical mechanism may be the following: in some cases of bisphosphonate-induced secondary hyperparathyroidism there may be a relatively smaller reduction of bone turnover caused by the bisphosphonate because of the antagonistic effect of the high parathyroid hormone (PTH). In such cases, bone turnover higher than expected from treatment with a bisphosphonate unopposed by a high PTH may lead to relatively higher bone uptake of bisphosphonate and higher than average concentration of the drug in the bone microenvironment. This in turn may result in a localized, relatively increased bisphosphonate-induced production of interleukin-6 and other proinflammatory cytokines, and an inflammatory reaction confined to bones [91].

Ocular Side Effects

The eyes have a relatively small mass and rich blood supply and are therefore highly susceptible to toxic substances. Nearly every drug can cause changes to ocular tissues through a variety of mechanisms such as the production of arachidonic acid derivatives, the liberation of free radicals, and the disruption of blood–aqueous and blood–retinal barriers [92].

The most common ocular side effect of bisphosphonates is nonspecific conjunctivitis, which usually improves without specific therapy and despite continuing treatment with a bisphosphonate. Rarely, treatment with a non-steroidal anti-inflammatory eye drop is required [93].

Other ocular side effects such as ptosis, eyelid edema, optic or retrobulbar neuritis, periorbital edema, cranial nerve palsy, and eyelid edema have been reported. Uveitis and scleritis are the most serious ocular side effects of bisphosphonate therapy and require the discontinuation of BPs treatment [94, 95].

Fraunfelder et al., after reviewing the inflammatory ocular side effects caused by intravenous pamidronate disodium, remarked that scleritis patients had an associated acute-phase reaction [93]. Therefore, ocular inflammation may represent a localized manifestation of a systemic adverse reaction to the drug.

So far with this type of BP, one case of posterior scleritis [94, 95] and two cases of anterior uveitis [96, 97] complicating treatment with zoledronic acid have been reported. In two patients, ALN was associated with scleritis [98, 99], which was confirmed by a positive rechallenge test. Three more cases of scleritis and one case of anterior uveitis [100, 101] occurred during treatment with ALN, and one case of uveitis with clodronate [102].

In several studies the number of patients developing ocular inflammation was higher in patients treated with zoledronate when compared with placebo [95, 103–107]. Inflammation appeared within the first 15 days after the infusion. This complication can resolve after a short course of corticosteroid treatment. Prevention includes the avoidance of bisphosphonates or caution in their use (especially IV) for those with a history of inflammatory eye disease or uveitis.

No recommendation can safely be made about the use of biphosphonates in patients with uveitis and/or scleritis, but discontinuation of biphosphonate treatment should be considered [93].

Esophageal Cancer

Since the launch of the first oral amino-containing bisphosphonate in 1995 and up through 2008, the FDA received reports of 23 cases of patients diagnosed with esophageal cancer, with ALN as the suspect drug in 21 of them, and the concomitant drug in the other two cases [108].

Thirty-one patients from Europe and Japan were also reported with esophageal cancer, with ALN as the suspect drug in 21 of the patients, risedronate, ibandronate, or etidronate in six cases, and bisphosphonates as concomitant drugs in four cases. The median time from ALN use to diagnosis was 2.1 years in the USA cases and 1.3 years in the European and Japanese cases. The relatively short time from drug exposure to diagnosis seems more consistent with the drugs acting as tumor promoters rather than as tumor initiators. Risk factors were not described for all patients, but some were taking oral bisphosphonates despite absolute contraindications such as Barrett esophagus [108, 109]. Wysowski recommends that practitioners should avoid prescribing bisphosphonates to patients with Barrett's esophagus [108].

However, there is no plausible mechanism by which such cancers are likely to be truly associated with bisphosphonate therapy. In contrast, it has been suggested that bisphosphonates may be associated with a reduced rather than increased risk of certain cancers, for example, breast cancers or colorectal cancers. In the first case such an association might be explained in part by low endogenous estrogen levels being associated both with breast cancer and with osteoporosis [110].

Moreover, Rennert et al. showed that the use of bisphosphonates for more than 1 year before diagnosis was associated with a significantly reduced relative risk of colorectal cancer. This association remained statistically significant after adjustment in a model for vegetable consumption, sports activity, family history of colorectal cancer, body mass index, and use of low-dose aspirin, statins, vitamin D, and postmenopausal hormones [111].

Atrial Fibrillation

Atrial fibrillation (AF) came to attention as a possible side effect of BPs in a recent large international randomized trial (The HORIZON Pivotal Fracture Trial) in which there was a higher risk of serious AF (defined as fatal, life-threatening, or resulting in hospitalization or disability) in those participants taking zoledronate than in recipients of placebo. The vast majority of AF events were reported more than 30 days after infusion, by which time zoledronate would be at extremely low concentrations in the circulation. A total of 50 patients in the zoledronate group had serious AF (1.3%), compared with 20 patients (0.5%) in the placebo group ($P<0.001$), whereas the incidence of any AF events was similar between treatment and placebo groups (2.4% vs. 1.9%; $P=0.12$) [103].

Re-analysis of a previous trial also reported a "trend" towards an increased risk of serious AF events among patients treated with oral ALN compared with placebo (1.5% vs. 1.0%; $P=0.07$) and a recent population-based case–control study found that "ever-use" of oral ALN was associated with an increased risk of incident [112, 113].

For this reason the FDA began reviewing placebo-controlled clinical trial information of BP to explore the potential risk of AF in patients treated with these agents. The review included data on 19,687 bisphosphonate-treated patients and 18,358 placebo-treated patients who were followed for 6 months to 3 years. In 2008, the FDA issued a statement indicating that their review revealed no clear association between overall BP exposure and the incidence of AF. It was also reported that increasing the dose or duration of therapy was not linked to an increased risk of AF [114]. Nevertheless, information on the incidence of AF was included in the Summary of Product Characteristics for zoledronate [115].

Grosso et al. reported an observational study investigating the association of oral BP and AF involving over 2,000 exposed patients with one or more recorded episode of AF or atrial flutter. They found no evidence of an overall long-term increased risk of AF or atrial flutter associated with alendronic acid or risedronate sodium [116].

However, these data conflict with results from other studies.

Two recent case–control studies investigated whether patients who presented with atrial fibrillation were treated more often with bisphosphonates [113, 117] compared to control patients without atrial fibrillation. In the first study, a group of 719 women with atrial fibrillation was compared to a group of 966 without atrial fibrillation. In the atrial fibrillation group 6.5% of the women were treated with alendromate at any time, compared to 4.1% of the women without atrial fibrillation ($P=0.03$). Thus, the risk of atrial fibrillation was greater among the ALN users than among the women who never used this drug (odds ratio 1.86, 95% CI 1.09–3.15). In contrast, the second study found that the use of bisphosphonates does not increase the risk of AF; the records of 13,586 women with a history of atrial fibrillation or flutter were surveyed and compared with 68,054 control women without these cardiac complications. In this study, 3.2% of the women with atrial fibrillation or flutter and 2.9% of the control women were taking a BPs (etidronate or ALN). The relative risk of atrial fibrillation was 0.95.

The mechanism that explains this side effect remains unknown. Since AF is associated with an altered sarcoendoplasmic reticulum calcium load, Kemeny-Suss et al. studied how ALN affects cardiomyocyte calcium homeostasis and protein isoprenylation in vitro. Acute and long-term (48 h) treatment of atrial and ventricular cardiomyocytes with ALN (10(−8)-10(−6)M) was performed. Changes in calcium dynamics were determined by both fluorescence measurement of cytosolic free Ca^{2+} concentration and western blot analysis of calcium-regulating proteins. Finally, effect of ALN on protein farnesylation was also identified. In both atrial and ventricular cardiomyocytes, ALN treatment delayed and diminished calcium responses to caffeine. Only in atrial cells, long-term exposure to ALN induced transitory calcium oscillations and led to the development of oscillatory component in calcium responses to caffeine. Changes in calcium dynamics were accompanied by changes in expression of proteins controlling sarcoendoplasmic reticulum calcium. Treatment of atrial cardiomyocytes with ALN induced abnormalities in calcium dynamics consistent with induction of a self-stimulatory, pacemaker-like behavior, which may contribute to the development of cardiac side effects associated with these drugs [118].

Atypical Fractures of the Femoral Diaphysis

A possible association between prolonged BP therapy and the occurrence of atypical subtrochanteric or diaphyseal fractures is a further example of adverse effects coming to attention. These fractures have been postulated to be the result of excessive reduction of bone remodeling [119–122]. They often have a characteristic radiographic pattern, defined as a simple transverse fracture with a unicortical break in an area of cortical hypertrophy.

Findings from a recent registry-based national cohort study from Denmark involving nearly 16,000 patients suggest that subtrochanteric fractures are likely to occur as a result of osteoporosis, and not as a complication of ALN therapy [123].

However several authors reported patients on long-term (3–8 years) treatment with ALN who suffered unusual, low-energy, non-spinal fractures [124, 125].

It has been hypothesized that the unique radiographic pattern and prodromal pains suggest that the complete femoral diaphyseal fracture in bisphosphonate users may result from propagation of a stress fracture unable to heal, because of the excessive bisphosphonate bone turnover suppression [119, 120].

Other mechanisms could explain these fractures. Indeed, in a study of a patient on ALN who sustained bilateral fractures at the level of the femoral diaphysis, although bone formation was lower at the site of the fracture than in the iliac crest, the number of osteoclasts in the femoral cortex was sixfold higher than in the iliac cancellous bone [126]. Moreover, these osteoclasts had all the morphologic characteristics of normal active osteoclasts, and no resemblance to the giant osteoclasts described previously in patients treated with ALN [127].

Hypocalcemia

Symptomatic hypocalcemia and hypomagnesemia occurred in 8% of patients with various malignancies treated with zoledronic acid despite prophylactic administration of vitamin D and calcium supplements [128]. Risk factors for severe bisphosphonate-induced hypocalcemia are preexisting hypoparathyroidism, parathyroid dysfunction during thyroidectomy in a patient receiving chronic bisphosphonate therapy, vitamin D deficiency, and renal failure. The bisphosphonate-induced hypocalcemia can be avoided by the administration of adequate vitamin D and calcium supplements, starting about 2 weeks before the administration of the bisphosphonate. Berruti et al. emphasized the need to treat effectively the secondary hyperparathyroidism in cancer patients because of the ability of high PTH to promote tumor growth [129].

Cutaneous Reactions

A few cases of bisphosphonate-associated cutaneous reactions have been reported, such as urticaria to pamidronate and ALN [130, 131]. Remarkably, ALN challenge did not provoke a reaction in two of the patients with a previous urticarial reaction to pamidronate, while one patient was subsequently treated with clodronate without an untoward reaction. Rare cases of rash/pruritus have been reported with ALN, risedronate, and etidronate [132]. Reported skin reactions to ALN also include lichen planus, superficial gyrate erythema, or perivascular spongiotic dermatitis, and papulo-petechial skin eruption [133–135]. A case of risedronate-associated cutaneous vasculitis has also been reported [136, 137]. Some cases were confirmed by a positive rechallenge test. Probably, skin reactions to bisphosphonates are not a class effect [135].

Teratogenic Risk

The use of BP in pre-menopausal women prior to conception may also pose a teratogenic risk because bisphosphonates remain in mineralized bone for several years. Consequently a clinical dilemma exists in treating and preventing glucocorticoid-induced osteoporosis in pre-menopausal women. McNicholl and Heaney reviewed the evidence regarding the use of bisphosphonates in pre-menopausal woman [138]. Bisphosphonate treatment of metastatic hypercalcemia during pregnancy has not demonstrated adverse fetal events. Cases of preconception bisphosphonate exposure have failed to describe developmental or bone density abnormalities; however neonatal hypocalcemia has occurred. Lower birth weight, lower gestational age at birth, and higher rate of spontaneous abortion in mothers with preconception and first trimester bisphosphonate exposure have been described, but notable confounding factors and lack of adequate sample size make extrapolation of this data difficult. Human preconception and first trimester bisphosphonate use has to date not been associated with the same adverse effects evident in animals. However, larger well-controlled studies in pre-menopausal women with preconception and prenatal bisphosphonate exposure are required to confirm the safe use throughout pregnancy to both mother and fetus [138].

Hepatitis

A few cases of hepatitis developing several months or years after starting bisphosphonate therapy and resolving several months after discontinuing bisphosphonates have been reported [139–142]. Liver biopsy revealed lesions suggestive of a drug effect.

The mechanism by which ALN may cause liver damage is not known. A possibility is that ALN, like other aminobisphosphonates, inhibits the synthesis of cholesterol in the liver, which may alter liver function, since this pathway is essential in the maturation of Ras-related G proteins [143].

Osteonecrosis of the Jaw

In late 2003 a preliminary report described 36 cases of avascular necrosis of the jaws in patients receiving intravenous bisphosphonates. With the exception of one subject being treated for osteoporosis, all patients were receiving BP as an adjunct to cancer therapy [40].

Subsequently, Ruggiero et al. reported on 63 cases of ONJ. In this series 55 patients were under treatment for cancer (28 multiple myeloma, 20 breast cancer) and 7 patients were being treated for osteoporosis [144]. Later several other cases were reported [145, 146].

The American Association of Oral and Maxillofacial Surgeons (AAOMS) defines ONJ as exposure of portions of the jawbone in patients who have been exposed to bisphosphonates that has persisted for more than 8 weeks with no history of radiation therapy to the jaws. The Australian and New Zealand Bone and Mineral Society, Osteoporosis Australia, Medical Oncology Group of Australia, and the Australian Dental Association define ONJ as an exposed area of bone present for more than 6 weeks. The American Society of Bone and Mineral Research uses the term "suspected" for ONJ in which bone has been exposed for less than 8 weeks, and at 8 weeks or longer, it uses the term "confirmed" [147–150].

Clinically, the lesions appear as exposed bone, which is painful in 60–69% of cases [151]. AAOMS designates exposed bone with absence of pain and infection as Stage 1 ONJ [152].

Pain may be associated with suppuration or infection and intraoral sinus track formation (Stage 2 ONJ). If there is also extraoral fistula, exposed necrotic bone extending beyond the region of the alveolar bone, pathological fracture, osteolysis extending to the inferior border or oroantral communication, the disease is designated as Stage 3 ONJ. Other staging systems, however, have also been proposed. More recently, Mawardi and colleagues suggested that ONJ might occur even in the absence of exposed bone (Stage 0) [153–156] (Table 4).

Several definitions for ONJ have recently been proposed and a breech in the oral mucosa is an absolute requirement for such a definition to apply [154].

Table 4 Possible ONJ staging

Stage 0 (?)	Sinus tracts formation
Stage I	Exposed bone with absence of pain and infection
Stage II	Exposed bone with suppuration, or infection
Stage III	Extraoral fistula

Suspected ONJ: the bone has been exposed for less than 8 weeks
Confirmed ONJ: the bone has been exposed for 8 weeks or longer

However, at the present time, it is not definitively known whether the ONJ lesion, in fact, initiates in the bone, or whether it may originate in the epithelial cells of the mucosa [157].

Toxicity of bisphosphonates to gastric mucosal cells has been previously documented. Several authors have shown that a number of bisphosphonates, including pamidronate, are cytotoxic to human intestinal epithelial cells [158–161].

Landesberg et al. showed that pamidronate suppresses cellular proliferation and delays wound healing of oral mucosal cells. Furthermore, the effects of pamidronate are observed at doses that would theoretically be expected in the maxillofacial bone and soft tissues of patients receiving the drug by the intravenous route. These effects however, were not the result of apoptotic cell death [162].

In any case, we have to bear in mind that immediate contact of the BP tablets with the oral mucosa, for example sucking the tablets instead of swallowing them, was the cause of contact stomatitis with oral ulcerations in several cases [163–167]. Probably, some of these cases are remarkable in that they could be misdiagnosed as cases of osteonecrosis of the jaw and raise the question that habitual contact of the bisphosphonate tablet with the gingival mucosa could possibly lead to osteonecrosis of the jaw. Diagnosis of ONJ should be mainly based on clinical and radiographic criteria. Tissue biopsy is not always necessary and should be performed only if metastatic disease is suspected. At microscopic examination, the specimen shows areas of partially or completely necrotic bone and debris, and fibrinous exudates accompanied by an inflammatory infiltrate formed by neutrophils, histiocytes, eosinophils, and plasma cells. The non-vital bone exhibits loss of osteocytes from the lacunae, peripheral resorption, and bacterial colonization. Scattered sequestra and pockets of abscess formation are common [168].

Epidemiology

Until 2006, of 368 cases of ONJ reported in the literature, only 15 (4.1%) had received oral bisphosphonates for osteoporosis [169]. A German study of 780,000 people with osteoporosis found only three cases of ONJ, yielding a risk of less than 1 in 100,000 patient treatment years [170]. An Australian study identified 36 cases of ONJ in people with osteoporosis, giving an overall risk ranging between 0.01 and 0.04%, rising to between 0.09 and 0.34% after dental extraction [171].

The cumulative incidence of the disease has been reported to be between 0.8 and 12% in patients using intravenous bisphosphonates [154]. Hoff reported on 4,000 cancer patients where 33 developed ONJ. This retrospective study found the overall occurrence rate of ONJ was 0.8%, with 2.8% occurring in multiple myeloma patients and 1.2% in breast cancer patients [172].

The incidence of ONJ in breast cancer patients seems to be lower than in patients with prostate cancer or patients with multiple myeloma, with reported incidence rates ranging from 3.8 to 11.0%. These observations might suggest that not only co-medication but also the underlying malignant disease may influence the occurrence of ONJ [173].

Several studies suggest that the frequency of ONJ may be different according to geographical regions [174–183]. Most of the published studies are from the United States, Italy, Poland, and Greece. The frequency of ONJ in multiple myeloma patients seems to be lower in the United States, Poland, and Italy (1.2–3.5%) and higher in Greece (4–11%). In breast cancer patients, the frequency varies from 0.6% to the 1.2% observed in this study in the United States, 2.9% in Greece [175], and 6.2% in a study from Italy [180]. These differences raise interesting questions regarding potential influence of genetic backgrounds in the development of ONJ.

Obesity and smoking were associated significantly with ONJ development [184].

Pathoetiology of ONJ

Despite considerable effort, it has proved very difficult to produce lesions that might be relevant to human ONJ in animal models. None of the animal models to date fully simulate the human condition. The absence of appropriate animal models has also hampered efforts to search for plausible mechanisms.

Several theories have been proposed regarding the pathoetiology of ONJ with most speculation focusing on origin from the bone (Table 5). For instance, many authors have hypothesized that ONJ is related to an over-suppression of bone turnover by bisphosphonates. Both human and animal studies [185–189] show that bisphosphonate doses used for osteoporosis treatment significantly suppress tissue-level bone remodeling (measured using fluorochrome-labeled bone histomorphometry). By suppressing turnover, regions of nonviable osteocytes are inevitably removed at a slower rate. This produces localized tissue necrosis, containing regions of nonviable osteocytes or, if allowed sufficient time, regions where the osteocytes and their canaliculi fill with mineral [190]. Presence of matrix necrosis in the alveolar bone of a tooth that undergoes dental extraction is likely to contribute to a disrupted healing/remodeling response.

This theory, however, fails to explain why ONJ occurs almost exclusively in the maxillofacial region.

Another hypothesis is that ONJ is a result of the inhibition of angiogenesis by the BP. While several reports have shown that BP reduce angiogenesis, almost all of

Table 5 Hypothetical mechanisms of ONJ pathoetiology

Inhibition of bone remodeling	Compromised bone remodeling and repair
Functional effects	Inhibition of human osteoblasts adhesion and migration
Vascular	Antiangiogenic affects delaying wound healing and/or affecting micro-infarction in bone and/or soft tissues
Action on different cells	Effect on mesenchymal stem cells
Infection	Microorganisms of the oral cavity promoting cell death in the bone and/or oral soft tissues
Vitamin D pathway	Hypovitaminosis D may contribute to both the obstruction of bone repair and soft tissue destruction through an immune response that involves $\gamma\delta$ T cells
Genetic predisposition	Genetic polymorphisms affecting drug metabolism, excretion, or drug targets within pathways of bone metabolism and/or wound healing
Drug interactions	Drug interactions between chemotherapy and bisphosphonate selectively promoting cell death

these studies have only been performed in vitro and in transgenic animal models [191–196].

Allegra et al. demonstrated a modified angiogenesis and a decrease in circulating endothelial progenitor cells (EPCs) in patients with ONJ compared to control subjects [197, 198].

Moreover they found an increase in endothelial cell apoptosis in Multiple Myeloma patients after BP administration and in osteonecrosis subjects [199].

Of interest are data from patients treated with bisphosphonates and antiangiogenic drugs.

In a retrospective analysis, ONJ occurred in 2% of patients treated with bevacizumab and BP compared with 1.1% of patients receiving BP alone [200], and Estilo et al. reported the first case of ONJ related to bevacizumab [201]. Given the mechanism of action of these drugs, an additive toxic effect of antiangiogenic drugs in patients receiving BP may explain the cases of ONJ observed [201–203].

Recent studies have indicated toxic effects of BP on different cell types, apart from osteoclast inhibition, and a possible interference of bisphosphonates on cytokine production [204].

It has been showed that multipotent stem cells activity plays an important role in the processes of wound healing and bone regeneration, which seem to be especially impaired in the jaws of patients receiving high doses of N-BP. Otto et al. studied the effect of zoledronate, ibandronate, and clodronate on immortalized mesenchymal stem cells tested at different concentrations, reflecting 1, 3, and 6 months and 1, 3, 5, and 10 years of exposure to standard oncology doses of the 2 N-BP and equimolar concentrations of clodronate. Both types of BP revealed remarkable differences. Zoledronate and ibandronate showed a dose- and pH-dependent cellular toxicity. Increasing concentrations of both N-BP and an acidic milieu led to a significant decrease in cell viability and activity, with more pronounced effects for zoledronate. Equimolar concentrations of clodronate did not affect the cell survival or activity significantly, apart from the effect of pH reduction itself, which was also detectable in the patients in the control group who did not receive bisphosphonates [205].

Moreover, BP seem to inhibit human osteoblast adhesion, and migration, and ONJ could be enhanced by an inhibition of osteoblast adhesion and migration. The gene expression results, however, suggest a cell line dependent effect of bisphosphonates, which could explain the interindividual differences of ONJ incidences [206].

Interestingly, the vitamin D pathway has been proposed as one possible biologic pathway leading to ONJ in humans. Ardine et al. [207], after studying patients with bone metastatic breast cancer on long-term zoledronate treatment, observed that low serum calcium and elevated levels of PTH may be significantly correlated with ONJ development. These results are consistent with data from Berruti et al., who suggested that hypercalcemia may predispose prostate cancer patients to ONJ development, whereas secondary parathyroidism may play a significant role as well [208, 209]. The coexistence of hypovitaminosis D may also contribute to both the obstruction of bone repair and soft tissue destruction through an immune response that involves γδ T cells [210]. As said, genetic predisposition may also play an important role in the development of ONJ. Genetic polymorphisms that might lead to hypercalcemia or hyperparathyroidism, for example, in the vitamin D receptor gene (VDR) or in the parathyroid gene (PTH1), may explain how the vitamin D pathway is involved in ONJ. Interestingly, certain VDR genotypes have been implicated in the metabolic syndrome and in obesity via interactions with the insulin-like growth factor system, which may also explain the epidemiologic findings associating obesity with ONJ in humans [184]. Therefore, tailoring vitamin D and calcium supplementation might prove helpful in reducing the frequency of ONJ.

Finally several observations suggested that bisphosphonate use facilitates actinomycotic infection, resulting in osteonecrosis. Osteonecrosis due solely to bisphosphonates has not been described in animals or humans with normal bone, and the concomitant occurrence of actinomyces infection with ONJ has been seen in an increasing number of cases. Histological studies have shown a near-universal presence of the actinomyces-like organisms from affected bone. Furthermore, actinomyces can instigate bone resorption, and histologically, in the setting of ONJ, sulfur granules have been regularly noted to occur at the location of osteonecrotic bone [211–218].

In any case, Subramanian et al. [219] proposed that ONJ pathogenesis is multifactorial and is the combined result of attenuated osteoblastic activity (owing to the patient's underlying disease, e.g., osteoporosis or multiple myeloma), BP-mediated osteoclast toxicity, and the resultant compromised osteoblast–osteoclast interactions during bone remodeling. Consequently, a vicious cycle of ineffective local remodeling results in the persistence of defective bone, compromised tissue perfusion, and if unresolved, ultimately leads to necrosis.

Table 6 Suggested ONJ treatments

Prevention
Diet supplemented with vitamin D and calcium
Antibiotic prophylaxis
Antibiotic therapy
Surgical intervention
Bone resection and autologous platelet-rich plasma
Teriparatide treatment
Infusion of mesenchymal stem cells
Hyperbaric oxygen therapy

Prevention and Therapy of ONJ

There is no definitive treatment for ONJ at this time, and the management of ONJ remains an important clinical challenge in cancer patients and requires a multidisciplinary approach involving dentists, medical oncologists, oral and maxillofacial surgeons, and others [220] (Table 6).

Prevention remains the best care. Dimopolos et al. found a significant, almost threefold, decrease of zoledronic acid-related ONJ since they implemented preventive measures [221].

In the prevention of ONJ, optimal dental health during treatment is essential, and all patients should be informed of the importance of good oral hygiene and the use of chlorhexidine rinses. In addition, regular visual inspections by the treating physician and routine assessments (at least annually) by a dental specialist are warranted.

Consensus-based prevention guidelines recommend a panoramic radiograph to help in the diagnosis of caries and periodontal disease, and the identification of metastatic cancer and other bony pathology [222].

If BP therapy can be delayed, preventive surgery to eliminate potential sites of infection should be performed. Only after the mucosa has completely healed should BP therapy be started. Finally, dentures should be adjusted to relieve pressure on mucosal surfaces.

Whenever dental treatment is necessary, the less invasive endodontic techniques with preservation of the dental root are preferred over total tooth extraction. Only when teeth have a mobility score = 3 should extraction be considered, and it should be performed as atraumatically as possible [223]. To prevent local infection, the use of antibiotics may be indicated.

Some authors suggest the withdrawal of bisphosphonate therapy for 1–3 months before performing invasive dental procedures [223]. Considering the long half-life of bisphosphonates in bone, however, it remains doubtful if adequate osteoclast recovery is realistic [224].

The use of serum CTX-I has recently been proposed to predict the risk for ONJ after an invasive dental procedure in patients receiving oral BP therapy. Patients with morning fasting serum CTX-I levels >150 pg/mL are considered to have adequate residual bone remodeling capacity and to be only at minimal risk of developing ONJ [225].

However, the technique is subject to considerable variability, and most patients on BP therapy will not have had a baseline CTX-I measurement. Moreover, known sources of CTX-I variability include, amongst others, the circadian cycle, gender, age, food intake, renal and liver function, and concomitant medication use [226], while in patients with skeletal metastases, the theoretical merit of bone turnover markers is lacking, and they should thus not be used to guide clinical management decisions [227].

Diet supplemented with vitamin D and calcium has also been proven to be beneficial because it maintains proper levels of PTH during therapy.

A reduction of duration of BP treatment could be relevant in the prevention of ONJ. Palaska et al. calculated the weighted mean time to event, as well as the minimum reported time and dose for zoledronate, pamidronate, and oral bisphosphonates. The mean time to ONJ after zoledronate treatment was calculated at 1.8 years and the minimum was 10 months; after pamidronate, the mean time was 2.8 years and the minimum was 1.5 years; and after oral BP therapy, the mean time was 4.6 years and the minimum was 3 years [228].

Prophylactic antibiotic coverage is recommended for any invasive dental procedure. Penicillins remain the drug of choice. For individuals with penicillin allergy, quinolones and metronidazole or erythromycin and metronidazole are good second choices. When osteonecrosis is present a possible treatment is long-term therapy with penicillin V-K 500 mg four times a day and 0.12% chlorohexidine. In refractory cases, metronidazole 500 mg per os three times a day could be added. In more symptomatic cases, ampicillin 1 g with clavulanate 500 mg intravenously every 8 h is recommended [145].

As far as surgical therapy, a major goal in the prevention of ONJ is to limit the possibility of extensive or multifocal involvement. Although there is no evidence to support a conservative clinical approach, it may be prudent to proceed conservatively in some cases. On the other hand, periapical pathoses, sinus tracts, purulent periodontal pockets, severe periodontitis, and active abscesses that already involve the medullary bone, all may exacerbate osteonecrosis, and these areas should be treated immediately even if multiple quadrants are involved.

Depending on the stage of the disease, therapy ranges from mouth rinses, debridement, to huge resections of the affected area with a broad influence on the quality of life for these patients.

Treatment of refractory ONJ with a combination of bone resection and autologous platelet-rich plasma could be considered an alternative treatment modality for management of advanced cases [229].

Cheung et al. suggested Teriparatide therapy for the treatment of Alendronate-Associated ONJ. Teriparatide in fact may counteract alendronnate action by stimulating bone remodeling. An increased number of remodeling units and increased bone formation within each unit may promote healing and the removal of damaged bone. In addition, bone necrosis and infection trigger T-cell immune responses, stimulating the secretion of Wnt10b and Wnt-mediated bone formation locally.

Teriparatide may augment this process by stimulating marrow CD8+ T cells to produce Wnt10b.3. Additional stimulation of Wnt signaling through the action of teriparatide on sclerostin may further enhance this reparative pathway [230].

However, Teriparatide is contraindicated in patients with ostcosarcoma or metastatic bone disease, the group in which BP-ONJ is most commonly observed.

Hyperbaric oxygen therapy is a way of treatment which increases the oxygen content of the blood and, consequently, the amount of oxygen delivered to tissues. Hyperbaric oxygen therapy has also been suggested in the treatment of ONJ. The proposed rationale behind the beneficial effects of hyperbaric oxygen therapy in ONJ is increased wound healing, reduction of edema and inflammation, stem cell mobilization, and moderation of the suppression of bone turnover by BPs. It is a concern for some physicians that hyperbaric oxygen may have a cancer-promoting effect or may lead to metastatic growth. However, Feldmeier et al. reviewed all available data and concluded that there is no evidence to support the idea that hyperbaric oxygen therapy may enhance tumor growth or metastases in cancer patients [231–238].

Very recently, a new treatment modality was introduced elucidating one possible causative factor for ONJ. Kikuiri et al. infused mesenchymal stem cells in ONJ-like mice [239]. The stem cells modulated the immune system, and may prevent and cure ONJ-like disease [240–244].

Conclusions

It is estimated that there have been more than 190 million prescriptions in the United States, for oral ALN, risedronate, and ibandronate, and more than six million patients treated with IV bisphosphonates for cancer, worldwide (1.9 million with pamidronate, 1.2 million with zoledronate, and 3.1 million with both). When compared to many other therapies, especially in the cancer setting, the severity of adverse events related to BPs is generally mild and, with the exception of the AFR causing transient fever, myalgia, and bone pain, side effects are infrequent. Thus, the benefits of treatment with BPs almost always outweigh the risks [245]. However, when making the decision to use a bisphosphonate, the balance between benefit and potential risks according to clinical circumstances of each patient should be considered, and with the increased use of antiangiogenic agents in patients with metastatic breast, colon, lung, or kidney cancer, physicians should be aware of the potential for development of ONJ.

All this is relevant, and in the future the use of BPs will be even more widespread as emerging data suggest that BPs may have different antitumor activities. In fact, it is becoming increasingly clear that bisphosphonates—especially the more potent nitrogen-containing bisphosphonates and, in particular, zoledronic acid—show a multitude of antitumor effects in the laboratory. These drugs have been shown to induce apoptosis in a variety of cancer cells and enhance the antitumor effects of a variety of treatments including steroids, chemotherapeutic agents and hormonal therapies as well as many classes of new anticancer drugs. In addition, the agents inhibit vasculogenesis and angiogenesis. The N-BP also show immunostimulatory effects that have been shown to have potent antitumor effects. They also block tumor cell adhesion making tumor cells more susceptible to the cytotoxic effects of chemotherapy as well as migration and invasion of tumor cells [246–251].

Prospective studies are required to determine the degree of risk for BP side effects and how best this risk can be minimized.

References

1. Fleisch H, Russell RG, Bisaz S, Casey PA, Mühlbauer RC. The influence of pyrophosphate analogues (diphosphonates) on the precipitation and dissolution of calcium phosphate in vitro and in vivo. Calcif Tissue Res. 1968;2(Suppl):10–10a.
2. Fleisch H, Russell RG, Francis MD. Diphosphonates inhibit hydroxyapatite dissolution in vitro and bone resorption in tissue culture and in vivo. Science. 1969;165:1262–4.
3. Van den Wyngaert T, Huizing MT, Vermorken JB. Disambiguating the bisphosphonates. Ann Oncol. 2008;19:1357–9.
4. Bassett CA, Donath A, Macagno F, Preisig R, Fleisch H, Francis MD. Diphosphonates in the treatment of myositis ossificans. Lancet. 1969;2:845.
5. Rosen LS, Gordon D, Kaminski M, Howell A, Belch A, Mackey J, et al. Zoledronic acid versus pamidronate in the treatment of skeletal metastases in patients with breast cancer or osteolytic lesions of multiple myeloma: a phase III, double-blind, comparative trial. Cancer J. 2001;7:377–87.
6. Saad F, Gleason DM, Murray R, Tchekmedyian S, Venner P, Lacombe L, et al., Zoledronic Acid Prostate Cancer Study Group. A randomized, placebo-controlled trial of zoledronic acid in patients with hormone-refractory metastatic prostate carcinoma. J Natl Cancer Inst. 2002;94:1458–68.
7. Major P. The use of zoledronic acid, a novel, highly potent bisphosphonate, for the treatment of hypercalcemia of malignancy. Oncologist. 2002;7:481–91.
8. Van Poznak CH. The use of bisphosphonates in patients with breast cancer. Cancer Control. 2002;9:480–9.
9. Amin D, Cornell SA, Gustafson SK, Needle SJ, Ullrich JW, Bilder GE, et al. Bisphosphonates used for the treatment of bone disorders inhibit squalene synthase and cholesterol biosynthesis. J Lipid Res. 1992;33:1657–63.
10. Rogers MJ, Gordon S, Benford HL, Coxon FP, Luckman SP, Monkkonen J, et al. Cellular and molecular mechanisms of action of bisphosphonates. Cancer. 2000;88:2961–78.
11. Roelofs AJ, Thompson K, Gordon S, Rogers MJ. Molecular mechanisms of action of bisphosphonates: current status. Clin Cancer Res. 2006;12:6222s–30.
12. Weinstein RS, Roberston PK, Manolagas SC. Giant osteoclast formation and long-term oral bisphosphonate therapy. N Engl J Med. 2009;360:53–62.
13. Papapoulos SE. Pharmacodynamics of bisphosphonates in man: implications for treatment. In: Bijvoet OLM, Fleisch H, Canfield RE, Russell RGG, editors. Bisphosphonate therapy in acute and chronic bone loss. Amsterdam: Elsevier; 1995. p. 231–45.
14. Russell RGG, Rogers M. Bisphosphonates: from the laboratory to the clinic and back again. Bone. 1999;25:97–106.
15. Mitchell DY, Barr WH, Eusebio RA, Stevens KA, Duke FP, Russell DA, et al. Risedronate pharmacokinetics and intra- and inter-subject variability upon single dose intravenous and oral administration. Pharm Res. 2001;18:166–70.
16. Kahn SA, Kanis JA, Vasikaran S, Kline WF, Matuszewski BK, McCloskey EV, et al. Elimination and biochemical responses to intravenous alendronate in postmenopausal osteoporosis. J Bone Miner Res. 1997;12:1700–7.

17. Luhe A, Kunkele KP, Schad K, Haiker M, Zihlmann C, Bauss F, et al. Preclinical evidence for nitrogen-containing bisphosphonate inhibition of farnesyl diphosphate (FPP) synthase in the kidney: implications for renal safety. Toxicol In Vitro. 2008;22:899–909.
18. Troehler U, Bonjour JP, Fleisch H. Renal secretion of diphosphonates in rats. Kidney Int. 1975;8:6–13.
19. Lasseter KC, Porras AG, Denker A, Santhanagopal A, Daifotis A. Pharmacokinetic considerations in determining the terminal elimination half-lives of bisphosphonates. Clin Drug Investig. 2005;25(2):107–14.
20. Mitchell DY, Eusebio RA, Sacco-Gibson NA, Pallone KA, Kelly SC, Nesbitt JD, et al. Dose proportional pharmacokinetics of risedronate on single-dose oral administration to healthy volunteers. J Clin Pharmacol. 2000;40:258–65.
21. Body JJ, Pfister T, Bauss F. Preclinical perspectives on bisphosphonate renal safety. Oncologist. 2005;19 Suppl 1:3–7.
22. Dumon JC, Magritte A, Body JJ. Efficacy and safety of the bisphosphonate tiludronate for the treatment of tumor-associated hypercalcemia. Bone Miner. 1991;15:257–66.
23. O'Sullivan TL, Akbari A, Cadnapaphornchai P. Acute renal failure associated with the administration of parenteral etidronate. Ren Fail. 1994;16:767–73.
24. Kanis JA, Preston CJ, Yates AJ, Percival RC, Mundy KI, Russell RG. Effects of intravenous diphosphonates on renal function. Lancet. 1983;1:1328.
25. Nancollas GH, Tang R, Phipps RJ, Henneman Z, Gulde S, Wu W, et al. Novel insights into actions of bisphosphonates on bone: differences in interactions with hydroxyapatite. Bone. 2006;38:617–27.
26. Lawson MA, Xia Z, Barnett BL, Triffitt JT, Phipps RJ, Dunford JE, et al. Differences between bisphosphonates in binding affinities for hydroxyapatite. J Biomed Mater Res B Appl Biomater. 2010;92(1):149–55.
27. Dunford JE, Thompson K, Coxon FP, Luckman SP, Hahn FM, Poulter CD, et al. Structure-activity relationships for inhibition of farnesyl diphosphate synthase in vitro and inhibition of bone resorption in vivo by nitrogen-containing bisphosphonates. J Pharmacol Exp Ther. 2001;296:235–42.
28. Dunford JE, Kwaasi AA, Rogers MJ, Barnett BL, Ebetino FH, Russell RG, et al. Structure-activity relationships among the nitrogen containing bisphosphonates in clinical use and other analogues: time-dependent inhibition of human farnesyl pyrophosphate synthase. J Med Chem. 2008;51(7):2187–95.
29. Russell RGG, Watts NB, Ebetino FH, Rogers MJ. Mechanisms of action of bisphosphonates: similarities and differences and their potential influence on clinical efficacy. Osteoporos Int. 2008;19:733–59.
30. IMS Health. NPA Plus™. Norwalk, CT: IMS Health Inc.; 2006.
31. Black DM, Cummings SR, Karpf DB, Cauley JA, Thompson DE, Nevitt MC, et al. Randomised trial of effect of alendronate on risk of fracture in women with existing vertebral fractures. Fracture Intervention Trial Research Group. Lancet. 1996;348:1535–41.
32. Cummings SR, Black DM, Thompson DE, Applegate WB, Barrett-Connor E, Musliner TA, et al. Effect of alendronate on risk of fracture in women with low bone density but without vertebral fractures: results from the Fracture Intervention Trial. JAMA. 1998;280:2077–82.
33. Pols HA, Felsenberg D, Hanley DA, Stepán J, Muñoz-Torres M, Wilkin TJ, et al. Multinational, placebo-controlled, randomized trial of the effects of alendronate on bone density and fracture risk in postmenopausal women with low bone mass: results of the FOSIT study. Fosamax International Trial Study Group. Osteoporos Int. 1999;9:461–8.
34. Harris ST, Watts NB, Genant HK, McKeever CD, Hangartner T, Keller M, et al. Effects of risedronate treatment on vertebral and nonvertebral fractures in women with postmenopausal osteoporosis: a randomized controlled trial. Vertebral Efficacy with Risedronate Therapy (VERT) Study Group. JAMA. 1999;282:1344–52.
35. Reginster J, Minne HW, Sorensen OH, Hooper M, Roux C, Brandi ML, et al. Randomized trial of the effects of risedronate on vertebral fractures in women with established postmeno-

pausal osteoporosis. Vertebral Efficacy with Risedronate Therapy (VERT) Study Group. Osteoporos Int. 2000;11:83–91.
36. McClung MR, Geusens P, Miller PD, Bensen WG, Roux C, Adami S, et al.; Hip Intervention Program Study Group for Hip Intervention Program Study Group. Effect of risedronate on the risk of hip fracture in elderly women. N Engl J Med. 2001;344:333–40.
37. Chesnut CH III, Skag A, Christiansen C, Recker R, Stakkestad JA, Hoiseth A, et al.; Oral Ibandronate Osteoporosis Vertebral Fracture Trial in North America and Europe (BONE). Effects of oral ibandronate administered daily or intermittently on fracture risk in postmenopausal osteoporosis. J Bone Miner Res. 2004;19:1241–9.
38. Lewiecki EM. Safety of long-term bisphosphonate therapy for the management of osteoporosis. Drugs. 2011;71(6):791–814.
39. Diel IJ, Bergner R, Grotz KA. Adverse effects of bisphosphonates: current issues. J Support Oncol. 2007;5:475–82.
40. Marx RE. Pamidronate (Aredia) and zoledronate (Zometa) induced avascular necrosis of jaws: a growing epidemic. J Oral Maxillofac Surg. 2003;61:1115–7.
41. de Groen PC, Lubbe DF, Hirsch LJ, Daifotis A, Stephenson W, Freedholm D, et al. Esophagitis associated with the use of alendronate. N Engl J Med. 1996;335:1016–21.
42. Bauer DC, Black D, Ensrud K, Thompson D, Hochberg M, Nevitt M, et al. Upper gastrointestinal tract safety profile of alendronate. The fracture intervention trial. Arch Intern Med. 2000;160:517–25.
43. Cryer B, Miller P, Petruschke RA, Chen E, Geba GP, De Papp AE. Upper gastrointestinal tolerability of once weekly alendronate 70 mg with concomitant non-steroidal anti-inflammatory drug use. Aliment Pharmacol Ther. 2005;21:599–607.
44. Eisman JA, Rizzoli R, Roman-Ivorra J, Lipschitz S, Verbruggen N, Gaines KA, et al. Upper gastrointestinal and overall tolerability of alendronate once weekly in patients with osteoporosis: results of a randomized, double-blind, placebo-controlled study. Curr Med Res Opin. 2004;20:699–705.
45. Greenspan S, Field-Munves E, Tonino R, Smith M, Petruschke R, Wang L, et al. Tolerability of once-weekly alendromate in patients with osteoporosis: a randomized, double-blind, placebo-controlled study. Mayo Clin Proc. 2002;77:1044–52.
46. Cryer B, Bauer DC. Oral bisphosphonates and upper gastrointestinal tract problems: what is the evidence? Mayo Clin Proc. 2002;77:1031–43.
47. Bobba RS, Beattie K, Parkinson B, Kumbhare D, Adachi JD. Tolerability of different dosing regimens of bisphosphonates for the treatment of osteoporosis and malignant bone disease. Drug Saf. 2006;29:1133–52.
48. Epstein S, Geusens P, Fishcr JE. Disintegration and esophageal irritation profiles of alendronate formulations: implications for clinical safety and efficacy. J Appl Res. 2005;5(2):253–65.
49. Epstein S, Cryer B, Ragi S, Zanchetta JR, Walliser J, Chow J, et al. Disintegration/dissolution profiles of copies of Fosamax (alendronate). Curr Med Res Opin. 2003;19:781–9.
50. Dansereau RJ, Crail DJ, Perkins AC. In vitro disintegration and dissolution studies of once-weekly copies of alendronate sodium tablets (70 mg) and in vivo implications. Curr Med Res Opin. 2008;24:1137–45.
51. Dansereau RJ, Crail DJ, Perkins AC. In vitro disintegration studies of weekly generic alendronate sodium tablets (70 mg) available in the US. Curr Med Res Opin. 2009;25:449–52.
52. Miller PD, Miday R, Klemes A, Ramsey D. Renal safety across a wide range of dosing regimens of risedronate. J Bone Miner Res. 2008;23 Suppl 1:SA402.
53. Jamal SA, Bauer DC, Ensrud KE, Cauley JA, Hochberg M, Ishani A, et al. Alendronate treatment in women with normal to severely impaired renal function: an analysis of the Fracture Intervention Trial. J Bone Miner Res. 2007;22:503–8.
54. Miller PD. The kidney and bisphosphonates. Bone. 2011;49:77–81.
55. Miller PD, Roux C, Boonen S, Barton IP, Dunlap LE, Burgio DE. Safety and efficacy of risedronate in patients with age-related reduced renal function as estimated by the Cockcroft and Gaukt method: a pooled analysis of nine clinical trials. J Bone Miner Res. 2005;20:2105–15.

56. Miura N, Mizuno N, Aoyama R. Massive proteinuria and acute renal failure after oral bisphosphonate (alendronate) administration in a patient with segmental glomerulosclerosis. Clin Exp Nephrol. 2009;13:85–8.
57. Choudhury D, Ahmed Z. Drug-associated renal dysfunction and injury. Nat Clin Pract Nephrol. 2006;2:80–91.
58. Chen T, Berenson J, Vescio R, Swift R, Gilchick A, Goodin S, et al. Pharmacokinetics and pharmacodynamics of zolendronic acid in cancer patients with bone metastases. J Clin Pharmacol. 2002;42:1228–36.
59. Phipps R, Christiansen C, Burgio D, Sun L, Russell D, Keck B. Comparison of risedronate and alendronate pharmacokinetics at clinical doses. Osteoporos Int. 2003;14 Suppl 7:38.
60. Buemi M, Fazio MR, Bolignano D, Coppolino G, Donato V, Lacquaniti A, et al. Renal complications in oncohematologic patients. J Investig Med. 2009;57(8):892–901.
61. Boonen S, Sellmeyer DE, Lippuner K, Orlov-Morozov A, Abrams K, Mesenbrink P, et al. Renal safety of annual zoledronic acid infusions in osteoporotic postmenopausal women. Kidney Int. 2008;74:641–8.
62. Perazella MA, Markowitz GS. Bisphosphonate nephrotoxicity. Kidney Int. 2008;74:1385–93.
63. Markowitz GS, Appel GB, Fine PL, Fenves AZ, Loon NR, Jagannath S, et al. Collapsing focal segmental glomerulosclerosis following treatment with high-dose pamidronate. J Am Soc Nephrol. 2001;12:1164–72.
64. Bergner R, Diel IJ, Henrich D, Hoffmann M, Uppenkamp M. Differences in nephrotoxicity of intravenous bisphosphonates for the treatment of malignancy-related bone disease. Onkologie. 2006;29:534–40.
65. Balla J. The issue of renal safety of zoledronic acid from a nephrologist's point of view. Oncologist. 2005;10:306–8.
66. Hillner BE, Ingle JN, Chlebowski RT, Gralow J, Yee GC, Janjan NA, et al.; American Society of Clinical Oncology. American society of clinical oncology 2003 update on the role of bisphosphonates and bone health issues in women with breast cancer. J Clin Oncol. 2003;21:4042–57.
67. Body JJ. The risk of cumulative renal effects of intravenous bisphosphonates. Support Cancer Ther. 2006;3:77–83.
68. Adami S, Zamberlan N. Adverse effects of bisphosphonates: a comparative review. Drug Saf. 1996;14(3):158–70.
69. Novartis Pharmaceuticals Corporation. Aredia (pamidronate) [injection prescribing information]. East Hanover, NJ: Novartis Pharmaceuticals Corporation; 2007.
70. Rosen LS, Gordon D, Kaminski M, Howell A, Belch A, Mackey J, et al. Long-term efficacy and safety of zoledronic acid compared with pamidronate disodium in the treatment of skeletal complications in patients with advanced multiple myeloma or breast carcinoma: a randomized, double-blind, multicenter, comparative trial. Cancer. 2003;98:1735–44.
71. Chang JT, Green L, Beitz J. Renal failure with the use of zoledronic acid. N Engl J Med. 2003;349:1676–9.
72. F. Hoffmann-La Roche Ltd. Bondronat (ibandronate acid). Summary of product characteristics. Basel: F. Hoffmann-La Roche; 2003.
73. Bergner R, Nauth B, Henrich DM, Hoffmann M, Ullmann M, Honecker A, et al. Renal safety and pharmacokinetics of ibandronate in multiple myeloma patients with pre-existing renal insufficiency. In: 31 st ESMO congress, Istanbul, Sept 29 to Oct 3, 2006.
74. Weide R, Koppler H, Antras L, Smith M, Chang E, Green J, et al. Renal toxicity in patients with multiple myeloma receiving zoledronic acid vs. ibandronate: a retrospective medical records review. J Cancer Res Ther. 2010;6:31–5.
75. Novartis Pharmaceuticals Corporation. Zometa (zoledronic acid). US summary of product characteristics. East Hanover, NJ: Novartis Pharmaceuticals Corporation; 2004.
76. Guarneri V, Donati S, Nicolini M. Renal safety of zoledronic acid in patients with bone metastases from breast cancer or other tumors treated with intravenous bisphosphonates for up to ten years. Ann Oncol. 2004;15:811.

77. Rosen LS, Gordon D, Tchekmedyan S. Zoledronic acid versus placebo in the treatment of skeletal metastases in patients with lung cancer and other solid tumors: a phase III, double-blind, randomized trial. J Clin Oncol. 2003;21:3150–7.
78. Hillner BE, Ingle JN, Chlebowski RT, Gralow J, Yee GC, Janjan NA, et al. American Society of Clinical Oncology 2003 update on the role of bisphosphonates and bone heath issues in women with breast cancer. J Clin Oncol. 2004;21:1342–57.
79. Adami S, Bhalla AK, Dorizzi R. The acute-phase response after bisphosphonate administration. Calcif Tissue Int. 1987;41:326–31.
80. Olson K, Van Poznak C. Significance and impact of bisphosphonate-induced acute phase responses. J Oncol Pharm Pract. 2007;13:223–9.
81. Thompson K, Rogers MJ. Statins prevent bisphosphonate-induced γ, δ-T-cell proliferation and activation in vitro. J Bone Miner Res. 2004;19:278–88.
82. Coxon FP, Thompson K, Rogers MJ. Recent advances in understanding the mechanism of action of bisphosphonates. Curr Opin Pharmacol. 2006;6(3):307–12.
83. Hewitt RE, Lissina A, Green AE, Slay ES, Price DA, Sewell AK. The bisphosphonate acute phase response: rapid and copious production of proinflammatory cytokines by peripheral blood gd T cells in response to aminobisphosphonates is inhibited by statins. Clin Exp Immunol. 2005;139(1):101–11.
84. Roelofs AJ, Jauhiainen M, Mönkkönen H, Rogers MJ, Mönkkönen J, Thompson K. Peripheral blood monocytes are responsible for gammadelta T cell activation induced by zoledronic acid through accumulation of IPP/DMAPP. Br J Haematol. 2009;144(2):245–50.
85. Schweitzer DH, Oostendorp-vande Ruit M, van der Pluijm G, Lowik CWGM, Papapoulos SE. Interleukin-6 and the acute phase response during treatment of patients with Paget's disease with the nitrogen-containing bisphosphonate dimethylaminohydroxypropylidene bisphosphonate. J Bone Miner Res. 1995;10:956–62.
86. Srivastava T, Haney CJ, Alon US. Atorvastatin may have no effect on acute phase reaction in children following intravenous bisphosphonate infusion. J Bone Miner Res. 2009;24:334–7.
87. Wark JD, Bensen W, Recknor C, Ryabitseva O, Chiodo III J, Mesenbrink P, et al. Treatment with acetaminophen/paracetamol or ibuprofen alleviates post-dose symptoms related to intravenous infusion with zoledronic acid 5 mg. Osteoporos Int. 2012;23(2):503–12.
88. http://www.fda.gov/cder/drug/infopage/bisphosphonates/default.htm
89. Bock O, Boerst H, Thomasius FE, Degner C, Stephan-Oelkers M, Valentine SM, et al. Common musculoskeletal adverse effects of oral treatment with once weekly alendronate and risedronate in patients with osteoporosis and ways for their prevention. J Musculoskelet Neuronal Interact. 2007;7:144–8.
90. Wysowski DK, Chang JT. Alendronate and risedronate: reports of severe bone, joint, and muscle pain. Arch Intern Med. 2005;165:346–7.
91. Nakchbandi IA, Grey A, Masiukewicz U, Mitnick M, Insogna K. Cytokines in primary hyperparathyroidism. In: Bilezikian JP, Marcus R, Levine M, editors. The parathyroids, basic and clinical concepts. San Diego, CA: Academic; 2001. p. 411–21.
92. Peponis V, Kyttaris VC, Chalkiadakis SE, Bonovas S, Sitaras NM. Ocular side effects of anti-rheumatic medications: what a rheumatologist should know. Lupus. 2010;19(6):675–82.
93. Fraunfelder FW, Fraunfelder FT, Jensvold B. Scleritis and other ocular side effects associated with pamidronate disodium. Am J Ophthalmol. 2003;135:219–22.
94. French DD, Margo CE. Postmarketing surveillance of uveitis and scleritis with bisphosphonates among a national veteran cohort. Retina. 2008;28:889–93.
95. Benderson D, Karakunnel J, Kathuria S, Badros A. Scleritis complicating zoledronic acid infusion. Clin Lymphoma Myeloma. 2006;7:145–7.
96. El Saghir NS, Otrock ZK, Bleik JH. Unilateral anterior uveitis complicating zoledronic acid therapy in breast cancer. BMC Cancer. 2005;5:156.
97. Kilickap S, Ozdamar Y, Altundag MK, Dizdar O. A case report: zoledronic acid-induced uveitis. Med Oncol. 2008;25:238–40.

98. Leung S, Ashar BH, Miller RG. Bisphosphonate-associated scleritis: a case report and review. South Med J. 2005;98:733–5.
99. Tabbara KF. Nodular scleritis following alendronate therapy. Ocul Immunol Inflamm. 2008;16:99–101.
100. Mbekeani JN, Slamivits TL, Schwartz BH, Sauer HL. Ocular inflammation associated with alendronate therapy. Arch Ophthalmol. 1999;117:837–8.
101. Malik AR, Campbell SH, Toma NMG. Bilateral acute anterior uveitis after alendronate. Br J Ophthalmol. 2002;86:1443.
102. Fietta P, Manganelli P, Lodigiani L. Clodronate induced uveitis. Ann Rheum Dis. 2003;62:378.
103. Black DM, Delmas PD, Eastell R, Reid IR, Boonen S, Cauley JA, et al. for HORIZON Pivotal Fracture Trial. Once-yearly zoledronic acid for treatment of postmenopausal osteoporosis. N Engl J Med. 2007;356:1809–22.
104. Lyles KW, Colon-Emeric CS, Magaziner JS, Adachi JD, Pieper CF, Mautalen C, et al. Zoledronic acid and clinical fractures and mortality after hip fracture. N Engl J Med. 2007;357:1799–809.
105. Moore MM, Beith JM. Acute unilateral anterior uveitis and scleritis following a single infusion of zoledronate for metastatic breast cancer. Med J Aust. 2008;188:370–1.
106. Banal F, Briot K, Ayoub G, Dougados M, Roux C. Unilateral anterior uveitis complicating zoledronic acid therapy in prostate cancer. J Rheumatol. 2008;35(12):2458–9.
107. Sharma NS, Ooi JL, Masselos K, Hooper MJ, Francis IC. Zoledronic acid infusion and orbital inflammatory disease. N Engl J Med. 2008;359(13):1410–1.
108. Wysowski DK. Reports of esophageal cancer with oral bisphosphonate use. N Engl J Med. 2009;360:89–90.
109. Papapetrou PD. Bisphosphonate-associated adverse events. Hormones. 2009;8(2):96–110.
110. Rennert G, Pinchev M, Rennert HS. Use of bisphosphonates and risk of postmenopausal breast cancer. J Clin Oncol. 2010;28(22):3577–81.
111. Rennert G, Pinchev M, Rennert HS, Gruber SB. Use of bisphosphonates and reduced risk of colorectal cancer. J Clin Oncol. 2011;29(9):1146–50.
112. Cummings SR, Schwartz AV, Black DM. Alendronate and atrial fibrillation. N Engl J Med. 2007;356(18):1895–6.
113. Heckbert SR, Li G, Cummings SR, Smith NL, Psaty BM. Use of alendronate and risk of incident atrial fibrillation in women. Arch Intern Med. 2008;168(8):826–31.
114. US Food and Drug Administration (FDA). Update of safety review follow-up to the Oct 1, 2007 Early communication about the ongoing safety review of bisphosphonates. Silver Spring, MD: FDA; 2008. http://www.fda.gov/Drugs/DrugSafety/PostmarketDrugSafetyInformationforPatientsandProviders/DrugSafetyInformationforHeathcareProfessionals/ucm136201.htm. Accessed 24 March 2010.
115. European Medicines Agency (EMEA). London: EMEA; July 8, 2009. Aclasta (zoledronic acid): European Public Assessment Report, revision 9. http://www.emea.europa.eu/humandocs/Humans/EPAR/aclasta/aclasta.htm. Accessed 24 March 2010.
116. Grosso A, Douglas I, Hingorani A, MacAllister R, Smeeth L. Oral bisphosphonates and risk of atrial fibrillation and flutter in women: a self-controlled case-series safety analysis. PLoS One. 2009;4(3):e4720.
117. Sorensen HT, Christensen S, Mehnert F, Pedersen L, Chapurlat RD, Cummings SR, et al. Use of bisphosphonates among women and risk of atrial fibrillation and flutter: population based case–control study. BMJ. 2008;336:813–6.
118. Kemeny-Suss N, Kasneci A, Rivas D, Afilalo J, Komarova SV, Chalifour LE, et al. Alendronate affects calcium dynamics in cardiomyocytes in vitro. Vascul Pharmacol. 2009;51(5–6):350–8.
119. Goh SK, Yang KY, Koh JS, Wong MK, Chua SY, Chua DT, et al. Subtrochanteric insufficiency fractures in patients on alendronate therapy: a caution. J Bone Joint Surg Br. 2007;89:349–53.
120. Lenart BA, Lorich DG, Lane JM. Atypical fractures of the femoral diaphysis in postmenopausal women taking alendronate. N Engl J Med. 2008;358:1304–6.
121. Kwek EBK, Goh SK, Koh JSB, Png MA, Howe TS. An emerging pattern of subtrochanteric stress fractures: a long-term complication of alendronate therapy. Injury. 2008;39:224–31.

122. Neviaser AS, Lane JM, Lenart BA, Edobor-Osula F, Lorich DG. Low-energy femoral shaft fractures associated with alendronate use. J Orthop Trauma. 2008;22:346–50.
123. Abrahamsen B, Eiken P, Eastell R. Subtrochanteric and diaphyseal femur fractures in patients treated with alendronate: a register-based national cohort study. J Bone Miner Res. 2009;24(6):1095–102.
124. Odvina CV, Zerwekh JE, Rao DS, Maalouf N, Gottschalk FA, Pak CY. Severely suppressed bone turnover: a potential complication of alendronate therapy. J Clin Endocrinol Metab. 2005;90:1294–301.
125. Visekruna M, Wilson D, McKienan FE. Severely suppressed bone turnover and atypical skeletal fragility. J Clin Endocrinol Metab. 2008;93:2948–52.
126. Somford MP, Draijer FW, Thomassen BJ, Chavassieux PM, Boivin G, Papapoulos SE. Bilateral fractures of the femur diaphysis in a patient with rheumatoid arthritis on long-term treatment with alendronate: clues to the mechanism of increased bone fragility. J Bone Miner Res. 2009;24(10):1736–40.
127. Weinstein RS, Roberson PK, Manolagas SC. Giant osteoclast formation and long-term oral bisphosphonate therapy. N Engl J Med. 2009;360(1):53–62.
128. Chennuru S, Koduri J, Bauman MA. Risk factors for symptomatic hypocalcemia complicating treatment with zoledronic acid. Intern Med J. 2008;38:635–7.
129. Berruti A, Tucci M, Generali D, Mosca A, Ardine M, Vana F, et al. Management of the side-effects of intravenous bisphosphonates: targeting the serum parathyroid hormone elevation. Ann Oncol. 2006;17:1854–5.
130. Kontoleon P, Ilias I, Stavropoulos PG, Papapetrou PD. Urticaria after administration of alendronate. Acta Derm Venereol. 2000;80:398.
131. Phillips E, Knowles S, Weber E, Shear NH. Skin reactions associated with bisphosphonates: a report of 3 cases and an approach to management. J Allergy Clin Immunol. 1998;102 (4 Pt 1):697–8.
132. Anastasakis AD, Goulis DG, Kita M, Avramidis A. Oral bisphosphonate adverse effects in 849 patients with metabolic bone disease. Hormones (Athens). 2007;6:233–41.
133. Lazarov A, Moss K, Plosk N, Cordova M, Baitelman L. Alendronate-induced lichen planus. Isr Med Assoc J. 2002;4:389–90.
134. High WA, Cohen JB, Wetherington W, Cockerell CJ. Superficial gyrate erythema as a cutaneous reaction to alendronate for osteoporosis. J Am Acad Dermatol. 2003;48:945–6.
135. Kimura M, Kawada A, Murayama Y, Murayama M. Drug eruption due to alendronate sodium hydrate. Contact Dermatitis. 2003;48:116.
136. Brinkmeier T, Kügler K, Lepoittevin JP, Frosch PJ. Adverse cutaneous drug reaction to alendronate. Contact Dermatitis. 2007;57:123–5.
137. Belhadjali H, Slim R, Aouam K, Youssef M, Zilli J. Cutaneous vasculitis induced by risedronate. Allergy. 2008;63:1405.
138. McNicholl DM, Heaney LG. The safety of bisphosphonate use in pre-menopausal women on corticosteroids. Curr Drug Saf. 2010;5(2):182–7.
139. Halabe A, Lifschitz BM, Azuri J. Liver damage due to alendronate. N Engl J Med. 2000;343:365–6.
140. Carrére C, Duvaal JL, Godard B, De Jaureguiberry JP, Ciribilli JM. Severe acute hepatitis induced by alendronate. Gastroenterol Clin Biol. 2002;16:179–80.
141. Yanik B, Turkay C, Atalar H. Hepatotoxicity induced by alendronate therapy. Osteoporos Int. 2007;18:829–31.
142. Phillips MB. Risedronate-induced hepatitis. Am J Med. 2007;120:e1–2.
143. Russell RG, Rogers MJ, Frith JC, Luckman SP, Coxon FP, Benford HL, et al. The pharmacology of bisphosphonates and new insights into their mechanisms of action. J Bone Miner Res. 1999;14 Suppl 2:53–65.
144. Ruggiero SL, Mehrotra B, Rosenberg TJ, Engroff SL. Osteonecrosis of the jaws associated with the use of bisphosphonates: a review of 63 cases. J Oral Maxillofac Surg. 2004;62(5):527–34.
145. Nastro E, Musolino C, Allegra A, Oteri G, Cicciù M, Alonci A, et al. Bisphosphonate-associated osteonecrosis of the jaw in patients with multiple myeloma and breast cancer. Acta Haematol. 2007;117(3):181–7.

146. Nastro E, Allegra A, Oteri G, Ferlito S, Alonci A, Bellomo G, et al. Avascular necrosis of bone in leukemia and osteonecrosis of jaw by bisphosphonates. J Oral Maxillofac Surg. 2009;67(12):2701–3.
147. Ruggiero SL, Dodson TB, Assael LA, Landesberg R, Marx RE, Mehrotra B. American Association of Oral and Maxillofacial Surgeons position paper on bisphosphonate-related osteonecrosis of the jaws: 2009 update. J Oral Maxillofac Surg. 2009;67(5 Suppl):2–12.
148. Sambrook P, Olver I, Goss A. Bisphosphonates and osteonecrosis of the jaw. Aust Fam Physician. 2006;35(10):801–3.
149. Khosla S, Burr D, Cauley J, Dempster DW, Ebeling PR, Felsenberg D, et al.; American Society for Bone and Mineral Research. Bisphosphonate-associated osteonecrosis of the jaw: report of a task force of the American Society for Bone and Mineral Research. J Bone Miner Res. 2007;22(10):1479–1.
150. Soulafa A, Almazrooa BDS, Woo SB. Bisphosphonate and nonbisphosphonate-associated osteonecrosis of the jaw. J Am Dent Assoc. 2009;140:864–75.
151. Ruggiero SL, Woo SB. Biophosphonate-related osteonecrosis of the jaws. Dent Clin North Am. 2008;52(1):111–28.
152. McMahon RE, Bouquot JE, Glueck CJ, Griep JA, Adams WR, Spolnik KJ, et al. Staging bisphosphonate-related osteonecrosis of the jaw should include early stages of disease. J Oral Maxillofac Surg. 2007;65(9):1899–900.
153. Mawardi H, Treister N, Richardson P, Anderson K, Munshi N, Faiella RA, et al. Sinus tracts: an early sign of bisphosphonate-associated osteonecrosis of the jaws? J Oral Maxillofac Surg. 2009;67(3):593–601.
154. Advisory Task Force on Bisphosphonate-Related Ostenonecrosis of the Jaws, American Association of Oral and Maxillofacial Surgeons. American Association of Oral and Maxillofacial Surgeons position paper on bisphosphonate-related osteonecrosis of the jaws. J Oral Maxillofac Surg. 2007;65:369–76.
155. Kyle RA, Yee GC, Somerfield MR, Flynn PJ, Halabi S, Jagannath S, et al., American Society of Clinical Oncology. American Society of Clinical Oncology 2007 clinical practice guideline update on the role of bisphosphonates in multiple myeloma. J Clin Oncol. 2007;25:2464–72.
156. Ruggiero S, Gralow J, Marx RE, Hoff AO, Schubert MM, Huryn JM, et al. Practical guidelines for the prevention, diagnosis, and treatment of osteonecrosis of the jaw in patients with cancer. J Oncol Pract. 2006;1:7–14.
157. Nastro Siniscalchi E, Cutroneo G, Catalfamo L, Santoro G, Allegra A, Oteri G, et al. Immunohistochemical evaluation of sarcoglycans and integrins in gingival epithelium of multiple myeloma patients with bisphosphonate-induced osteonecrosis of the jaw. Oncol Rep. 2010;24(1):129–34.
158. Twiss IM, de Water R, den Hartigh J, Sparidans R, Ramp-Koopmanschap W, Brill H, et al. Cytotoxic effects of pamidronate on monolayers of human intestinal epithelial (Caco-2) cells and its epithelial transport. J Pharm Sci. 1994;83:699–703.
159. Twiss IM, Pas O, Ramp-Koopmanschap W, Den Hartigh J, Vermeij P. The effects of nitrogen-containing bisphosphonates on human epithelial (Caco-2) cells, an in vitro model for intestinal epithelium. J Bone Miner Res. 1999;14:784–91.
160. Wallace JL, Dicay M, McKnight W, Bastaki S, Blank MA. N-bisphosphonates cause gastric epithelial injury independent of effects on the microcirculation. Aliment Pharmacol Ther. 1999;13:1675–82.
161. Suri S, Mönkkönen J, Taskinen M, Pesonen J, Blank MA, Phipps RJ, et al. Nitrogen-containing bisphosphonates induce apoptosis of Caco-2 cells in vitro by inhibiting the mevalonate pathway: a model of bisphosphonate-induced gastrointestinal toxicity. Bone. 2001;29:336–43.
162. Landesberg R, Cozin M, Cremers S, Woo V, Kousteni S, Sinha S, et al. Inhibition of oral mucosal cell wound healing by bisphosphonates. J Oral Maxillofac Surg. 2008;66(5):839–47.
163. Demerjian N, Bolla G, Spreux A. Severe oral ulcerations induced by alendronate. Clin Rheumatol. 2000;18:349–50.
164. Gonzalez-Moles MA, Bagan-Sebastian JV. Alendronate-related oral mucosa ulcerations. J Oral Pathol Med. 2000;29:514–8.

165. Krasagakis K, Krüger-Krasagakis S, Ioannidou D, Tosca A. Chronic erosive and ulcerative oral lesions caused by incorrect administration of alendronate. J Am Acad Dermatol. 2004;50:651–2.
166. Schmut JL, Barbaud A, Trechot P. Alendronate et ulcerations buccales: d'abord bien interroger le patient. Ann Dermatol Venereol. 2005;132:930.
167. Rubegni P, Fimiani M. Images in clinical medicine. bisphosphonate-associated contact stomatitis. N Engl J Med. 2006;355:e25.
168. Ficarra G, Beninati F. Bisphosphonate-related osteonecrosis of the jaws: an update on clinical, pathological and management aspects. Head Neck Pathol. 2007;1(2):132–40.
169. Woo SB, Hellstein JW, Kalmer JR. Narrative review: bisphosphonates and osteonecrosis of the jaws. Ann Intern Med. 2006;144:753–61.
170. Felsenberg D, Hoffmeister B, Amling M, Mundlos S, Seibel MJ, Fratzl P. Onkologie: Kiefernekrosen nach hoch dosierter bisphosphonattherapic. Dtsch Arztebl. 2006;103:3078–81.
171. Mavrokokki T, Cheng A, Stein B, Goss A. Nature and frequency of bisphosphonate-associated osteonecrosis of the jaws in Australia. J Oral Maxillofac Surg. 2007;65:415–23.
172. Hoff A. Practical guidelines for the prevention, diagnosis, and treatment of osteonecrosis of the jaw in patients with cancer. Proc Am Soc Clin Oncol. 2006;24:475s. Abstract 8528.
173. Walter C, Al-Nawas B, du Bois A, Buch L, Harter P, Grötz KA. Incidence of bisphosphonate-associated osteonecrosis of the jaws in breast cancer patients. Cancer. 2009;115(8):1631–7.
174. Durie BG, Katz M, Crowley J. Osteonecrosis of the jaw and bisphosphonates. N Engl J Med. 2005;353:99–100.
175. Bamias A, Kastritis E, Bamia C, Moulopoulos LA, Melakopoulos I, Bozas G, et al. Osteonecrosis of the jaw in cancer after treatment with bisphosphonates: incidence and risk factors. J Clin Oncol. 2005;23:8580–7.
176. Badros A, Weikel D, Salama A, Goloubeva O, Schneider A, Rapoport A, et al. Osteonecrosis of the jaw in multiple myeloma patients: clinical features and risk factors. J Clin Oncol. 2006;24:945–52.
177. Van Poznak CH, Estilo CL, Sauter NP, Hudis C, Tunick S, Huryn JM, et al. Osteonecrosis of the jaw in patients with metastatic breast cancer. Breast Cancer Res Treat. 2004;88 Suppl 1:S131.
178. Tosi P, Zamagni E, Cangini D, Tachetti P, DiRaimondo F, Catalano L, et al. Osteonecrosis of the jaws in newly diagnosed multiple myeloma patients treated with zoledronic acid and thalidomide-dexamethasone. Blood. 2006;108:3951–2.
179. Kraj M, Ryszard P, Stanislaw M, Owczarska K. The incidence of jaw osteonecrosis in multiple myeloma patients treated with bisphosphonates. Acta Pol Pharm. 2006;63:450–2.
180. Sanna G, Zampino MG, Pelosi G, Nole F, Goldhirsh A. Jaw avascular bone necrosis associated with long-term use of bisphosphonates. Ann Oncol. 2005;16:1207–8.
181. Pozzi S, Marcheselli R, Sacchi S, Baldini L, Angrili F, Pennese E, et al. Bisphosphonate-associated osteonecrosis of the jaw: a review of 35 cases and an evaluation of its frequency in multiple myeloma patients. Leuk Lymphoma. 2007;48:56–64.
182. Zervas K, Verrou E, Teleioudis Z, Vahtsevanos K, Banti A, Mihou D, et al. Incidence, risk factors and management of osteonecrosis of the jaw in patients with multiple myeloma: a single-centre experience in 303 patients. Br J Haematol. 2006;134:620–3.
183. Dimopoulos MA, Kastritis E, Anagnostopoulos A, Melakopoulos I, Gika D, Moulopoulos LA, et al. Osteonecrosis of the jaw in patients with multiple myeloma treated with bisphosphonates: evidence of increased risk after treatment with zoledronic acid. Haematologica. 2006;91:968–71.
184. Wessel JH, Dodson TB, Zavras AI. Zoledronate, smoking, and obesity are strong risk factors for osteonecrosis of the jaw: a case–control study. J Oral Maxillofac Surg. 2008;66(4):625–31.
185. Chavassieux PM, Arlot ME, Reda C, Wei L, Yates AJ, Meunier PJ. Histomorphometric assessment of the long-term effects of alendronate on bone quality and remodeling in patients with osteoporosis. J Clin Invest. 1997;100:1475–80.
186. Eriksen EF, Melsen F, Sod E, Barton I, Chines A. Effects of long-term risedronate on bone quality and bone turnover in women with postmenopausal osteoporosis. Bone. 2002;31:620–5.
187. Recker RR, Boonen S, Garcia P, Supronik J, Peichl P, Black D, et al. The effect of annual treatment with zoledronic acid 5 mg on bone remodeling: bone histomorphometry results from the HORIZON-PFT. J Bone Miner Res. 2006;21:S290.

188. Balena R, Toolan BC, Shea M, Markatos A, Myers ER, Lee SC, et al. The effects of 2-year treatment with the aminobisphosphonate alendronate on bone metabolism, bone histomorphometry, and bone strength in ovariectomized nonhuman primates. J Clin Invest. 1993;92:2577–86.
189. Allen MR, Iwata K, Phipps R, Burr DB. Alterations in canine vertebral bone turnover, microdamage accumulation, and biomechanical properties following 1-year treatment with clinical treatment doses of risedronate or alendronate. Bone. 2006;39:872–9.
190. Remaggi F, Ferretti M, Cane V, Zaffe D. Histomorphological and chemico-physical analyses of the mineral matrix of micropetrotic human bone. Ann Anat. 1996;178:223–7.
191. Wood J, Bonjean K, Ruetz S, Bellahcène A, Devy L, Foidart JM, et al. Novel antiangiogenic effects of the bisphosphonate compound zoledronic acid. J Pharmacol Exp Ther. 2002;302:1055–61.
192. Bezzi M, Hasmim M, Bieler G, Dormond O, Rüegg C. Zoledronate sensitizes endothelial cells to tumor necrosis factor-induced programmed cell death: evidence for the suppression of sustained activation of focal adhesion kinase and protein kinase B/Akt. J Biol Chem. 2003;31(278):43603–14.
193. Hamma-Kourbali Y, Di Benedetto M, Ledoux D, Oudar O, Leroux Y, Lecouvey M, et al. A novel non-containing-nitrogen bisphosphonate inhibits both in vitro and in vivo angiogenesis. Biochem Biophys Res Commun. 2003;310:816–23.
194. Giraudo E, Inoue M, Hanahan D. An amino-bisphosphonate targets MMP-9-expressing macrophages and angiogenesis to impair cervical carcinogenesis. J Clin Invest. 2004;114:623–33.
195. Okamoto T, Yamagishi S, Inagaki Y, Amano S, Takeuchi M, Kikuchi S, et al. Incadronate disodium inhibits advanced glycation end products-induced angiogenesis in vitro. Biochem Biophys Res Commun. 2002;297:419–24.
196. Santini D, Vincenzi B, Avvisati G, Dicuonzo G, Battistoni F, Gavasci M, et al. Pamidronate induces modifications of circulating angiogenetic factors in cancer patients. Clin Cancer Res. 2002;8:1080–4.
197. Allegra A, Oteri G, Nastro E, Alonci A, Bellomo G, Del Fabro V, et al. Patients with bisphosphonates-associated osteonecrosis of the jaw have reduced circulating endothelial cells. Hematol Oncol. 2007;25:164–9.
198. Alonci A, Allegra A, Bellomo G, Quartarone E, Oteri G, Nastro E, et al. Patients with bisphosphonate-associated osteonecrosis of the jaw have unmodified levels of soluble vascular endothelial growth factor receptor 1. Leuk Lymphoma. 2007;48(9):1852–4.
199. Allegra A, Alonci A, Penna G, Granata A, Nastro Siniscalchi E, Oteri G, et al. Bisphosphonates induce apoptosis of circulating endothelial cells in multiple myeloma patients and in subjects with bisphosphonate-induced osteonecrosis of the jaws. Acta Haematol. 2010;124(2):79–85.
200. McArthur HL, Estilo C, Huryn J. Osteonecrosis of the jaw among intravenous bisphosphonate—and/or bevacizumab—treated patients at Memorial Sloan-Kettering Cancer Center. J Clin Oncol. 2008;26:Abstr 9588.
201. Estilo CL, Fornier M, Farooki A, Carlson D, Bohle III G, Huryn JM. Osteonecrosis of the jaw related to bevacizumab. J Clin Oncol. 2008;26(24):4037–8.
202. Santini D, Schiavon G, Angeletti S, Vincenzi B, Gasparro S, Grilli C, et al. Last generation of aminobisphosphonates (N-BPs) and cancer angiogenesis: a new role for these drugs? Recent Pat Anticancer Drug Discov. 2006;1(3):383–96.
203. Ayllon J, Launay-Vacher V, Medioni J, Cros C, Spano JP, Oudard S. Osteonecrosis of the jaw under bisphosphonate and antiangiogenic therapies: cumulative toxicity profile? Ann Oncol. 2009;20(3):600–1.
204. Oteri G, Allegra A, Bellomo G, Alonci A, Nastro E, Penna G, et al. Reduced serum levels of interleukin 17 in patients with osteonecrosis of the jaw and in multiple myeloma subjects after bisphosphonates administration. Cytokine. 2008;43(2):103–4.
205. Otto S, Pautke C, Opelz C, Westphal I, Drosse I, Schwager J, et al. Osteonecrosis of the jaw: effect of bisphosphonate type, local concentration, and acidic milieu on the pathomechanism. J Oral Maxillofac Surg. 2010;68(11):2837–45.
206. Koch FP, Wunsch A, Merkel C, Ziebart T, Pabst A, Yekta SS, et al. The influence of bisphosphonates on human osteoblast migration and integrin aVb3/tenascin C gene expression in vitro. Head Face Med. 2011;7(1):4.

207. Ardine M, Generali D, Donadio M, Bonardi S, Scoletta M, Vandone AM, et al. Could the long-term persistence of low serum calcium levels and high serum parathyroid hormone levels during bisphosphonate treatment predispose metastatic breast cancer patients to undergo osteonecrosis of the jaw? Ann Oncol. 2006;17:1336–7.
208. Berruti A, Ortega C, Fusco V. Re: Christan Walter, Bilal Al-Nawas, Knut A. Grötz (2009) Prevalence and risk factors of bisphosphonate-associated osteonecrosis of the jaw in prostate cancer patients with advanced disease treated with zoledronate. Eur Urol 2008;54:1066–1072. Eur Urol. 2009;55:e61–e2.
209. Berruti A, Dogliotti L, Tampellini M. Effect of zoledronic acid (Z) treatment based on serum parathyroid hormone (PTH) levels in patients (pts) with malignant bone disease. J Clin Oncol. 2006;24(18 Suppl):8610.
210. Chen L, Cencioni MT, Angelini DF, Borsellino G, Battistini L, Brosnan CF. Transcriptional profiling of gamma delta T cells identifies a role for vitamin D in the immunoregulation of the V gamma 9V delta 2 response to phosphate-containing ligands. J Immunol. 2005;174:6144–52.
211. Biasotto M, Chiandussi S, Dore F, Cavalli F, Cova MA, Di Lenarda R. Clinical aspects and management of bisphosphonates-associated osteonecrosis of the jaws. Acta Odontol Scand. 2006;64:348–54.
212. Bisdas S, Chambron Pinho N, Smolarz A, Sader R, Vogl TJ, Mack MG. Biphosphonate-induced osteonecrosis of the jaws: CT and MRI spectrum of findings in 32 patients. Clin Radiol. 2008;63:71–7.
213. Hansen T, Kunkel M, Springer E, Walter C, Weber A, Siegel E, et al. Actinomycosis of the jaws—histopathological study of 45 patients shows significant involvement in bisphosphonate-associated osteonecrosis and infected osteoradionecrosis. Virchows Arch. 2007;451:1009–17.
214. Hansen T, Kunkel M, Weber A, James Kirkpatrick C. Osteonecrosis of the jaws in patients treated with bisphosphonates—histomorphologic analysis in comparison with infected osteoradionecrosis. J Oral Pathol Med. 2006;35:155–60.
215. Lazarovici TS, Yahalom R, Taicher S, Elad S, Hardan I, Yarom N. Bisphosphonate-related osteonecrosis of the jaws: a single-center study of 101 patients. J Oral Maxillofac Surg. 2009;67:850–5.
216. Lugassy G, Shaham R, Nemets A, Ben-Dor D, Nahlieli O. Severe osteomyelitis of the jaw in long-term survivors of multiple myeloma: a new clinical entity. Am J Med. 2004;117:440–1.
217. Dodson TB, Raje NS, Caruso PA, Rosenberg AE. Case records of the Massachusetts General Hospital: case 9-2008: a 65-year-old woman with a nonhealing ulcer of the jaw. N Engl J Med. 2008;358:1283–91.
218. Hansen T, Kunkel M, Kirkpatrick CJ, Weber A. Actinomyces in infected osteoradionecrosis—underestimated? Hum Pathol. 2006;37:61–7.
219. Subramanian G, Cohen HV, Quek SY. A model for the pathogenesis of bisphosphonate-associated osteonecrosis of the jaw and teriparatide's potential role in its resolution. Oral Surg Oral Med Oral Pathol Oral Radiol Endod. 2011;112(6):744–53.
220. Rayman S, Almas K, Dincer E. Bisphosphonate-related jaw necrosis: a team approach management and prevention. Int J Dent Hyg. 2009;7(2):90–5.
221. Dimopoulos MA, Kastritis E, Bamia C, Melakopoulos I, Gika D, Roussou M, et al. Reduction of osteonecrosis of the jaw (ONJ) after implementation of preventive measures in patients with multiple myeloma treated with zoledronic acid. Ann Oncol. 2009;20(1):117–20.
222. Migliorati CA, Casiglia J, Epstein J, Jacobsen PL, Siegel MA, Woo SB. Managing the care of patients with bisphosphonate-associated osteonecrosis: an American Academy of Oral Medicine position paper. J Am Dent Assoc. 2005;136:1658–68.
223. Thakkar SG, Isada C, Smith J, Karam MA, Reed J, Tomford JW, et al. Jaw complications associated with bisphosphonate use in patients with plasma cell dyscrasias. Med Oncol. 2006;23:51–6.
224. Papapoulos SE, Cremers SC. Prolonged bisphosphonate release after treatment in children. N Engl J Med. 2007;356:1075–6.

225. Marx RE, Cillo Jr JE, Ulloa JJ. Oral bisphosphonate-induced osteonecrosis: risk factors, prediction of risk using serum CTX testing, prevention, and treatment. J Oral Maxillofac Surg. 2007;65:2397–410.
226. Seibel MJ. Biochemical markers of bone turnover: part I: biochemistry and variability. Clin Biochem Rev. 2005;26:97–122.
227. Van den Wyngaert T, Huizing MT, Fossion E, Vermorken JB. Bisphosphonates in oncology: rising stars or fallen heroes. Oncologist. 2009;14:181–91.
228. Palaska PK, Cartsos V, Zavras AI. Bisphosphonates and time to osteonecrosis development. Oncologist. 2009;14(11):1154–66.
229. Curi MM, Cossolin GS, Koga DH, Zardetto C, Christianini S, Feher O, et al. Bisphosphonate-related osteonecrosis of the jaws-an initial case series report of treatment combining partial bone resection and autologous platelet-rich plasma. J Oral Maxillofac Surg. 2011;69(9):2465–72.
230. Cheung A, Seeman E. Teriparatide therapy for alendronate-associated osteonecrosis of the jaw. N Engl J Med. 2010;363:2473–4.
231. Freiberger J, Padila-Burgos R, Chhoêu AH, Kraft KH, Boneta O, Moon RE, et al. Hyperbaric oxygen treatment and bisphosphonate-induced osteonecrosis of the jaw: a case series. J Oral Maxillofac Surg. 2007;65(7):1321–7.
232. Shimura K, Shimazaki C, Taniguchi K, Akamatsu S, Okamoto M, Uchida R, et al. Hyperbaric oxygen in addition to antibiotic therapy is effective for bisphosphonate-induced osteonecrosis of the jaw in a patient with multiple myeloma. Int J Hematol. 2006;84(4):343–5.
233. Lee CY, David T, Nishime M. Use of platelet-rich plasma in the management of oral bisphosphonate-associated osteonecrosis of the jaw: a report of 2 cases. J Oral Implantol. 2007;33(6):371–82.
234. Magopoulos C, Karakinaris G, Telioudis Z, Vahtsevanos K, Dimitrakopoulos I, Antoniadis K, et al. Osteonecrosis of the jaws due to bisphosphonate use. A review of 60 cases and treatment proposals. Am J Otolaryngol. 2007;28(3):158–63.
235. Freiberger JJ. Utility of hyperbaric oxygen in treatment of bisphosphonate-related osteonecrosis of the jaws. J Oral Maxillofac Surg. 2009;67(5 Suppl):96–106.
236. Feldmeier JJ, Heimbach RD, Davolt DA, Brakora MJ. Hyperbaric oxygen and the cancer patient: a survey of practice patterns. Undersea Hyperb Med. 1993;20(4):337–45.
237. Feldmeier J, Carl U, Hartmann K, Sminia P. Hyperbaric oxygen: does it promote growth or recurrence of malignancy? Undersea Hyperb Med. 2003;30(1):1–18.
238. Erkan M, Bilgi O, Mutluoglu M, Uzun G. Bisphosphonate-related osteonecrosis of the jaw in cancer patients and hyperbaric oxygen therapy. JOP. 2009;10(5):579–80.
239. Kikuiri T, Kim I, Yamaza T, Akiyama K, Zhang Q, Li Y, et al. Cell-based immunotherapy with mesenchymal stem cells cures bisphosphonate-related osteonecrosis of the jaw-like disease in mice. J Bone Miner Res. 2010;25:1668–79.
240. Handschel J, Naujoks C, Langenbach F, Berr K, Depprich RA, Ommerborn MA, et al. Comparison of ectopic bone formation of embryonic stem cells and cord blood stem cells in vivo. Tissue Eng Part A. 2010;16:2475–83.
241. Luo JZ, Xiong F, Al-Homsi AS, Roy T, Luo LG. Human BM stem cells initiate angiogenesis in human islets in vitro. Bone Marrow Transplant. 2010;46(8):1128–37.
242. Meyer U, Meyer T, Handschel J, Wiesmann HP. Fundamentals of tissue engineering and regenerative medicine. Berlin: Springer; 2009.
243. Handschel J, Berr K, Depprich RA, Kubler NR, Naujoks C, Wiesmann HP, et al. Induction of osteogenic markers in differentially treated cultures of embryonic stem cells. Head Face Med. 2008;4:10.
244. Handschel J, Meyer U. Infection, vascularization, remodelling—are stem cells the answers for bone diseases of the jaws? Head Face Med. 2011;7:5.
245. Coleman R, Burkinshaw R, Winter M, Neville-Webbe H, Lester J, Woodward E, et al. Zoledronic acid. Expert Opin Drug Saf. 2011;10(1):133–45.

246. Di Salvatore M, Orlandi A, Bagalà C, Quirino M, Cassano A, Astone A, et al. Anti-tumour and anti-angiogenetic effects of zoledronic acid on human non-small-cell lung cancer cell line. Cell Prolif. 2011;44(2):139–46.
247. Tamura T, Shomori K, Nakabayashi M, Fujii N, Ryoke K, Ito H. Zoledronic acid, a third-generation bisphosphonate, inhibits cellular growth and induces apoptosis in oral carcinoma cell lines. Oncol Rep. 2011;25(4):1139–43.
248. Gnant M, Eidtmann H. The anti-tumor effect of bisphosphonates ABCSG-12, ZO-FAST and more…. Crit Rev Oncol Hematol. 2010;74 Suppl 1:S2–6.
249. Coleman RE, Winter MC, Cameron D, Bell R, Dodwell D, Keane MM, et al.; AZURE (BIG01/04) Investigators. The effects of adding zoledronic acid to neoadjuvant chemotherapy on tumour response: exploratory evidence for direct anti-tumour activity in breast cancer. Br J Cancer. 2010;102(7):1099–105.
250. Berenson JR. Antitumor effects of bisphosphonates: from the laboratory to the clinic. Curr Opin Support Palliat Care. 2011;5(3):233–40.
251. Meraviglia S, Eberl M, Vermijlen D, Todaro M, Buccheri S, Cicero G, et al. In vivo manipulation of Vgamma9Vdelta2 T cells with zoledronate and low-dose interleukin-2 for immunotherapy of advanced breast cancer patients. Clin Exp Immunol. 2010;161(2):290–7.

Part III
Conclusions and Therapeutic Opportunities

Conclusion

Stephen T. Sonis and Dorothy M. Keefe

Toxicities have been an accepted consequence of radiation- and drug-based cancer therapies since the times of Curie and Farber. Inevitably patients' anxieties were not only about the prognosis associated with their cancer diagnosis but also about the treatment-related hell they would be expected to endure. The pitch that treatment side effects however horrible were a small price to pay for effective cancer interventions was preached by caregivers and tolerated by patients. Supportive cancer care was reactive, rudimentary, and without scientific rationale. Dose de-escalation or breaks in radiation were primary strategies to minimize toxicity, but had an adverse effect on tumor control.

In the past two decades the inevitability of regimen-related toxicities has been challenged. Research in the area shifted from being solely descriptive in nature to investigations that were comprehensive critical epidemiologic analyses or focused on the pathobiological underpinnings of toxicities. The concept of clustering—the fact that cancer patients tend to develop groups of toxicities that share similar biological causality—was recognized and pursued. These efforts were rewarded by the identification of mechanistic targets, opportunities for pharmacological intervention, and the beginnings of a therapeutic portfolio for supportive cancer care. "Beginnings" is the key word in the previous sentence as we still have a long way to go before most patients have a treatment course that is free of regimen-driven side effects.

S.T. Sonis, DMD, DMSc (✉)
Clinical Professor of Oral Medicine, Harvard School of Dental Medicine,
Chief, Divisions of Oral Medicine, Brigham and Women's Hospital and the Dana-Farber Cancer Institute, Chief Scientific Officer, Biomodels, LLC 75 Francis Street Boston, Boston, MA 02115, USA
e-mail: ssonis@partners.org

D.M. Keefe, MD
Service Director, South Australian Cancer Service, Professor of Cancer Medicine, Faculty of Health Sciences, University of Adelaide, Adjunct Professor, Sansom Institute, University of South Australia, Senior Consultant Medical Oncologist, Royal Adelaide Hospital Cancer Centre, Adelaide, Australia
e-mail: Dorothy.keefe@health.sa.gov.au

S.T. Sonis and D.M. Keefe (eds.), *Pathobiology of Cancer Regimen-Related Toxicities*,
DOI 10.1007/978-1-4614-5438-0_15, © Springer Science+Business Media New York 2013

The preceding chapters describe the current state of the science of regimen-related toxicities. From the introductory chapters which discuss general concepts of normal tissue injury to those dealing with specific treatment complications, the reader will, no doubt, recognize common themes. While radiation- or chemotherapy-induced injury contributes to normal tissue damage, it is secondary biological mediators which create the most havoc. The finding that pro-inflammatory cytokines play roles in both the breakdown of intestinal mucosa and in the genesis of fatigue is illustrative.

But the story is far from complete and more granularity is being added at a rapid rate.

Furthermore, the development of oncologic drugs is ongoing. While each new class brings optimism for clinicians and their patients for effective tumor amelioration, each also comes with a new collection of toxicities which impact treatment tolerability and noncancer health outcomes. Clearly any discussion of regimen-related toxicities is therefore a moving target and compounded by the use of multi-agent regimens. Thus it is naïve to believe that the contents of this book represent anything more than the first reel in a movie that will play out in the years to come. While we cannot tell how long the movie will be, we are nonetheless convinced that with continued research and development, we can anticipate a happy ending.

Index

S.T. Sonis and D.M. Keefe (eds.), *Pathobiology of Cancer Regimen-Related Toxicities*,
DOI 10.1007/978-1-4614-5438-0, © Springer Science+Business Media New York 2013